AF412568

Human Adenoviruses

From Villains to Vectors

Human Adenoviruses

From Villains to Vectors

Jane Flint
Princeton University, USA

Glen Nemerow
The Scripps Research Institute, USA

World Scientific

NEW JERSEY · LONDON · SINGAPORE · BEIJING · SHANGHAI · HONG KONG · TAIPEI · CHENNAI · TOKYO

Published by

World Scientific Publishing Co. Pte. Ltd.

5 Toh Tuck Link, Singapore 596224

USA office: 27 Warren Street, Suite 401-402, Hackensack, NJ 07601

UK office: 57 Shelton Street, Covent Garden, London WC2H 9HE

Library of Congress Cataloging-in-Publication Data

Names: Flint, S. Jane, author. | Nemerow, Glen R. (Glen Robert), author.

Title: Human adenoviruses : from villains to vectors / Jane Flint, Glen R. Nemerow.

Description: New Jersey : World Scientific, 2016. |

    Includes bibliographical references and index.

Identifiers: LCCN 2016027777 | ISBN 9789813109797 (hardcover : alk. paper)

Subjects: | MESH: Adenoviruses, Human

Classification: LCC QR396 | NLM QW 165.5.A3 | DDC 579.2/443--dc23

LC record available at https://lccn.loc.gov/2016027777

British Library Cataloguing-in-Publication Data

A catalogue record for this book is available from the British Library.

Typeset by Stallion Press

Email: enquiries@stallionpress.com

Printed in Singapore

We dedicate this book to our former and present students, postdoctoral
fellows and collaborators to whom we are ever in their debt.
We also dedicate this to our families who inspired
and supported our efforts:
Jonn, Gethyn and Amy Leedham
Laureen Bloomer, Brian, Carolyn, Brittney and Wylder Nemerow

"Reading furnishes the mind only with materials of knowledge;
it is thinking that makes what we read ours."
John Locke, English philosopher

Preface

Adenoviruses are significant pathogens in humans, especially immuno-compromised patients, prompting new approaches to the development of antiviral drugs and other therapies. Furthermore, by 2015, vectors derived from them were the most used in clinical trials worldwide (Abedi, 2013), and one oncolytic adenovirus is in clinical use in China. The quests for adenovirus-specific antivirals and vectors tailored to optimize gene transfer, vaccine delivery or tumor cell-specific cytotoxicity, build on the extensive knowledge of adenovirus molecular and structural biology collected since the first adenoviruses were described in the winter of 1952–53. They have in turn inspired renewed scrutiny of the anti-viral defenses mounted by individual cells and the host immune system, leading to new strategies for the development of antiviral drugs and vectors.

Descriptions of the fundamentals of adenovirus biology are published regularly, for example in Fields Virology, and adenovirus vectors are the subjects of numerous book chapters and reviews, for example, >1,500 reviews matching the terms 'adenovirus vector review' in PubMed. In contrast, no up-to-date book or monograph devoted entirely to adenovirus is available. This book was conceived to fill this void by providing a comprehensive discussion of human adenoviruses and their interactions with the host, in the context of pathogenesis, specific anti-adenovirus drugs, and vector development. Numerous recent advances in elucidation of adenovirus–host interactions and increasing success have made this a timely and exciting endeavor—one that we believe is long overdue.

In assembling this monograph, we tried to avoid presenting this work as a collection of individual chapters written by one or another of us, but rather sought to integrate our knowledge in specific areas with input from both authors to produce a joint endeavor. The research discussed in this monograph was conducted in many laboratories worldwide and the field is indebted to their tireless efforts. In this regard, we would like to thank members of our own laboratories who throughout the past ~35 years have initiated and performed some of the research described in several of these chapters. We apologize for not including reference to every research publication, an omission imposed by the enormous adenovirus research endeavor and space limitations. Our special thanks to Drs. Vijay Reddy and Phoebe Stewart, who pioneered the X-ray diffraction and cryoEM structural analyses of intact adenovirus particles, respectively, during the past few decades and who provided some of the high-resolution structure figures in Chapter 2. It has been a great privilege of one of us (G.R.N.) to collaborate with them in these endeavors. We are grateful to Drs. Marlene Bouvier, Steven Cusack, Walter Mangel, Clodagh O'Shea, Jon Rux, Carmen San Martin, and Thilo Stehle for making or providing images for this book. We also thank Dr. Balazs Harrach who provided input on special issues such as virus evolution and nomenclature. We are also indebted to Lorraine Lathrop (the Scripps Research Institute) and Ellen Brindle-Clark (Princeton University) for excellent technical assistance, with special thanks to Ellen for compiling and formatting the text.

As with any textbook, some of the research findings discussed herein are likely to be overtaken by new discoveries in the coming months or years. We have tried to anticipate this certainty at the end of each chapter ('What's Next?'). Nonetheless, viruses such as adenoviruses continue to surprise us even in the most subtle of ways. We trust that this monograph will provide a knowledge base for virologists, but most of all, we hope that it will ignite new research efforts on this important human pathogen and vector.

Jane Flint
Glen Nemerow

Contents

List of Tables

List of Figures

A Short History of Adenovirus   1

1.1 Discovery and Initial Characterization

The viruses we now know as adenoviruses were first described over 60 years ago, when the medium from cultures of adenoids of young children in Washington D.C. that underwent degeneration within 1–4 weeks after isolation was shown to contain a filterable agent (i.e. a virus) that transmitted the cytopathic effect to established human cells lines, such as HeLa cells (Rowe *et al.*, 1953). An agent that induced very similar cytopathic responses was isolated the same winter from the throat washings of a military recruit diagnosed with primary atypical pneumonia (Hilleman *et al.*, 1954). This respiratory illness (R1) and the adenoid-degenerating (AD) agents were found by neutralization and complementation fixation assays to be antigenically related. Over the next few years, additional viruses that share a group-specific complement fixation antigen were isolated from multiple human subjects, including apparently healthy individuals (Fox *et al.*, 1969; Fox *et al.*, 1977; Wold & Ison, 2013), from other mammals, such as monkeys (Hull *et al.*, 1958), dogs (Kapsenberg, 1959) and mice (Hartley & Rowe, 1960), and from birds (Yates & Fry, 1957). The family name adenovirus (*Adenoviridae*) was chosen in 1956, in part because of its echo of the tissue of origin of the prototype strain.

Adenoviruses are non-enveloped virus particles with diameters of 70–90 nm, with a distinctive morphology first visualized by electron microscopy of a species C human adenovirus (HAdV-C5) (Horne *et al.*,

1959). In contrast to many viral capsids with icosahedral symmetry, adenovirus particles actually have an icosahedral appearance. Furthermore, they carry distinctive fibers projecting from each of the 12 vertices of the particle (one, two, or three fibers/vertex, depending on the genus). Adenoviral genomes are linear, double-stranded molecules of DNA that carry inverted terminal repetitions (ITR), a packaging signal (ψ), and a viral protein, the terminal protein (TP), covalently linked to the 5′ end of each strand. Despite these common properties, and a conserved core of protein coding sequence, adenoviral genomes differ in size (ranging from ~26 kbp to >48 kbp), GC content and arrangement, and sequences of non-conserved genes (see Chapter 5).

1.2 Classification and Evolution

Adenoviruses have been identified only in vertebrates, from fish to mammals. Viruses isolated from humans dominate the family, and only small members of piscine and reptilian viruses have been reported to date, a bias that almost certainly reflects intensity of study, rather than the actual distribution of members of this family. The ~170 distinct members recognized by the International Committee on the Taxonomy of Viruses (ICTV) have been classified in five genera (Harrach *et al.*, 2012). The criteria for classification have included the species of the host, serological properties, AT content of the genome, sequence relationships among genomes, and organization and coding capacity of the genome. For example, some structural proteins (core protein V and cement protein IX) are unique to the *Mastadenovirus*. Similarly, the genes that encode viral regulatory proteins differ considerably among the genera. Phylogenetic comparisons of sequences encoding specific proteins, typically those of hexons, have identified five discrete clades that correspond to the five genera (Fig. 1.1).

In general, adenoviruses appear to have co-evolved with their hosts; phylogenetic trees for the viral protease and host mitochondrial rRNA are similar, as are the evolutionary distances between the viruses within a genus and their hosts (Benko & Harrach, 2003; Harrach *et al.*, 2012; Kovacs *et al.*, 2003). Nevertheless, these relationships are by no means absolute. For example, adenoviruses that infect birds or mammals can belong to different genera (Fig. 1.1). In the absence of any adenovirus

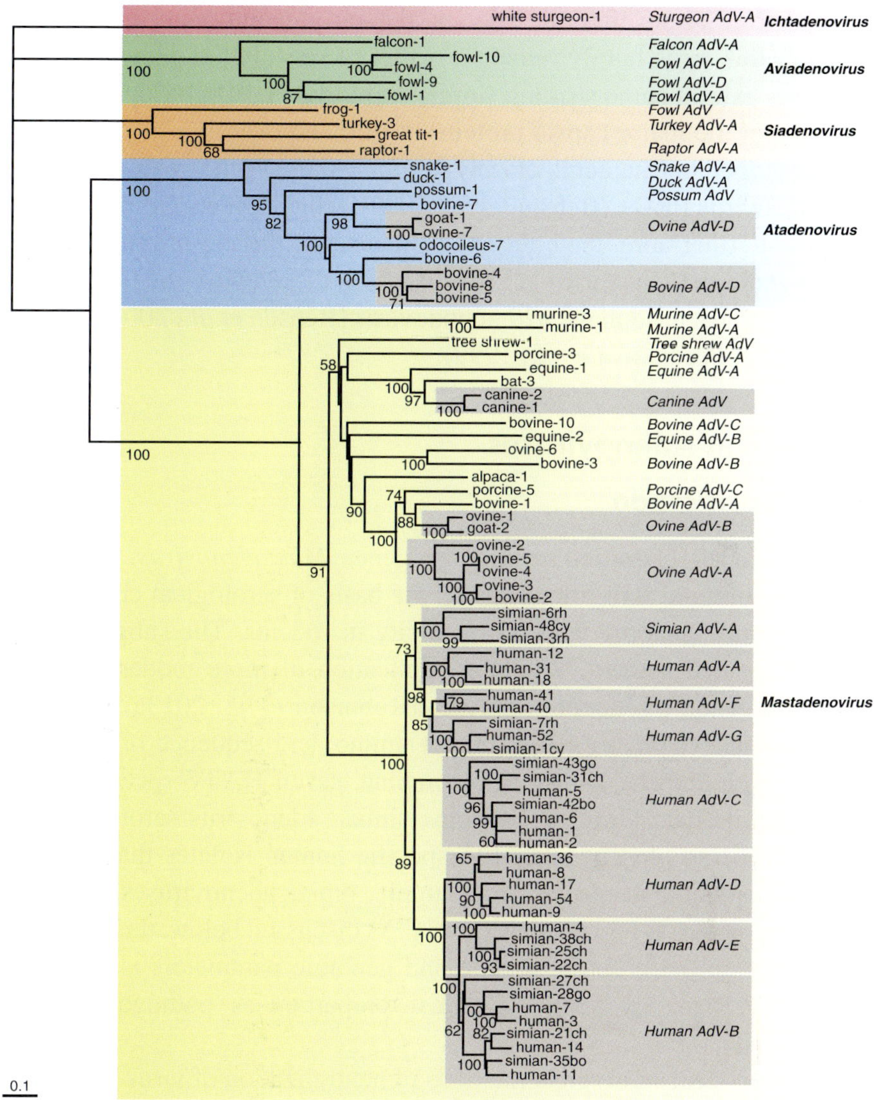

Figure 1.1. Phylogenetic tree of adenovirus. This tree is based on analysis of hexon amino acid sequences, using only some members of primate adenovirus species with more than three members and excluding the sequences of primate adenovirus hexons known to have arisen by homologous recombination. This unrooted tree was generated with the ichtaadenovirus white sturgeon virus 1 chosen as the outgroup. Bo, bonobo; ch, chimpanzee; cy, cynomologous macaque; go, gorilla; rh, rhesus macaque. Adapted from Harrach *et al.*, 2012, with permission.

isolated from an invertebrate species, the early evolutionary history and origin of this virus family remain a matter of speculation. However, adenoviruses exhibit some striking similarities to various bacteriophages; the topology of the major capsid protein (hexon) and overall capsid architecture, described in Chapter 2, closely resemble those of the double-stranded DNA tectivirus PRD1 (Benson *et al.*, 1999), which infects Gram-negative bacteria. Furthermore, the unusual mechanism of protein priming of viral DNA synthesis (Chapter 6) is also shared by these two virus families, as well as with bacteriophage ø29, a podovirus (Benson *et al.*, 2003; de Jong *et al.*, 2003; Pecenkova *et al.*, 1999).

1.3 Human Adenoviruses

1.3.1 *Classification*

HAdVs are all classified within the genus *Mastadenovirus*, and were originally ordered into subgroups on the basis of serological criteria, GC content of the genome, and oncogenicity in rodents. The subgroups are now designated species. With the application of more modern criteria, including genome sequence analyses (Robinson *et al.*, 2013a), phylogenetic distance (typically based on the amino acid sequence of the DNA polymerase), and the ability to recombine, seven HAdV species (A–G) have been defined (Table 1.1). Some simian adenovirus serotypes have proved to be so closely related to specific human isolates that they are included with human species; for example, simian adenoviruses 21 and 31 in human species B and C, respectively. As discussed below, this grouping of HAdVs on the basis of genetic and genomic parameters is mirrored, at least in part, by such biological properties as pathogenesis and oncogenicity.

The classical serological methods of neutralization of virus infectivity by serum antibodies (serum neutralization) and inhibition of hemagglunitation activity initially distinguished 51 HAdV serotypes (Lion, 2014; Wold & Horwitz, 2006). Hemagglutination is a property of the fiber (Nicklin *et al.*, 2005), whereas the majority of epitopes recognized by neutralizing antibodies lie within the seven hypervariable loops of the hexon (the major capsid protein) (Roberts *et al.*, 2006). These classic assays are relatively time-consuming, require the availability of

Table 1.1. Human adenoviruses (HAdVs).

Species	Types[1]	Genome GC content	Oncogenic potential	Infections
A	12, 18, 31, *61*	48–49	High	Respiratory, urinary and gastrointestinal
B	3, 7, 11, 14, 16, 21, 34, 35, 50, *51, 66*	50–52	Moderate	Respiratory, urinary,[2] and gastrointestinal; keratoconjunctivitis
C	1, 2, 5, 6, *7, 57* (SiAdV-31, BoAdV-9)	57–59	None	Upper respiratory, urinary,[2] and gastrointestinal
D	8–10, 13, 15, 17, 19, 20, 22–30, 32, 33, 36–39, 41–49, 51, *53, 54, 56, 56–60, 63–67, 70*	56	None	Gastrointestinal; keratoconjunctivitis
E	4 (SiAdV 22-25)	57–61	None	Keratoconjunctivitis; respiratory
F	40, 41	57–59	None	Gastrointestinal
G	52 (SiAdV 1, 2, 7, 11, 12, 15)	55	ND	Gastrointestinal

[1]HAdV types listed in italics have been identified by genome sequencing and computational analysis.
[2]More common in transplant patients.
BoAdV, bovine adenovirus; SiAdV, simian adenovirus; ND, not determined.

appropriate sera and report on only a small fraction of the genetic diversity among HAdV isolates. Consequently, they have been increasingly replaced by typing methods based on sequencing of segments of the genome that specify type-specific epitopes of the hexon or fiber (Lion, 2014; McCarthy *et al.*, 2009) and complete genome sequencing (Seto *et al.*, 2011). With decreasing cost and increasing power, full genome sequencing is now quite feasible, and obviously provides far more information than can be obtained by focus on type-specific epitopes and the regions that encode them. Full genome sequencing has resolved apparently anomalous results obtained by serotyping (e.g. Liu *et al.*, 2012; Walsh *et al.*, 2010). This approach also facilitates characterization of newly arising recombinants, which are particularly prevalent within species D, the largest group (Robinson *et al.*, 2013a).

The designation serotype has been replaced by the term type, and there is general agreement that genome sequence information should be the basis for the identification of new HAdV types. Indeed, the types numbered from 52 were detected and characterized by sequencing methods (see Lion, 2014). Nevertheless, there is ongoing debate among members of the adenovirus research community about how much sequence information should be collected, the contribution of serological methods, and how new HAdV types should be named and by whom (Aoki *et al.*, 2011; Kajon *et al.*, 2013; Lion, 2014; Seto *et al.*, 2011; Seto *et al.*, 2013).

As of this writing, June, 2016, more than 70 HAdV types have been reported. Their distribution among species is highly skewed, with but one representative of both species E and G, and over 40 types in species D (Table 1.1). The sole member of species E, HAdV-E4, was one of the first HAdVs to be isolated (Hilleman *et al.*, 1954), and was originally placed in species B. However, recent computational analyses exploiting the availability of genome sequences of all human prototypes and various SAdVs (e.g. Robinson *et al.*, 2013b) established that this isolate is unique, and arose by recombination and zoonosis; the HAdV-E4 genome contains sequences for hypervariable hexon loops L1 and L2 from HAdV-B16 (hence its original classification) in the backbone of a simian adenovirus (Dehghan *et al.*, 2013). Similarly, although both species F and G types are associated with gastroenteritis (Table 1.1), phylogenetic analyses of genome sequences have established that HAdV-G52 is quite distinct from the members of species F (Jones *et al.*, 2007).

Homologous recombination between highly related sequences in the genomes of members of the same species has been recognized and exploited for decades (e.g. Sambrook *et al.*, 1975; Takemori, 1972). Recombination within the coding sequences for hexon, penton, and fiber is common (Lion, 2014; Lukashev *et al.*, 2008; Robinson *et al.*, 2011b; Robinson *et al.*, 2013b), but has also been observed in the E3 region (Singh *et al.*, 2013). Species D HAdV types appear to be especially prone to recombination; the more recently identified members of this species have been shown by sequencing and bioinformatic analyses to have arisen by recombination between the genomes of two or more HAdV-D types

(Kajon *et al.*, 2014; Kaneko *et al.*, 2011a; Matsushima *et al.*, 2013; Robinson *et al.*, 2013b). This property implies that co-infection of a single cell by two (or more) HAdV types, a pre-requisite for recombination, is not infrequent, perhaps because these viruses can persist in immuno-suppressed patients. Duplication and divergence of sequences, as well as deletion, contribute to the evolution of genetic diversity in HAdVs (Davison *et al.*, 2003; Jones *et al.*, 2011) and substitutions may be important in some cases (Robinson *et al.*, 2013b). However, it is clear that recombination is the predominant mechanism for the generation of new HAdV types (see Lion, 2014; Robinson *et al.*, 2013b), which are associated, in many cases, with distinct pathogenic properties.

1.4 Prevalence and Pathogenesis

Members of the *Adenoviridae* are widespread in the human population and present in all parts of the world, with different types of predominating in particular geographic regions or at different times (see Lion, 2014; Wold & Ison, 2013). They are associated with a variety of acute diseases in normal individuals.

Infection of epithelial surfaces by HAdVs causes respiratory disease, epidemic keratoconjuctivitis, and gastroenteritis, and is associated with a variety of other syndromes (Chapter 8). There is some correlation of the pathogenicity of these viruses with the other properties on which classification is based. For example, species B and C HAdVs cause acute respiratory disease, and pneumonia and pharyngitis, respectively. In contrast, epidemic keratoconjunctivitis is associated largely with species D and E types, and gastroenteritis with the members of species F and G. At present, there is no explanation for such differences in tropism and pathogenicity, although it is clear that they cannot be ascribed to a single interaction with the host, such as binding of virus particles to different, cell type-specific receptors (Chapter 3). Consequently, it is more likely that HAdV tissue tropism and disease are determined by multiple aspects of virus-host interactions.

As described in Chapter 8, HAdV infections can be endemic; for example, several species C types in young children, or epidemic, such as outbreaks of keratoconjunctivitis or of acute respiratory disease in the military. Infection by these viruses can lead to significant morbidity, a situation that prompted the U.S. Department of Defense to support the development and redeployment of a vaccine against HAdV-4 and HAdV-7 (Gaydos *et al.*, 1995; Gray *et al.* 2000; Radin *et al.*, 2014). However, HAdVs are a far more serious problem in immunocompromised individuals, such as transplant recipients and those infected with immunodeficiency virus type 1. In such populations, morbidity is markedly increased, and death is a not infrequent outcome of infection (Echavarria, 2008; Lion, 2014; Wold & Ison, 2013).

1.5 Lessons from Adenovirus

1.5.1 *Insights into fundamental cellular processes*

The demonstration that HAdV-C2/5 genes, like cellular protein-coding genes, are transcribed by host cell RNA polymerase II (Price & Penman, 1972; Weinman, 1974) prompted a cohort of investigators to turn to adenovirus-infected cells to study the mechanisms by which mRNA is synthesized in mammalian cells. Indeed within just a few years, intensive characterization of viral mRNAs and the sequence that encode them in the species C HAdV genome solved the long-standing conundrum of how mRNAs could be produced from considerably longer precursors (pre-mRNA) while retaining both the 5′ caps and the 3′ poly(A) tails. Abundant adenoviral mRNAs were shown by elegant electron microscopy to be spliced from multiple segments that are not contiguous in pre-mRNA and in the viral genome (Berget *et al.*, 1977; Chow *et al.*, 1977). Shortly thereafter, it was established that mammalian pre-mRNAs, notably those encoding murine β-globin (Tilghman *et al.*, 1978a; Tilghman *et al.*, 1978b), an immunoglobulin light chain (Rabbitts, 1978) and chicken ovalbumin (Breathnach *et al.*, 1978), are spliced, a modification that turned out to be the rule for mammalian pre-mRNAs. The discovery of RNA splicing by Phillip Sharp, Richard Roberts, and their colleagues was recognized by the 1993 Nobel Prize in physiology or medicine.

　　The discovery of RNA splicing represents the best-known contribution to fundamental molecular biology of studies of adenovirus-infected cells. However, this system also provided the first, and still one of the most dramatic examples, of polyadenylation of a single pre-mRNA at multiple, alternative sites (Nevins & Darnell, 1978a; Nevins & Darnell, 1978b). Furthermore, the precise mapping of the site of initiation of transcription of HAdV-C2 major late (ML) transcription unit (Ziff *et al.*, 1978) allowed the development of the first *in vitro* assay for specific initiation of RNA polymerase II transcription (Weil *et al.*, 1979). This advance paved the way for identification and characterization of the elaborate suite of proteins required to confer specificity and efficiency upon this process. These proteins include sequence-specific transcriptional activators, some of which were first recognized by virtue by their binding to adenoviral promoters. An important example is E2F, which binds to the viral E2 early (E2E) promoter. This protein proved to be a crucial target of mechanisms that regulate cell cycle progression (see 'Mechanisms of transformation' below).

1.5.2 *Insights into virology*

Investigation of the reproduction of HAdVs and their interactions with host cells has also established a number of paradigms in virology. These include the discovery of small viral RNAs, virus-associated (VA) RNAs, in HAdV-C2 infected cells (Forget & Weissman, 1967; Reich *et al.*, 1966), which are synthesized by RNA polymerase III (Price *et al.*, 1972; Weinmann *et al.*, 1974). All HAdV genomes contain at least one VA gene, and analogous RNA polymerase III genes are present in the genomes of various herpesviruses (Rosa *et al.*, 1981). VA RNAs contribute to blocking antiviral defense mechanisms, and also serve as precursors for production of viral microRNAs (Chapters 5 and 9).

It is now well established that the initial steps in a viral infectious cycle — attachment to and entry into the host cell — frequently require binding of virus particles to multiple cell surface proteins. This receptor–coreceptor paradigm was first established in studies of HAdV-C5 entry, where the cellular integrins $\alpha_v\beta_3$ or $\alpha_v\beta_5$ were shown to be essential for internalization of virus particles, but not for their initial attachment to the host cell (Wickham *et al.*, 1993; see also Chapters 3 and 4).

Adenovirus-infected cells also provided the first example of another feature common to cells infected by many different types of virus, disruption of antigen presentation by major histocompatibility (MHC) proteins. In this case, a viral glycoprotein encoded in the E3 gene (E3 gp19K) binds to MHC class I proteins to retain them in the endoplasmic reticulum (ER) (Anderson *et al.*, 1985; Burgert & Kvist, 1985).

As DNA-dependent DNA polymerases cannot initiate DNA synthesis, all require a primer with a free 3′ OH group. During cellular genome replication, short RNAs synthesized by DNA polymerase α/primase serve as these primers (see Waga & Stillman, 1998). However, one of the daughter DNA strands (the lagging strand) must be made discontinuously from multiple DNA primers, because RNA polymerases make new strands in only the 5′ to 3′ direction. Investigation of the mechanism of species C HAdV replication identified for the first time an elegant mechanism that allows continuous copying of both strands of the genome, priming from a dedicated viral protein that associates with the viral DNA polymerase at the terminal origins of replication (Challberg *et al.*, 1980; Lichy *et al.*, 1982; Pincus *et al.*, 1981). It is now known that protein priming initiates replication of the genomes of a variety of bacteriophages, such as ø29 and PRDI (see Salas, 1991), picornaviruses such as poliovirus (Morrow & Dasgupta, 1983) as well as hepadnavirus reverse transcription (see Ganem *et al.*, 1994). The first *in vitro* system capable of initiation of replication of an exongenous template was also developed from adenovirus-infected cells (Challberg & Kelly, 1979).

The HAdV-C5 E1A proteins, the first to be synthesized in the infected cells, have been paradigms for studying mechanisms of transcriptional regulation. The two major E1A proteins are translated from alternatively spliced mRNAs, but share all sequences expect for an internal domain of 46 amino acids unique to the larger protein (see Chapter 5). This unique domain confers the ability to activate transcription of the other viral genes expressed during the early phase of infection and hence promotes progression through the infectious cycle (Berk *et al.*, 1979; Jones & Shenk, 1979; Nevins, 1981). This E1A protein stimulates transcription without binding to specific promoter or enhancer sequences, and was the first of a sizeable class of viral transcriptional regulators to be identified (see Flint & Shenk, 1989). As discussed in the next section, common molecular properties of the two E1A proteins led to important insight into mechanisms of transformation and the regulation of cell-cycle progression.

1.5.3 *Mechanisms of transformation*

One of the very first exciting developments in adenovirus research was the finding that HAdV-A12 could induce tumors in rodents (Trentin *et al.*, 1962). Within a decade or so of this report, two species C HAdV genes were shown, using a variety of experimental approaches, to be necessary and sufficient for transformation of rodent cells in culture (see Bishop, 1985; Branton *et al.*, 1985). The E1A gene encodes mitogenic proteins that can immortalize primary cells, while E1B proteins allow stable transformation. Investigation of how these viral proteins reprogram cell proliferation and survival not only established the importance of subverting the activities of tumor suppressors, but also yielded groundbreaking insights into the molecular mechanisms that govern progression into and through the cell cycle.

The first cellular protein shown to interact with viral E1A proteins was the tumor suppressor RB (Harlow *et al.*, 1985). It was soon established that this interaction is critical for E1A-dependent transformation, as it allows reentry of quiescent cells in G0 into the cell cycle, and cell cycle progression through the G1 restriction point into S phase (Berk, 2005; DeCaprio, 2009; Dyson, 1994; Gallimore & Turnell, 2001; Moran, 1993; Nevins, 1995). Binding of RB to members of the E2F family of transcriptional activators recruits RB and associated repressive histone deacetylases to E2F-dependent promoters. This set includes the promoters of genes that encode the cyclins needed for the G1 to S phase transition and numerous genes expressed in S phase. Interaction of RB with specific E1A sequences sequesters the cellular protein from E2F, inducing activation of E2F-dependent genes, and progression through the cell cycle. Transforming proteins of polyomaviruses (LT) and oncogenic human papillomaviruses (E7) were soon also found to reverse the RB-mediated inhibition of transcription of E2F-regulated genes (DeCaprio, 2009; Dyson *et al.*, 1989; Felsani *et al.*, 2006), establishing both a general mechanism of transformation by these small DNA viruses and the critical regulatory role of RB.

Investigation of the molecular functions of E1B proteins reinforced the importance of inhibition of tumor suppressors for viral oncogenesis, and also focused attention on blocking apoptosis. Either E1B protein promotes survival of cells in which E1A proteins induce misregulation of cell-cycle progression. The E1B 19 kDa protein, a homolog of the cellular anti-apoptotic protein BCL-2, blocks E1A-induced programmed cell death (Rao *et al.*, 1992; Subramanian *et al.*, 1995). The E1B 55-kDa protein interacts with a second cellular tumor suppressor, p53 (Sarnow *et al.*, 1982a) to protect against consequences of p53 activation, such as G1 arrest or apoptosis (see Berk, 2005; Levine, 2009; White, 1995). In transformed cells in which only the viral E1A and E1B genes are expressed, the E1B–p53 association inhibits p53-dependent transcription and can sequester p53 in the cytoplasm (see Berk, 2005; Berk, 2013). It is clear from mutational analyses that the ability of the E1B 55-kDa protein to cooperate with E1A gene products to transform cells in culture correlates with repression of p53-dependent transcription (Endter *et al.*, 2005; Teodoro & Branton, 1994; Teodoro *et al.*, 1997; Yew *et al.*, 1990). Like the targeting of RB, blocking the activity of p53 proved to a mechanism common to oncogenic protein of other small DNA viruses (Levine, 2009; Levine *et al.*, 1991; Vousden, 1995).

1.6 Toward Clinical Applications

As of July 2015, adenovirus vectors remain among the most commonly used in clinical trials worldwide (Abedi, 2013). Replication defective derivatives of HAdV-C5 were the first to be used for therapeutic gene delivery (Lemarchand *et al.*, 1992; Rosenfeld *et al.*, 1994) and continue to be widely applied for this purpose in the laboratory. Investigation of therapeutic applications has spurred the development of alternative, less immunogenic vectors, methods to cloak virus particles and strategies to increase the efficiency or specificity of target cell transduction (see Chapter 11). Such replication-deficient adenoviruses are also components of a number of experimental vaccines, and have been used to deliver cytotoxic gene products to tumor cells. An alternative approach to the development of oncolytic adenoviruses relies on mutations that impair replication in normal, but not in tumor cells. A considerable variety of such viruses have been investigated in model systems, several are in clinical trials and one approved for use in the clinic in China (Chapter 12). These efforts to develop therapeutic adenoviruses have provided a major impetus for continued investigation of fundamental questions about adenovirus reproduction and the impact on host cells and organisms.

Adenovirus Composition, Structure, and Biophysical Properties

2

2.1 Overall Virus Composition

Human adenoviruses (HAdVs) are among the largest and most elaborate non-enveloped viruses, having a diameter of ~90 nanometers and a molecular mass of 150 megadaltons (Flint *et al.*, 2015). As determined by biochemical methods (Anderson, 1990; Mangel *et al.*, 2003; van Oostrum & Burnett, 1985) or by mass spectrometry-based proteomics (Benevento *et al.*, 2014; Chelius *et al.*, 2002), HAdV-C5 contains 13 distinct proteins, which make up the outer capsid shell and an inner core with a 36-kDa double-stranded (ds) DNA genome (Fig. 2.1).

The outer shell of the *pseudo-T=25* icosahedral capsid is primarily composed of 240 copies of the hexon trimer. Each of the 12 vertices (corners) of the particle contains a penton that comprises an extended fiber protein non-covalently attached to the penton base protein. Four distinct cement proteins, designated IIIa, IX, VI, and VIII are embedded in the outer and inner surfaces of the capsid shell. By means of their extensive associations with the hexons, these cement proteins provide stability for the relatively large virus particle. Underneath the vertex region, a segment of protein V associates with protein VI and links the inner surface of the capsid with the inner dsDNA core. This chapter discusses the growing

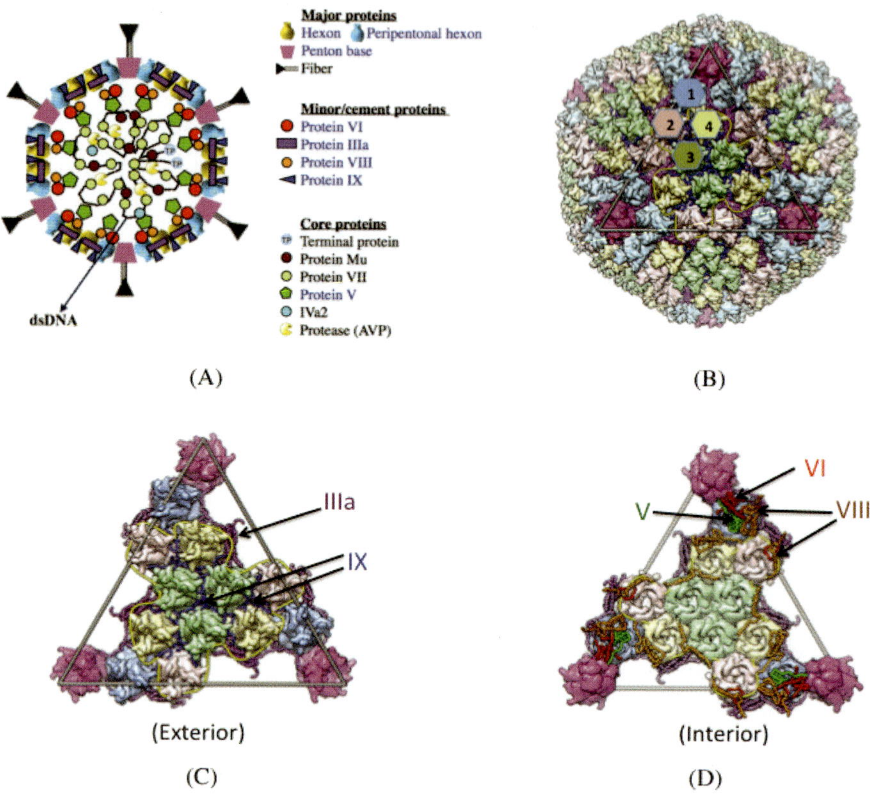

Figure 2.1. Crystal structure and organization of human adenovirus (HAdV). (**A**) A schematic illustration of the organization of capsid and core proteins in HAdV. The locations of various proteins are represented by different-colored symbols and the corresponding names are shown at the right. The indicated locations of the core proteins are approximate. Shown in blue lettering are the proteins whose structures have been identified in this crystal structure. (**B**) Overall organization of hexon and penton base subunits exhibiting *pseudo-T* = 25 icosahedral symmetry. Structurally unique hexons (1–4) are color-coded in light blue, pink, green, and khaki, respectively. Penton vertices are shown in magenta. Outer cement proteins IIIa and IX are shown in purple and blue, respectively. Fiber molecules associated with the penton base are disordered. The outline of the triangular icosahedral facet is shown as a gray triangle, whereas the border of the GON hexons is indicated by yellow-colored rope. (**C**) An exterior view of the triangular icosahedral facet that comprises 12 hexons along with penton base vertices shown in magenta. Color representations are the same as in B. (**D**) An interior view of the facet in C, with three minor proteins, V (green), VI (red), and VIII (orange). It is noteworthy that a copy of V, VI, and VIII forms a ternary complex beneath the vertices, whereas other VIII (orange) molecules are arranged as staples along the border (yellow-colored rope) of the GON hexons. Reprinted from Reddy & Nemerow (2014b), with permission.

number of structural analyses of adenovirus and adenovirus proteins, beginning nearly 50 years ago and continuing to the present date. These investigations have led to a more detailed understanding of adenovirus structure and provided insights into the processes of capsid assembly and disassembly. Undoubtedly, further investigations will be needed to illuminate more fully the stunningly complex architecture of adenovirus particles and to resolve ongoing controversies regarding the precise locations of some of the cement proteins.

2.2 Structures of Individual Adenovirus Capsid Proteins

2.2.1 *Hexon*

Although HAdVs are characterized by a large and complex macromolecular structure, several individual components of the virus capsid can be purified or synthesized as recombinant proteins, thereby facilitating structural analyses (Fig. 2.2). A prominent example is the HAdV-C2 hexon that comprises three identical subunits of 967 amino acids (Jornvall *et al.*, 1981). The hexon could be isolated from virus-infected cells and was the first animal virus protein to be crystallized (Franklin *et al.*, 1971; Macintyre *et al.*, 1969). The crystal structure of the hexon was subsequently determined by X-ray diffraction (Roberts *et al.*, 1986; Rux & Burnett, 2000). The bulk of the hexon monomer forms a tower-like structure built from two jellyroll domains. Each jellyroll is formed by an eight-stranded β (sandwich)-sheet, giving the resulting tertiary structure the appearance of a baker's delight. The jellyroll structural motif is common among many other structural proteins of non-enveloped and enveloped viruses (Klose & Rossmann, 2014; Veesler *et al.*, 2013). This conserved structural motif suggests a possible shared (ancient) common origin of adenovirus and bacteriophages such as PRD1. When the assembled hexon trimer is incorporated into the virus capsid, the six jellyrolls form a pseudo-hexonal shape. The top of the hexon tower contains seven distinct amino acid insertions that form extended loops (Rux & Burnett, 2000). These so-called hypervariable (HVR) loops are the major target of neutralizing (type-specific) antibody and are the most variable in amino acid sequences among the different adenovirus types (Crawford-Miksza & Schnurr, 1996). Interestingly, some of the HVR loops that were disordered and thus were not able to be seen in the isolated hexon, were visible in the

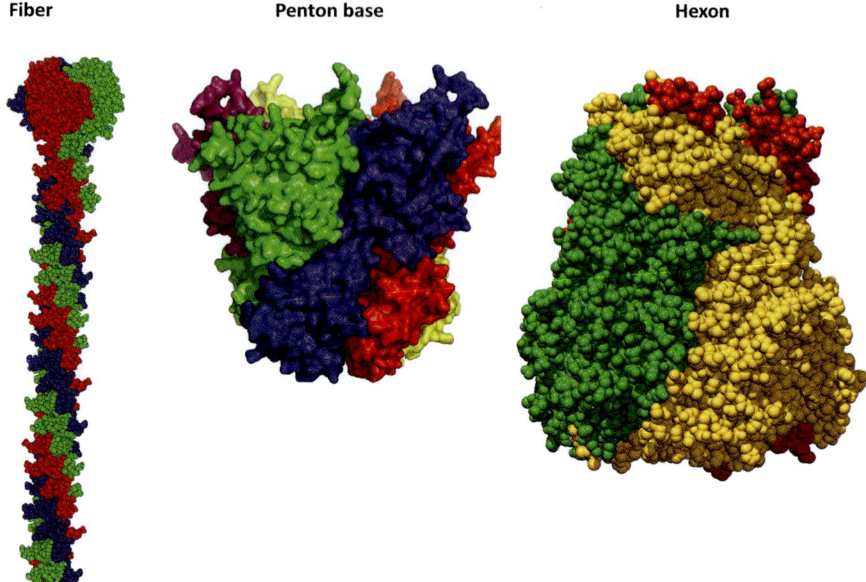

Figure 2.2. Surface representations of the crystal structures of the fiber, penton base, and hexon, not shown to scale relative to one another. (**Left**) A model of the HAdV-C2 fiber based on the 2.4 Å resolution crystal structure (reprinted from van Raaij *et al.*, 1999b, with permission). The fiber shaft forms a triple beta spiral, 265 Å in length, ending in the globular knob domain at the C-terminus of the protein. (**Middle**) A side view of the HAdV-C2 penton base protein (reprinted from Zubieta *et al.*, 2005, with permission), showing extensive outer loops on its upper surface that support integrin binding via multiple RGD sequences. (**Right**) A side view of the HAdV-C5 hexon trimer (reprinted from Rux & Burnett, 2000, with permission). Fiber and penton base, courtesy of S. Cusack, European Molecular Biology Laboratory, Grenoble, France; hexon courtesy of J. Rux, In Silico Molecular, LL., Blue Bell, PA.

crystal structure of intact adenovirus particles (Reddy *et al.*, 2010a). The inter-hexon contacts in the virus crystal lattice likely help stabilize the HVR loops. However, the largest HVR loop, HVR1, is disordered in the virus crystal structure and thus does not appear to have well-defined interactions with neighboring virus proteins in the crystal lattice.

2.2.2 *Fiber*

The trimeric HAdV-C2 fiber protein is an elongated molecule assembled from three identical polypeptides of 582 amino acids (Chroboczek *et al.*,

1995). A highly conserved sequence at the N-terminus of the fiber trimer mediates non-covalent interactions with the penton base protein (Cao *et al.*, 2012; Zubieta *et al.*, 2005). The C-terminal region of the fiber forms a globular head domain, known as the knob, that mediates attachment to specific cell receptors such as the Coxsackie and adenovirus receptor (CAR) (Bergelson *et al.*, 1997). The central domain of the fiber, known as the shaft, has an unusual triple β-spiral structure (van Raaij *et al.*, 1999b; see also Fig. 2.2) that is also present in the $\sigma 1$ protein of an unrelated RNA virus, reovirus (Mercier *et al.*, 2004; Stehle & Dermody, 2003). The β-helical structure of the fiber provides enhanced stability to the protein (Devaux *et al.*, 1987; Mitraki *et al.*, 1999). The fiber shafts of different adenovirus types contain a variable number of repeating β-sheet elements of 15–20 amino acids (Green *et al.*, 1983). Adenoviruses such as those belonging to species B1 (i.e. HAdV-B3) have only 5.5 shaft repeats whereas adenoviruses belonging to species A (i.e. HAdV-A12) have 22.5 repeats (Chiu *et al.*, 1999; Schoehn *et al.*, 1996). Importantly, the third repeat of most but not all adenovirus fiber shafts has a non-consensus sequence that does not conform to the folding pattern of the β-sheet repeats. This unusual third repeat corresponds to a region that provides flexibility to the otherwise rigid shaft domain (Wu & Nemerow, 2004). Earlier studies indicated that the length of the fiber, as determined by the number of repeat elements in the shaft, influences adenovirus infection (Shayakhmetov & Lieber, 2000). Subsequent studies employing HAdV-C5 vectors displaying chimeric fiber proteins demonstrated that the flexibility and length of the fiber protein determine optimal interactions with the receptor CAR (Wu *et al.*, 2003). Although the precise length that is optimal for CAR binding has not been determined, fibers with shaft domains containing 7.5 β-sheet repeats or fewer do not support binding.

At the C-terminal distal end of the fiber is the globular knob domain that engages in receptor interactions (Fig. 2.3). Many different HAdV trimeric fiber knobs can be overproduced as soluble recombinant proteins. The crystal structures of these receptor-binding domains have been determined by X-ray diffraction. The crystals structures of HAdV fiber head domains that bind to the CAR (van Raaij *et al.*, 1999a; Xia *et al.*, 1994), CD46 (Cupelli *et al.*, 2010; Persson *et al.*, 2007), sialic acid (Durmort *et al.*, 2001), or desmoglein 2 (Burmeister *et al.*, 2004; H. Wang *et al.*,

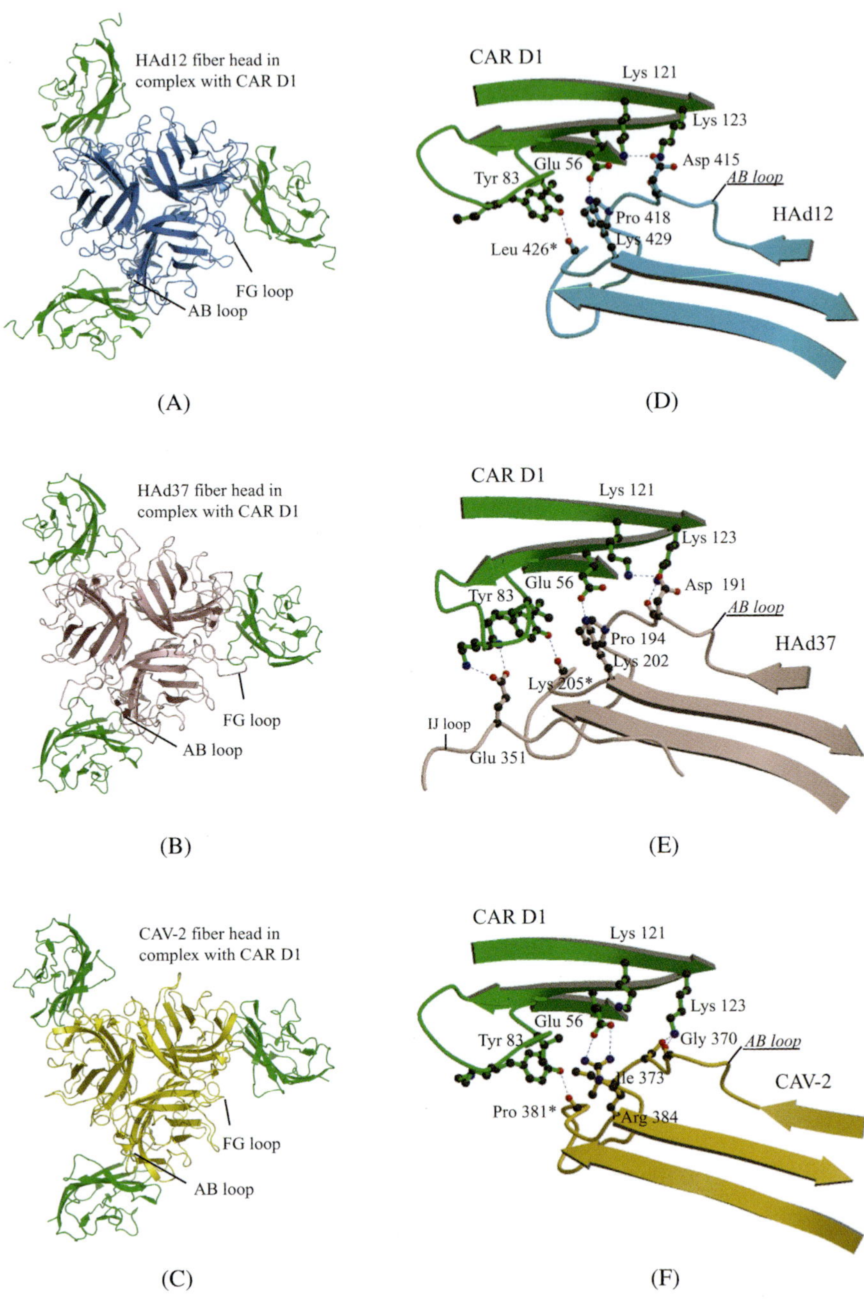
HAd12 fiber head in
complex with CAR D1
FG loop
AB loop
(A)
HAd37 fiber head in
complex with CAR D1
FG loop
AB loop
(B)
CAV-2 fiber head in
complex with CAR D1
FG loop
AB loop
(C)
CAR D1
Lys 121
Lys 123
Glu 56
Asp 415
Tyr 83
AB loop
HAd12
Pro 418
Lys 429
Leu 426*
(D)
CAR D1
Lys 121
Lys 123
Glu 56
Asp 191
Tyr 83
AB loop
HAd37
Pro 194
Lys 202
Lys 205*
IJ loop
Glu 351
(E)
CAR D1
Lys 121
Lys 123
Glu 56
Gly 370
AB loop
Tyr 83
Ile 373
CAV-2
Pro 381*
Arg 384
(F)

←

Figure 2.3. Models of adenovirus fiber knobs in complex with the Coxsackie and adenovirus receptor (CAR) D1. The fiber knobs of HAdV-A12, HAdV-D37, and CAV-2 are depicted in blue, pink, and yellow, respectively with CAR-D1 shown in green. (**A**) A top view of the HAdV-A12 fiber in complex with CAR-D1. (**B**) As **A**, but with the HAdV-D37 fiber knob. (**C**) As in **A** and **B** but with CAV-2 knob. (**D**) A segment of the binding interface between HAdV-A12 knob and CAR-D1 depicting key interacting residues. (**E**) An equivalent interface between HAdV-D37 knob and CAR-D1 illustrating the conservation of interactions and some additional unique associations provided by HAdV-D37 glutamine 351 in the IJ loop. (**F**) The same region of the CAV-2 fiber knob and CAR-D1 interface depicting equivalent associations made by different fiber residues. Hydrogen bonds are shown in dotted lines. Residues marked with an asterisk are not completely represented for clarity. Courtesy of S. Cusack, European Molecular Biology Laboratory, Grenoble, France.

2013) have all been determined by this method. Moreover, the crystal structures of non-HAdV fiber heads corresponding to canine (Schoehn *et al.*, 2008) and a snake adenovirus (A.K. Singh *et al.*, 2014) have also been determined at high resolution. The fiber knob domain (head) homotrimer has been made as a recombinant molecule in bacteria, usually in tandem with one or more terminal shaft repeats. The overall topology of the knob domain is quite similar for all fibers examined to date. Each polypeptide chain contains two antiparallel β-sheets that form a compact sandwich-like fold (Xia *et al.*, 1994). This core structure contains a number of extended loops on the side and top of the molecule that participate in interactions with the cell receptor, as discussed below.

2.2.3 *Cocrystal structures of fiber-receptor complexes*

The identification of CAR (Bergelson *et al.*, 1997), a member of the immunoglobulin superfamily, as a receptor for many adenovirus types (Roelvink *et al.*, 1998) provided a foundation for further structural studies. These early studies established that the adenovirus fiber protein binds to CAR to promote high affinity binding to cells prior to integrin-mediated internalization. The amino-terminal immunoglobulin-like domain of the CAR (D1) is recognized by type 2 (species C) and type 12 (species A) fibers (Freimuth *et al.*, 1999). Mutagenesis studies performed on the C-terminal fiber knob domain indicated that amino acid residues located in the AB loop on the lateral surface of the knob are critical for CAR binding

(Roelvink *et al.*, 1999). The crystal structures of different HAdV-type fiber knobs produced as recombinant proteins (Burmeister *et al.*, 2004; Durmort *et al.*, 2001; Henry *et al.*, 1994; Seiradake & Cusack, 2005; van Raaij *et al.*, 1999b; Xia *et al.*, 1994) as well as the CAR D1 domain (van Raaij *et al.*, 2000) revealed the tertiary structures of these molecules. Subsequently, the structure of the HAdV-A12 fiber knob bound to the CAR D1 domain confirmed that exposed AB loop residues on the side of the fiber knob interact with CAR (Fig. 2.3). This interaction allows a trivalent association with the receptor in which each fiber knob subunit binds to one CAR D1 (Bewley *et al.*, 1999). Importantly, the ability of the fiber knob to bind multiple CAR D1 receptor domains dramatically increases the avidity of the virus for this cell receptor with a dissociation constant of ~1 nM (Lortat-Jacob *et al.*, 2001). The complexes of canine adenovirus type 2 (CAdV-2) and human HAdV-D37 fiber knobs with the CAR D1 domain have also been solved by X-ray diffraction (Seiradake *et al.*, 2006). In these cases, the overall topology was similar to that of the HAdV-A12 fiber–CAR complex. However, there are some differences in the residues that form the HAdV-D37 and CAdV-2 fiber–CAR interfaces which could explain the variations noted in receptor affinity (Fig. 2.3).

Various microbial pathogens utilize membrane cofactor protein (MCP, CD46) as a receptor for host cell infection (Cattaneo, 2004), including species B and D HAdVs (Gaggar *et al.*, 2003; Segerman *et al.*, 2003; Sirena *et al.*, 2004; Wu *et al.*, 2004). CD46 binds to most adenoviruses via its two N-terminal short consensus repeats (SCR) that can be produced independently of the entire receptor (Casasnovas *et al.*, 1999). The cocrystal structure of the HAdV-B11 fiber knob bound to the first two SCRs of CD46 revealed the basis for virus–receptor interaction (Persson *et al.*, 2007; see also Fig. 2.4). CD46 binds on the outside of the fiber knobs with the HI and DG/FG loops of the fiber providing the major contact. An important association is provided by a salt bridge between arginine 280 of the fiber and glutamic acid 63 of CD46. Binding of the receptor can be abolished by a single amino acid substitution in the fiber (R279Q). An interesting consequence of CD46 association with the fiber knob is that the normally bent conformation of the two SCR modules is altered in such a way that they become more linear upon fiber association. An N-linked oligosaccharide present in SCR2 of CD46 is

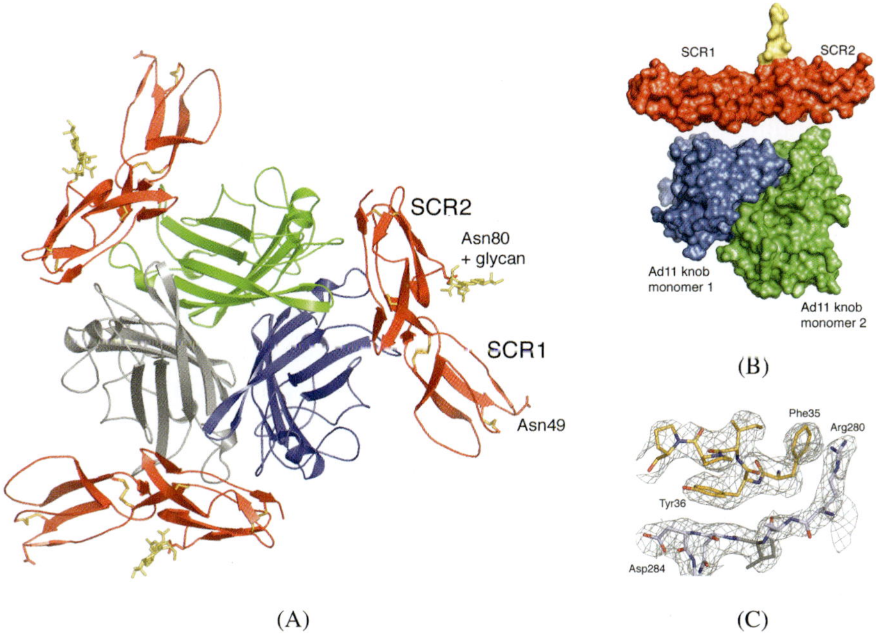

Figure 2.4. (**A**) Ribbon drawing of the complex between HAdV-B11 and CD46. The three subunits forming the fiber knob are blue, green, and gray; bound CD46 SCR1-SCR2 is red. Glycosylated residues Asn49 and Asn80 are red. The glycan attached to Asn80 is yellow. (**B**) Surface representation of a CD46 SCR1-SCR2 molecule bound to two Ad11 knob protomers, colored as in **A**. CD46 has been moved away from the two Ad11 knob protomers by about 5 Å to highlight shape complementarity. Views in **A** and **B** are from opposite directions. (**C**) $2F_\sigma - F_c$ electron density map, contoured at 1.0 σ and showing interactions between CD46 and the Ad11 knob. Reprinted from Persson *et al.* (2007), with permission.

positioned away from the fiber–receptor complex (Fig. 2.4). This mode of CD46–fiber association appears to be conserved among different HAdV-type fibers (Pache *et al.*, 2008a). However, certain species B types such as HAdV-B16 have relatively low affinity for CD46 (Pache *et al.*, 2008b). Biochemical and structure modeling studies indicated that two extra amino acids in the DG/FG loop of the HAdV-B16 fiber signifi-cantly reduce CD46 association, likely by causing steric hindrance. The cocrystal structure of CD46 bound to the HAdV-B21 fiber knob also showed that the protruding DG loop in this fiber knob prevents optimal

CD46 association (Cupelli *et al.*, 2010). Thus, differences in the extended fiber loops of distinct HAdV types influence receptor binding and thus the overall affinity for CD46.

Binding of certain species D HAdVs that are associated with severe ocular infections (D19, D8, and D37) to host cells is via sialic acid containing molecules (Arnberg *et al.*, 2000a; Arnberg *et al.*, 2000b; Arnberg *et al.*, 2002a). The cocrystal structures of HAdV-D37 and HAdV-D19 were obtained by soaking the fiber crystals with sialyl-lactose (Burmeister *et al.*, 2004). The location of sialic acid binding was revealed to be on top of the fiber knob trimer and it was observed that both $\alpha(2,3)$ and $\alpha(2,6)$ forms of the sialic acid-linked sugars can bind. The binding of sialic acid to the site on the HAdV-D37 fiber is mediated mostly by electrostatic interactions with highly basic residues on top of the fiber trimer and appears to be conserved among human species D adenoviruses. The binding constant for monomeric sialic acid was shown to be relatively low (HAdV-D37 $K_D = 5$ mM and K_D HAdV-D19 $= 7$ mM), suggesting that multivalent interactions with fibers on the intact virus are necessary to achieve increased avidity (Burmeister *et al.*, 2004). In support of this model, two sialic acid molecules were shown to bind to a single fiber knob when presented in the form of the GD1a glycan (Nilsson *et al.*, 2011; see also Fig. 2.5). Synthetic compounds displaying multivalent forms of sialic acid also show enhanced ability to restrict HAdV-D37 attachment and ocular infection (Aplander *et al.*, 2011; Spjut *et al.*, 2011).

2.2.4 *Penton base*

The crystal structure of a recombinant form of HAdV-C2 penton base lacking the first 49 amino acids was determined by Zubieta and colleagues in 2005. Each penton base monomer contains two distinct domains, a lower jellyroll-fold domain that faces the virus capsid and an outer exposed region displaying highly flexible RGD (arginine–glycine–aspartic acid) loops that facilitate cell integrin binding. Based on the crystal structure of intact virus as well as cryo-electron microscopy (cryoEM) HAdV structures (see below), both penton base domains make contact with peripentonal hexons, although there are relatively few of

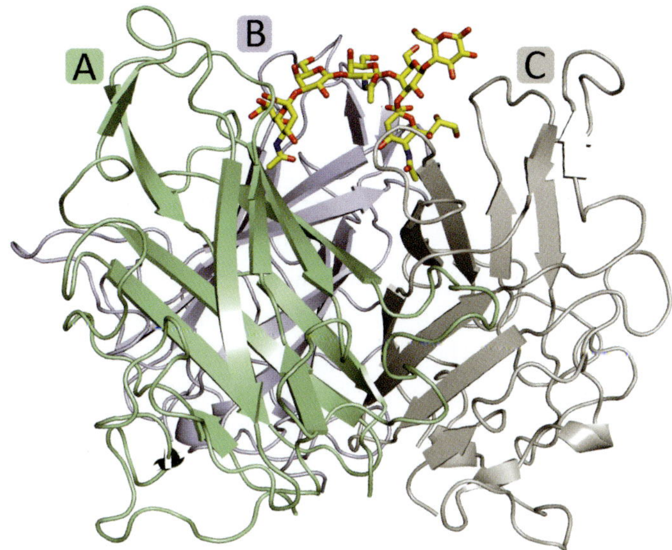

Figure 2.5. Structure of the HAdV-D37 fiber knob in complex with GD1a. The three HAdV-D37 fiber chains labeled A, B, and C are shown as ribbon tracings and colored green, blue, and gray. GD1a is bound on top of the knob and shown in stick representation, with carbons in yellow, oxygens in red and nitrogens in blue. Reprinted from Nilsson *et al.* (2011), with permission.

these associations compared to the extensive hexon–hexon interactions. The individual protein monomers form pentamers in solution as well as in the virus particle, primarily due to the sequestering of hydrophobic surfaces that bury 26% of the molecule. Several salt bridges also enhance multimerization of monomers. Interestingly, the amino terminal-truncated HAdV-C2 penton base as well as the full-length HAdV-B3 penton base (Schoehn *et al.*, 1996) are also capable of forming dodecahedrons. These highly regular multi-subunit structures have an internal cavity that is formed by interactions between loop regions of the jellyroll domain.

2.2.5 *Structure of the penton*

An initial mystery surrounding the adenovirus structure was how the five-fold symmetrical penton base associates with the three-fold symmetrical fiber protein. This enigma was addressed by analyzing cocrystals of the HAdV-C2 penton base with a 21-amino acid peptide corresponding to the fiber N-terminus (Zubieta *et al.*, 2005). The N-terminal fiber peptide was observed to lie in a narrow depression on the top of the penton base that is formed by adjacent monomers. A highly conserved amino acid sequence in the fiber tail, FNPVYPY, is the critical motif for penton base association. Three fiber tail peptides are able to dock at any of the five potential penton base interaction sites. However, as a result of steric hindrance, the fiber tails likely associate with alternating penton base depressions rather than with adjoining ones. The fiber peptide-binding site in the penton base is relatively hydrophobic with several hydrogen bonds helping to stabilize associations between the penton base and fiber. The binding of the fiber peptide to the penton base also induces a small conformational change in the penton base at residues 480–505. This change is restricted to a relatively small region and does not induce global alterations. Similar penton associations were observed in more recent cryoEM structural analyses of the HAdV-C5 penton base assembled with the HAdV-C5 (Liu *et al.*, 2011) or the HAdV-B35 fiber proteins (Cao *et al.*, 2012).

Not all penton base proteins encoded by HAdV genomes form a discrete pentameric assemblage. One interesting structural feature of adenovirus type 3 penton base is that this protein can assemble into a dodecahedron comprising 12 copies of the pentamer arranged on a $T=1$ icosahedral lattice in HAdV-B3-infected cells (Fuschiotti *et al.*, 2006; Schoehn *et al.*, 1996). It is highly likely that proteolysis of the N-terminal 37 amino acids of HAdV-B3 penton base is necessary to form the dodecahedrons, similar to that noted for recombinant HAdV-C2 penton base (Zubieta *et al.*, 2005). Some of the HAdV-B3 dodecahedrons produced during infection are equipped with fibers while others are not. Very few inter-penton contacts are involved in dodecahedron formation resulting in somewhat unstable macromolecular assemblies. These complexes also appear to contain a central cavity of 80 Å in diameter. The presence of a conserved RGD motif for integrin association on each pentamer increases the avidity for cell integrins, thereby improving dodecahedron cell surface association. This property has led researchers to explore the use of HAdV-B3 dodecahedrons to transfer various biomolecules of interest into host cells (Fender *et al.*, 1997; Fender *et al.*, 2003).

2.3 Structures of Adenovirus Capsids Analyzed by Cryo-Electron Microscopy

2.3.1 *Early structural studies of HAdV capsids at low to moderate resolution*

Structural analyses of the isolated hexon provided an important advance for investigating the structure of the entire adenovirus capsid using cryoEM. For cryoEM structural studies, purified adenovirus particles were flash frozen in vitreous ice, thereby preserving most of its native macromolecular structure. Hundreds to thousands of images of the ice-embedded HAdV particles, deposited on specialized EM grids, were collected using an electron microscope equipped with a CCD camera. The virus images were then computationally processed and used to reconstruct a three-dimensional image. The hexon crystal structure provided a pseudo-atomic model for interpreting the cryoEM density of the intact virus (Stewart *et al.*, 1991). Early cryoEM studies of adenovirus type 2 provided relatively low-resolution (25–50 Å) image virus reconstructions that nonetheless revealed the packing and associations of the 240 hexons as well as the penton base and fiber. Some hints about the locations of the cement proteins were also reported, although the precise folds and associations of these molecules could not be ascertained at this relatively low resolution. Subsequent improvements in cryoEM microscopes and CCD cameras led to higher resolution structures of adenovirus. A cryoEM reconstruction at 10 Å resolution showed further details of the cement protein densities on the outer and inner surfaces of the virus capsid (Fabry *et al.*, 2005). In this study, the investigators revealed a model in which cement proteins IIIa and IX were located on the outer surface of the capsid with protein IX appearing as a triskelelion. They also observed protein densities underneath the fivefold axis, but could not identify their origin with any certainty.

2.3.2 *Structural analyses of HAdV at moderate to high resolution*

Additional details of the adenovirus structure were revealed at 6 Å resolution by collecting and averaging thousands of images of particles of an HAdV-C5 vector displaying the short HAdV-B35 type fiber (Saban *et al.*,

2006). At this higher resolution, some secondary structure features including alpha helices were clearly visible in the hexon protein. Density could also be assigned to cement protein VIII on the inner surface of the capsid, consistent with the known copy number (120) of this molecule in the virus (Benevento *et al.*, 2014). Saban *et al.* (2006) tentatively assigned protein IIIa to a helical cluster on the inside of the capsid at the vertex region. Moreover, density corresponding to another cement protein was observed inside the hexon cavity and was tentatively assigned to protein VI.

With the more recent development of even more powerful electron microscopes, a cryoEM structure of adenovirus was obtained at 3.6 Å resolution (Liu *et al.*, 2010). This high-resolution structure revealed some of the HVR loops in the hexon protein that were missing in the crystal structure of the isolated protein, presumably because they were stabilized by inter-hexon associations. The networks of cement proteins IX, IIIa, and VIII on the inner and outer surfaces of the viral capsid could also be observed. Of particular interest, a prominent four-helix bundle was assigned to the C-termini of the protein IX molecules in the cryoEM model. Further, strong helical density underneath the vertex region was assigned to protein IIIa as in previous cryoEM studies. Currently, this assignment of IIIa on the inside of the capsid at the fivefold vertex is not in agreement with the assignments made from more recent X-ray diffraction analyses (see below).

2.3.3 *CryoEM structure of a hyperstable mutant adenovirus*

Propagation of a temperature-sensitive HAdV-C2 mutant, designated *ts1*, at the non-permissive temperature of ~39°C, results in the production of viral capsids that have six uncleaved precursor proteins: IIIa, VI, VII, VIII, TP, and mu (Rancourt *et al.* 1994). Moreover, these capsids are highly stable compared to those of normal, mature virions. This property renders *ts1* particles incapable of responding to the normal cellular cues (i.e. receptor engagement) that are needed to initiate capsid disassembly. Thus *ts1* mutants fail to penetrate the endosome following cell entry and are non-infectious (Imelli *et al.* 2009).

In order to gain insights into the structural consequences of the maturation cleavage reactions following adenovirus assembly, several investigators used cryoEM to analyze the structure of *ts1* particles at 10 Å resolution (Perez-Berna *et al.*, 2009; Silvestry *et al.*, 2009). Several differences with mature virus particles were identified. In particular, the internal core of *ts1* core showed a much greater connection to the outer capsid compared to normal virions. The density of preprotein VI inside *ts1* hexon trimers was also significantly greater than that seen with mature particles. These findings suggest that major conformational change normally occurs during maturation cleavage of one or more of the precursor cement proteins of the virus including preprotein VI.

2.3.4 *CryoEM structural analyses of non-HAdVs*

The very large collection of HAdVs and non-HAdVs has provided an opportunity to gain additional information on the diversity of viral capsid structural features. For example, canine type 2 adenovirus (CAdV-2), a virus vector that is being evaluated for gene transfer in the central nervous system (Cubizolle *et al.*, 2014), has been analyzed by cryoEM and its structure compared to that of human HAdV-C5 (Schoehn *et al.*, 2008). Overall, CAdV-2 capsids are similar to those of HAdVs with a few interesting exceptions. Many of the surface exposed antigenic loops on the hexon and penton base of CAdV-2 are shorter than those of HAdV, thereby giving CAdV-2 particles a smoother appearance. Furthermore, the CAdV-2 penton base does not have an integrin-binding RGD motif that is present in most human virus pentons. The fiber protein of CAdV-2 is highly flexible, with two predicted bends at repeats 4 and 10 in the shaft domain. Another difference in the CAdV-2 structure relates to cement protein IX. A rod-like density that is missing in HAdV-C2 is located above the triskelion of IX and may comprise the C-termini of IX. A similar organization of IX polypeptides was reported for bovine type 3 adenovirus (Cheng *et al.*, 2014).

2.4 X-Ray Diffraction Analyses of HAdVs

Numerous challenges have impeded the structural analyses of adenovirus by X-ray diffraction. These include the very large (910 Å diameter; 150 MDa) size of the virus particle and its tendency to undergo disassembly at non-neutral pH, the elongated fiber proteins that hinder uniform virus crystal packing, and the general insolubility of the virus at protein concentrations above 1 mg/ml. Most of these problems were recently overcome by using robotic crystallization screens and employing a powerful and highly focused synchrotron beam line (GM/CA CAT 23-ID-D, APS Chicago) to collect X-ray diffraction data of freshly frozen virus crystals (Reddy *et al.*, 2010b). In particular, conditions were discovered that allowed crystallization of a HAdV-C5-based vector displaying the short flexible fiber protein from HAdV-B35 (designated Ad35F) at close to neutral pH. Soaking the crystals in a cryo-protectant, Paratone N, allowed the freezing of Ad35F crystals so that multiple X-ray exposures could be obtained from each crystal. The extent of X-ray diffraction reached 3.5 Å resolution, the highest obtained for a macromolecular assembly of this size.

The crystal structures of the hexon and penton base proteins as well as the initial cryoEM structures described above provided valuable phasing models of the virus, thereby allowing determination of the Ad35F crystal structure at near-atomic resolution (3.5 Å) resolution (Reddy *et al.*, 2010b). This crystal structure revealed some of the HVR loop regions in the hexon that were invisible in the isolated hexon structure, likely because they engage in inter-hexon associations in the virus. Hexon associations with protein VIII, a cement protein on the inside of the virus capsid, were also visible in this structure. Somewhat disappointingly, the side-chain densities for residues comprising the other cement proteins including IX, IIIa, and VI, could not be accurately assigned, preventing their identification in the capsid and determination of their associations with hexons. An unusual association of the penton base and fiber was also revealed in these studies. In particular, the fiber was seen inserted into the central cavity of the penton base, a feature that had never been noted in previous cryoEM reconstructions of the same virus (Saban *et al.*, 2006). It is possible the crystallization conditions or poor resolution of this highly flexible region contributed to the unusual orientation of the penton complex.

Subsequently, a refined X-ray model of Ad35F was obtained by collecting additional diffraction data and by taking advantage of improvements in data processing procedures (Reddy & Nemerow, 2014a). This improved virus structure revealed significant new information on the locations, folding, and associations of the two cement proteins (VI and VIII) and the core bridging protein V on the inside and of two more cement proteins (IIIa and IX) on the outside of the capsid shell (Fig. 2.1). These cement proteins are important for their role in stabilizing the large adenovirus capsid and at least one of them (VI) has crucial roles in the viral infectious cycle. One of the most prominent features seen on the outer capsid surface in the X-ray structure is a cluster (bundle) of helices in a cement protein that mediate interactions among the hexons that comprise the group-of-12 hexons on each facet. This molecule was assigned to large segments of protein IIIa (amino acids 103–200 and 262–342). There are 60 IIIa molecules present within the intact virus capsid and six of these IIIa helices trace the border on the outer surface of each virus facet. The IIIa helices were previously assigned to four separate C-terminal extensions of protein IX based on cryoEM structural studies (H. Liu *et al.*, 2010; Saban *et al.*, 2006). However, these protein IX polypeptide assignments are not consistent with the rules of quasi-equivalence, because several of the proposed IX polypeptides occupied nearly the exact same location in the capsid and yet had distinct orientations. In the refined X-ray model, the helices appear to be connected to each other via loop extensions and therefore could not be the separate polypeptides corresponding to the C-termini of IX. The newly assigned IIIa structure also contained an N-terminal segment composed of amino acids 9–25 and 48–102 that extends between the peripentonal hexons and reaches all the way to the penton base, but does not appear to interact with the pentamer. A large segment of IIIa corresponding to the C-terminal region remains invisible in the structure, perhaps due to its lack of connections with other capsid proteins. Since the IIIa molecule spans nearly 170 Å across the outer capsid from the penton base to the twofold axis, it may be analogous to the tape-measure protein (P30) that is present in the bacteriophage, PRD1 (Abrescia *et al.*, 2004).

The other cement protein on the outer surface of the capsid is protein IX with 240 of these molecules per capsid. Protein IX forms a characteristic triskelion, containing a core of alanine and serine residues, that helps stabilize the hexons at the center of the icosahedral facet. Four structurally independent IX molecules with different length polypeptides (designated

P, Q, R, and S) were observed in the X-ray structure of the virus with the P molecule making contact with a peripentonal hexon. The X-ray model is also distinct from previous cryoEM models in that the direction of the IX polypeptide is reversed. This orientation of IX molecules allows the C-termini to be surface exposed similar to the orientation reported for the C-termini bundles of IX in non-HAdVs (Cheng *et al.*, 2014). However, these C-termini IX bundles that are present on non-HAdVs are not visible in HAdV types. The basis for this difference has yet to be determined.

Three separate proteins designated VI, VIII, and V form a multimeric complex that helps to stabilize the hexons on the inner capsid surface underneath the vertex region (Fig 2.6). Pre-protein VI is incorporated into the capsid via association of its 33 N-terminal residues with the hexon trimer (Snijder *et al.*, 2014). A large segment of preprotein VI (residues 1–21) was seen buried in the cavity of the peripentonal hexon (Reddy *et al.*, 2014b). This arrangement has been confirmed by mass spectrometry analyses in combination with hydrogen–deuterium exchange studies (Snijder *et al.*, 2014). Although three N-terminal VI peptides can bind to each trimer in solution, this mode of association is not possible in the virus particle due to steric hindrance imposed by the cement proteins V and VIII. Moreover, at neutral pH only one copy of VI peptide was observed to bind to each hexon trimer in solution. Another new feature that was visible in the virus crystal structure was the site of cleavage of protein VI between glycine 33 and alanine 34 that is recognized by the viral cysteine protease (AVP). This cleavage may allow more efficient release of the membrane lytic domain of protein VI (residues 34–54). Residues 34–79 and 87–157 of the VI molecule are visible in the virus particle, and contain a core α-helix, although the secondary structure is not as extensive as in the isolated recombinant protein (Moyer *et al.*, 2011). This difference suggests that VI undergoes conformational changes upon release from the virus particle during membrane association and endosome disruption. Overall, protein VI does not share any structural homology with known membrane lytic proteins and thus has a unique fold. Like the other cement proteins in the virus, proteins V and VIII, the majority of each protein VI molecule contributed to stabilizing the hexons underneath the virus capsid. One of the remaining puzzles that has not been solved is the location of all the copies of protein VI. Thus far, 180

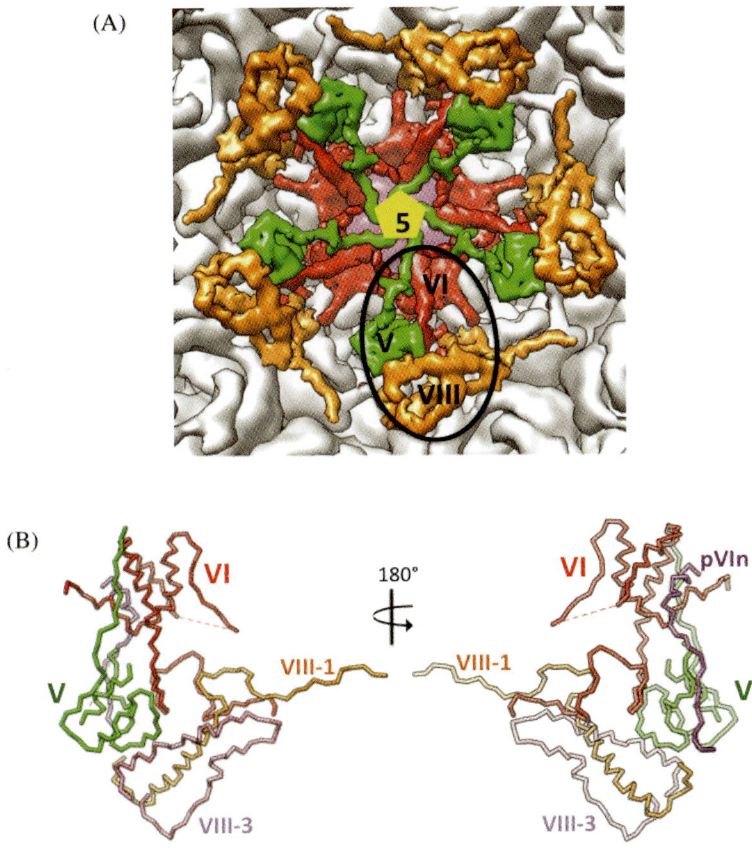

Figure 2.6. Organization of cement proteins underneath the vertex region. (**A**) Five copies of the ternary complex formed by proteins V (green), VI (red), and VIII (orange) that glue the peripentonal hexons (gray) to each other and connect them to the adjacent GONs. One of the copies is highlighted by a black oval. (**B**) A stick diagram showing the structure of the ternary complex. Color assignments of the proteins are as in **A**, with the exception that the propeptide of VI (pVIn) is shown in purple and the C-terminal fragment of VIII (VIII-3) is shown in orchid, whereas the N-terminal fragment of VIII (VIII-1) is shown in orange. Reprinted from Reddy & Nemerow (2014b), with permission.

molecules of VI can be accounted for in the crystal structure, but ~360 copies have been detected by mass spectrometry (Benevento *et al.*, 2014). One possibility is that the missing VI proteins reside inside the DNA core that is not highly ordered and hence has not been structurally characterized. In support of this possibility, it is known that protein VI binds to DNA during capsid maturation (Mangel & San Martin, 2014).

The two other cement proteins that form a ternary complex with protein VI are VIII and V. There are 120 copies of protein VIII per virion, with two distinct locations of this polypeptide on the inner capsid surface. Each of these separately located protein VIII molecules outlines the group of nine hexons and helps to stabilize hexons on the inside of the capsid. The protein VIII molecule is cleaved in two separate places producing three cleavage fragments. One of the three fragments (middle) is likely released during capsid assembly, as it is not visible in either of the two locations in the virus structure. Overall, the structure of VIII in the X-ray model agrees with that of VIII revealed in the cryoEM model with the exception of the third cleavage fragment whose fold is quite different. The remaining protein in the ternary complex, V, is generally considered a core protein of the virus as it has strong DNA-binding activity and participates in condensing the viral genome in the capsid (Chatterjee *et al.*, 1986b; Perez-Vargas *et al.*, 2014). There are 157 copies of protein V per virion and biochemical analyses performed in solution demonstrated that this molecule also binds to protein VI thereby acting as a bridge from the DNA inner core to the inner capsid shell (Perez-Vargas *et al.*, 2014). As seen in the crystal structure of Ad35F, the C-terminal region of V (residues 289–295) associates with protein VI underneath the vertex region. It is likely that the N-terminal region, of V, which is more basic and less ordered in the crystal structure, interacts with the viral nucleic acid inside the core. As part of the ternary complex, protein V also interacts with the third cleavage fragment of protein VIII and thereby assists in stabilizing the peripentonal hexons underneath the vertex.

2.5 Structural Analyses of Adenovirus Particles Bound to Immune or Coagulation Proteins

Over the past two decades there have been a number of structural studies of adenovirus bound to human immune or coagulation components. An early cryoEM structural study revealed that an anti-penton base antibody, designated DAV-1, recognizes the 22 Å highly flexible loops that comprise the integrin-binding RGD motif on the outer surface of the HAdV-C2 penton base (Stewart *et al.*, 1997). Because these integrin-binding sites are spaced only ~60 Å apart from each other, intact antibody molecules cannot fully saturate all of the five RGD sites on the penton base. In contrast, the 50 kDa Fab antibody fragments of DAV-1 can do so, thereby restricting virus internalization and infection. Consequently, the precise geometric display of the integrin binding RGD loops on the virus penton base serves to enhance multimeric integrin association while precluding antibody-mediated neutralization.

A completely different mode of antibody neutralization by an anti-hexon monoclonal antibody, designated 9C12 (Varghese *et al.*, 2004), was characterized by using structural and biochemical analyses. CryoEM structural studies of the 9C12 antibody bound to HAdV-C5 showed that the antibody enmeshed the virus in a well-ordered array of molecules that cross-linked with the outer hexon trimers. Nevertheless, this extensive network of antibodies over the surface of the particle did not prevent virus attachment or internalization, but instead interfered with a later step in the cell entry pathway. The 9C12 antibody was shown to block microtubule-dependent transfer of the virus to the microtubule-organizing center following endosome penetration (Smith *et al.*, 2008). However, the precise mechanism has not yet been elucidated. One possibility is that the Fc portion of the 9C12 antibody is recognized by a host protein known as TRIM21 (McEwan & James, 2015). This intracellular protein directs the ubiquitin- and proteasome-dependent breakdown of virus-antibody complexes and thereby neutralizes adenovirus infection (McEwan *et al.*, 2012; Chapter 9).

Adenoviruses are also susceptible to inactivation by other innate immune proteins such as alpha defensins (Smith *et al.*, 2008). These small cationic proteins are secreted from Paneth cells of the small intestine, neutrophils, or epithelial cells lining the mucosal surfaces. Human alpha defensins HD5 and HNP1 bind to most adenovirus types and prevent capsid disassembly in the early endosome. Thus, defensins restrict infection at an early step in the virus infection cycle (Nguyen *et al.* 2010). At relatively high concentrations of defensin HD5 (20 µM) that fully saturate and neutralize HAdV-C5, the outer capsid is well decorated with these innate immune molecules (Smith *et al.*, 2010). In particular, the strongest area of defensin binding was seen at a region spanning the base of the fiber protein and the RGD loops of the penton base. Further molecular genetic studies with HAdV chimeras that differ in their sensitivity to defensin indicated that this fiber-penton spanning region was crucial for virus neutralization. More recent molecular dynamics modeling, guided by cryoEM structural analyses, indicated that alpha defensin association stabilizes the RGD loops that are inherently disordered (Flatt *et al.* 2013). The stabilization of the vertex region of the capsid interferes with the disassembly process that is crucial for protein VI release during cell entry.

Many adenovirus types are also capable of binding soluble blood proteins such as coagulation factor (F) X (Doronin *et al.*, 2012; Waddington *et al.*, 2008). FX associates with the HAdV hexon via the γ-carboxyglutamic acid (Gla) domain of this clotting factor. CryoEM studies suggest that the FX–Gla domain binds on top of the hexon trimer in the central depression. However, these studies need to be confirmed by analyses at higher resolution. An important consequence of FX binding to the virus hexon is that this association directs the virus to the liver upon systemic administration, thereby bypassing other receptors such as CAR and integrins. FX binding to the virus may also shield the capsid from potentially neutralizing antibody and complement (Xu *et al.*, 2013). These findings have important implications for HAdV vector mediated gene transfer in animal models and perhaps, ultimately, for human gene delivery.

2.6 Structures of Adenovirus–Receptor Complexes

The fact that infection by many enveloped and non-enveloped viruses including adenoviruses, herpesviruses, reovirus, Echovirus 1, Coxsackie A9, and hantavirus depend on integrins suggests that integrins play more than just a fortuitous role in the cell entry process (Stewart & Nemerow, 2007). Consequently, there have been extensive efforts to reveal the structural basis of integrin–pathogen association. Structural models of adenoviruses associated with different cell receptors have emerged over the past several decades. Due to the complexity of these assemblies, most of these structures have been analyzed by cryoEM and single-particle image reconstructions. Particles of adenovirus type 2 and type 12 bound to a soluble recombinant integrin $\alpha_v\beta_5$ were analyzed by cryoEM at ~20 Å resolution (Chiu *et al.*, 1999). This moderate level of resolution combined with biochemical analyses established general features of the integrin interactions with the virus penton base. Up to four $\alpha_v\beta_5$ integrin molecules were capable of binding to each penton base with an affinity of $K_D = 70$ nM based on biosensor studies. Due to the large size of integrin heterodimer (~1,700 amino acid residues) as well as the protruding fiber protein, only four out of the five potential integrin-binding RGD loops can be occupied on top of each penton base. Compared to the larger RGD loop of HAdV-C2 (78 residues), the shorter HAdV-A12 RGD loop (17 residues) provided a more stable platform for integrin binding. This property enabled a more accurate assessment of the bound integrin receptor. The $\alpha_v\beta_5$ integrin consists of two separate domains, a globular region with an RGD-binding cleft of ~20 Å in diameter and a more distal domain with extended, flexible tails. These early studies provided valuable clues as to how adenovirus promotes clustering of α_v integrins on the cell surface to elicit a signaling response in host cells during virus internalization.

With further advances in cryoEM microscopes and CCD cameras, it has been possible to extend the resolution quite significantly (8–10 Å). Moreover, the available crystal structures of homologous integrins, $\alpha_v\beta_3$ and $\alpha_{II}\beta_3$ (Xiao et $al.$, 2004), were used to refine the model of the $\alpha_v\beta_5$ integrin–HAdV-A12 complex (Lindert et $al.$, 2009). These modeling studies combined with improved cryoEM structural analyses helped to alleviate the problems arising from the fact that the four $\alpha_v\beta_5$ integrin molecules do not bind to the penton base RGD loops in the exact same orientation— such heterogeneous integrin binding tends to "smear" the cryoEM density of the receptor. Heterogeneity of integrin binding was also reported in a recent cryoEM study of the HAdV-D9 penton bound to integrin $\alpha_v\beta_3$, in which four different receptor binding modes were observed (Veesler et $al.$, 2014). Interestingly, $\alpha_v\beta_5$ integrin binding to HAdV-A12 causes a subtle untwisting of the penton base (Lindert et $al.$, 2009). This change includes a displacement of the RGD loop by the bound integrin counter to the natural orientation of this loop. The combined movement of four RGD loops by multiple bound integrins could effectively cause a global conformational change in the penton pentamer such that contacts with adjacent peripentonal hexons could become loosened. This RGD loop conformational change in the penton base may represent one of the earliest steps in the disassembly process that leads to the loss (disassembly) of the vertex region of the capsid (Snijder *et al.*, 2013).

2.7 Physical Methods: Elastic Properties of Adenovirus Revealed by Atomic Force Microscopy

Recent developments in the realm of physical virology have provided additional methods for analyzing the structures of virus capsids. Among these is atomic force microscopy (AFM), which has gained increasing utility for assessing the tensile strength of these structures (Roos, 2011; Roos *et al.*, 2010). Briefly, the elastic properties of the viral capsid can be measured by exerting increased pressure on the virus capsid with an AFM cantilever probe of ~20 nm in diameter. This applied force process is referred to as nanoindentation and often results in the collapse or breaking of the viral capsid thereby allowing the determination of the breaking force. In initial AFM studies (Snijder *et al.*, 2013), adenovirus particles were immobilized on glass surfaces and then analyzed in the scanning mode by probing the surface of the virus without exerting force. This approach allowed identification of the two-, three-, and fivefold orientation of the particles on the glass surface. The AFM probe was then used in the nanoindentation mode to determine the breaking forces (and elastic properties) at each of these symmetry axes. The fivefold axis of the HAdV capsid was the weakest (most elastic) part of the virus whereas the twofold and threefold axes were stronger and less elastic. In further analyses, the effects of HD5 defensin or $\alpha_v\beta_5$ integrin binding on the elastic properties of the virus were evaluated. Binding of αHD5 to the virus selectively strengthened the fivefold axis, consistent with previous functional studies that showed that this defensin binds to the penton and prevents capsid disassembly. In contrast, binding of integrin $\alpha_v\beta_5$ to the capsid penton base loosened the fivefold axis and made the virus more susceptible to breaking, which was consistent with the observation that integrin binding to the virus causes a conformational change in the penton base to promote capsid disassembly.

In other AFM studies, the elastic properties of mature adenovirus particles were compared to those of the *ts1* mutant adenovirus that fails to undergo proper maturation cleavage (Perez-Berna *et al.*, 2012). Somewhat surprisingly, *ts1* capsids were actually more elastic (softer) than the mature viruses. The authors of these studies suggested that differences in the DNA core structure of wild type and *ts1* mutant capsids could explain the differences in the overall elasticity of the viruses. These findings suggest that maturation cleavage of adenovirus helps to stabilize the particle, consistent with the metastable state needed for efficient infection of host cells.

2.8 What Next?

While there have been significant advances in obtaining the structures of the outer capsid of HAdVs and non-HAdVs, we lack definitive information on the locations, folds, and interactions of the capsid cement proteins. In particular, only partial structures are available for most of these molecules. Certain regions of the cement proteins may not be highly organized because they are mobile. Segments of cement proteins that are not directly associated with the hexon or with other cement proteins could be particularly flexible. There also remains a lack of agreement on the assignments of some of the cement proteins such as IIIa (Reddy *et al.*, 2014a). Further structural analysis of recombinant adenoviruses that lack protein IIIa could help confirm the precise location of IIIa in native virions. Structural data derived from particles that lack IIIa may not be easy to interpret, however, as the absence of IIIa might alter the assembly of other cement proteins or the folding of these molecules.

In addition to unresolved structural questions related to the outer adenovirus capsid, the organization of the inner dsDNA core and associated proteins remains an even bigger mystery. Currently, we know that the adenovirus genome is not highly ordered like that of certain double stranded DNA-containing bacteriophages whose nucleic acid is spooled in a coaxial orientation (Jiang *et al.*, 2006; Lander *et al.*, 2006). As the HAdV core is not icosahedrally ordered, analyses of low resolution X-ray diffraction data (100–200 Å), similar to that used to define the structure of other viral nucleic acids (Venkataraman *et al.*, 2008), may reveal how the dsDNA is organized in adenovirus. Similarly, very little is known about the structures of the core proteins (V, VII, and mu) that are associated with the viral dsDNA genome. It might be feasible to determine the crystal structures of isolated core proteins produced as recombinant molecules (Perez-Vargas *et al.*, 2014). The structures of these core proteins could then be used in combination with X-ray data from the intact virus to obtain a more complete model of the HAdV inner core.

Another enigma is whether the adenovirus capsid contains a unique vertex or portal that is used for import of the dsDNA genome during virus assembly. Portals are present in an icosahedral dsDNA bacteriophage known as PRD1 (Stromsten *et al.*, 2003) and in human herpesvirus type 1 (Newcomb *et al.*, 2001). These dedicated subassemblies generally contain an ATPase that provides the biomechanical force necessary to drive dsDNA encapsidation. Interestingly, the viral genome does encode a sequence-specific DNA binding protein known as IVa2 that is essential for packing of the viral nucleic acid (Newcomb *et al.*, 2001). As most viral portal complexes are dodecamers and there are only ~five copies of IVa2 per virion (Benevento *et al.*, 2014), it is unlikely that IVa2 is the portal itself, but it could be one of the components. Protein IVa2 also contains the Walker A and B motifs characteristic of ATP binding proteins and has also been reported to have low affinity nucleotide binding activity (Ostapchuk & Hearing, 2008). The presence of IVa2 at one vertex of the virus has also been suggested by immunogold labeling experiments (Christensen *et al.*, 2008). CryoEM tomography analyses of single virus particles might uncover evidence for an adenovirus portal. Alternatively, adenovirus might not incorporate its nucleic acid through a unique vertex but could instead assemble the outer capsid around the pre-formed inner dsDNA core (Ostapchuk & Hearing, 2005).

A major incentive for acquiring further knowledge of adenovirus structure is that it could improve our understanding of virus assembly and disassembly. One potential avenue to increase this knowledge is to perform structural studies of immature adenovirus capsids that are presumably in the early stages of assembly. Immature particles are produced in excess during the normal course of cell infection and can be separated from mature particles on the basis of their lower densities. Such particles are mostly devoid of the viral nucleic acid but contain a putative scaffold assembly L1 protein known as 52/55K (Hasson *et al.*, 1992). Detailed structural analyses of such immature particles by cryoEM or X-ray diffraction analyses could provide valuable information on the process of virus assembly. In a similar vein, it might be possible to acquire further structural information on the process of disassembly by subjecting mature virus particles to heating or to receptor association in combination with low pH to simulate the process of disassembly in cells by release of the vertex region. The difficulty of these disassembly structural studies lies is preparing sufficiently homogenous samples for X-ray diffraction or cryoEM analyses. Nonetheless, the continuing developments in integrated structural, biochemical, and functional analyses may provide the means to address these remaining challenging issues that are central to an understanding of the adenovirus infectious cycle.

Cell Attachment and Entry 3

3.1 Receptors for Virus Attachment and Their Ligands

During the earliest phases of the adenovirus infection cycle, successful cell invasion requires that virus particles bind to and enter host cells. These cell invasion skills have contributed in part to the use of replication-defective and replication competent adenoviral vectors for gene or vaccine delivery. As is the case for many other human and nonhuman viruses, adenoviruses recognize specific cell surface proteins or carbohydrate molecules that provide ample binding affinity for virus attachment or internalization (Fig. 3.1). Two major outer coat proteins — the fiber and penton base — mediate human adenovirus (HAdV) attachment or internalization, respectively. Surprisingly, soluble blood proteins can also recognize adenovirus and circumvent normal receptor engagement and redirect virus particles to specific host tissues such as liver hepatocytes. This chapter will describe the different receptors and cell entry pathways used by adenovirus as well as the consequences for immune responses.

3.2 Cell Receptors for Adenovirus Attachment

3.2.1 *CAR and heparin sulfate proteoglycans*

In keeping with many other studies of HAdVs, the majority of virus-receptor binding analyses have been performed with species C types 2

Adenovirus Cell Receptors

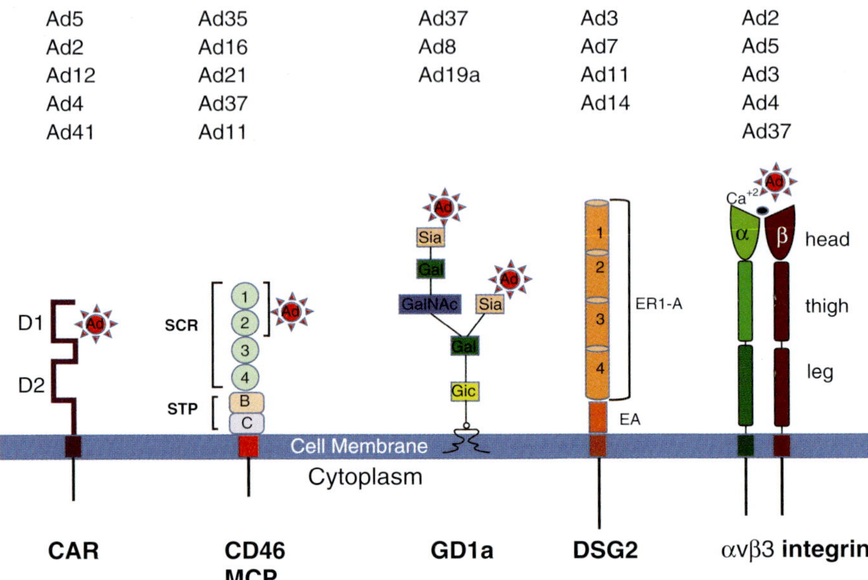

Figure 3.1. Cartoon depicting the cell receptors used by different adenovirus types. HAdV types that bind to each receptor are indicated above, using the short nomenclature. The virus-binding domains in each receptor are indicated by a virus symbol, except for desmoglein 2 (DSG2) whose virus-binding region has yet to be determined. Each protein receptor is anchored in the membrane by a single membrane-spanning domain whereas GD1a is membrane anchored via a phospholipid moiety. The extracellular domain of the Coxsackievirus and adenovirus receptor (CAR) contains two Ig-like domains (D1 and D2); CD46 has four short consensus repeats (SCRs); GD1a has multiple terminal sialic acid (Sia) residues; $\alpha_v\beta_3$ integrins are composed of an α and a β subunit with the N-terminal regions of both subunits contributing an RGD (Arg–Gly–Asp) binding cleft and a site for divalent metal cation association.

and 5. In the earliest studies of adenovirus host cell tropism, the elongated fiber proteins of species C adenoviruses were shown to mediate high affinity virus attachment to cells (Philipson, 1967; Philipson *et al.*, 1968). However, the molecular basis of this virus-host cell interaction remained an enigma for nearly three decades until a 46-kDa protein, known as CAR (Coxsackie and adenovirus receptor), was shown to be a high-affinity receptor for species C types 2 and 5 as well as other species A, D, and E adenoviruses (Bergelson *et al.*, 1997; Roelvink *et al.*, 1998; Tomko *et al.*, 1997). CAR on human cells, like several other virus receptors (e.g. CD155, poliovirus), is a member of the immunoglobulin (Ig) super family (Dermody *et al.*, 2009), containing two Ig-like domains in its extracellular region, followed by a single transmembrane anchor and a cytoplasmic tail of 107 amino acids. The cytoplasmic tail of CAR does not appear to be required for infection of cells in tissue culture as it can be replaced by a GPI-anchored form of the receptor that supports identical virus attachment and cell invasion (Bergelson *et al.*, 1998).

The role of CAR in adenovirus-mediated gene delivery is now well established. For example, a recombinant HAdV-C5 vector failed to transduce human airway epithelial cells during the treatment of cystic fibrosis patients (Stonebraker *et al.*, 2004). A major factor contributing to the failure of HAdV-mediated gene transfer in the human airway is that CAR is expressed on the basolateral surface of polarized epithelial cells in tight junctions (Cohen *et al.*, 2001a; Cohen *et al.*, 2001b; Walters *et al.*, 1999; Walters *et al.*, 2002) and is therefore inaccessible to virus receptor-mediated cell transduction. It is thought that homotypic CAR interactions help maintain the tight junctions that are crucial for normal tissue homeostasis (Lisewski *et al.*, 2008). These findings raised the question: how does adenovirus reach airway epithelial cells during a normal infection? One way this could happen is by the formation of transient breaks in the tight junctions allowing virus particles access to the basal lateral surface and CAR. It is possible that overproduction of the fiber protein in a limited number of cells could then outcompete for the normal homotypic CAR associations potentiate further enhancing basolateral surface infections (Walters *et al.*, 2002). Studies have also shown that one particular isoform of CAR, designated CAR-Ex8, co-localizes to the apical surface of polarized epithelial cells rather than the basolateral surface. This unusual receptor isoform may thus help the virus to bind to and infect these cells at the apical surface (Excoffon *et al.*, 2010). More recently, studies have indicated that expression of CAR-Ex8 on the apical epithelial surface can be enhanced by the host innate immune system and interleukin 8 (IL-8) in particular (Kotha *et al.*, 2015). This finding is consistent with an earlier report that the HAdV-C5 fiber is capable of activating inflammatory responses in airway epithelial cells (Tamanini *et al.*, 2006). These studies suggest that adenovirus infection induces a potent host immune response that could favor upregulation of receptor synthesis, thereby enhancing further virus spread. Although the cytoplasmic domain of CAR is not needed for infection, it may participate in a proinflammatory response involving activation of the MAP kinase, ERK (Farmer *et al.*, 2009).

The trimeric adenovirus fiber protein mediates CAR binding with relatively high affinity (K_D = 1 nM) (Lortat-Jacob *et al.*, 2001) due to its multivalent association with the receptor (Roelvink *et al.*, 1999). Extensive high-resolution structural analyses of different HAdV fiber knobs bound to CAR have been reported and are discussed in more detail in Chapter 2. Interestingly, it is not just the fiber knob domain that determines virus attachment to CAR on host cells. Certain structural features of the entire fiber, including its length and flexibility, also contribute to virus attachment (Shayakhmetov *et al.*, 2000; Smith *et al.*, 2003; Wu *et al.*, 2003). In particular, the fiber shaft domain has a unique triple β-spiral conformation (van Raaij *et al.* 1999) and comprises a variable number of β-repeating elements (5–22), depending on the virus type (Wu *et al.*, 2004). The longest adenovirus fiber (from HAdV-A12) projects the fiber knob domain ~340 Å away from the surface of the virus capsid, while the shortest, that of HAdV-B3, extends it only ~130 Å away. In addition to the different fiber lengths, the third β-repeat in most but not all virus fibers has a sequence that is distinct from the other repeats and is thought to confer flexibility to the fiber shaft domain (Chroboczek *et al.*, 1995). HAdV-D37 has a short (eight repeats), inflexible fiber shaft as a result of the absence of the conserved third shaft repeat. Consequently, HAdV-D37 lacks the ability to bind to CAR on host cells, even though its fiber exhibits this property in solid phase binding assays (Wu *et al.*, 2003). When the HAdV-D37 fiber knob domain was fused to the elongated and flexible HAdV-C5 fiber shaft on a recombinant HAdV-C5 virus, the virus chimera acquired the ability to bind to CAR on host cells. Thus, both the length and flexibility of the fiber shaft domain govern binding to CAR on host cells.

In some instances, CAR can serve as both an attachment and internalization receptor, as exemplified by canine adenovirus type 2 (CAV-2) entry into neuronal cells (Kremer & Nemerow, 2015). Such dual use of CAR allows CAV-2 to be taken up efficiently at the axon termini of neuronal cells and then be transported toward the soma (cell body) in a retrograde transport pathway (Salinas *et al.*, 2009; Soudais *et al.*, 2001; Soudais *et al.*, 2004). CAR-mediated CAV-2 internalization in neurons was shown to require lipid microdomain integrity and was actin and dynamin-dependent (Salinas *et al.*, 2014). These studies also showed that disruption of CAR homodimers with the CAV-2 fiber knob ligand triggered receptor internalization, a process that requires the CAR cytoplasmic tail. More recent studies indicate that the CAR intracellular domain plays a more important role in the internalization and infection of CAV-2 in multiple cell types (fibroblast and neurons) compared to the negligible role for this segment of CAR and human HAdV-C5 (Loustalot *et al.*, 2015). These unexpected findings illustrate the diversity of adenovirus receptor functions and may help lay the foundation for the use of CAR-binding adenovirus vectors to treat central nervous diseases (Cubizolle *et al.*, 2014).

The C-terminal fiber knob domain is not the only region of the HAdV-C5 fiber that possesses receptor-binding activity. Initial studies indicated that the third β-repeat element in the fiber shaft domain contains a KKTK motif that may promote association with heparin sulfate glycosaminoglycans based on the sensitivity of cell attachment to heparin (Dechecchi *et al.*, 2000; Dechecchi *et al.*, 2001; Smith *et al.*, 2003). Recombinant HAdV-C5 vectors were found to associate with bone marrow derived dendritic cells in a heparin-sulfate glycosaminoglycan dependent manner and this association led to potent immune responses *in vivo* (Cheng *et al.*, 2007). Thus, the HAdV-C5 fiber shaft region may well provide alternative modes of binding to cell surface molecules thereby expanding the tropism of this virus.

3.2.2 *Membrane cofactor protein/CD46*

Not all HAdV species use CAR as an attachment receptor. In particular, certain species B types including B16, B21, and B35 bind to CD46, also known as membrane cofactor protein (MCP), as a receptor for cell attachment (Gaggar *et al.*, 2003). CD46 is a widely produced type 1 membrane glycoprotein that is a member of the family of complement regulatory proteins (Hourcade *et al.*, 1989; Liszewski *et al.*, 1991). The ectodomain of CD46 contains four short consensus repeats (SCR) that mediate complement and microbial pathogen ligand binding interactions. The two amino terminal repeats, SCR1 and SCR2, comprise the HAdV fiber-binding site (Fleischli *et al.*, 2005; Gaggar *et al.*, 2005). Detailed structural information on the interactions of the species B fiber knobs with CD46 is discussed in Chapter 2. One of the striking features of CD46 is that this receptor undergoes a significant conformational change upon fiber binding. This entails a shape-complementarity association that creates a large continuous binding surface (Persson *et al.*, 2007).

During microbial interactions with CD46 on host cells, ligation and oligomerization of the receptor triggers macropinocytosis and downregulation of the molecule on the cell surface and consequently increases immune suppression and complement-dependent lysis (Crimeen-Irwin *et al.*, 2003). Type B adenoviruses also induce CD46-mediated macropinocytosis in non-polarized epithelial cells (Kalin *et al.*, 2010), a process that requires the participation of multiple signaling molecules including the p21-activated kinase (PAK) 1 (Amstutz *et al.*, 2008; Liberali *et al.*, 2008). Whether this cell signaling pathway is operative in polarized airway epithelial cells has yet to be discovered however.

The recognition of CD46 as a receptor by certain species B adenoviruses has important implications for gene transfer applications. Pre-existing antibodies to species C adenoviruses (HAdV-C5/2) are widely circulating and thus restrict efficient HAdV-C5 gene transfer applications. In contrast, species B adenovirus types 35 and 26 that bind CD46 (Li *et al.*, 2012) have low seroprevalance in the human population, and consequently vectors based on these HAdV types have shown promise in preclinical trials as vaccine vectors to protect against various pathogens, such as Ebolavirus (Geisbert *et al.* 2011; Chapter 11). A recombinant HAdV-B35 fiber knob with high affinity for CD46 has also been shown to sensitize CD46-expressing lymphoma cells to complement dependent cytotoxicity (Wang *et al.*, 2010). These findings suggest the potential clinical value of a HAdV-B35 component, in combination with tumor specific monoclonal antibodies, in cancer immunotherapy.

CD46 may also serve as the attachment receptor for the species D adenovirus, HAdV-D37 (Wu *et al.*, 2001; Wu *et al.*, 2004). This virus is associated with severe ocular infections known as epidemic keratoconjunctivitis (EKC). More recent studies have indicated that species

D viruses such as HAdV-D37 also bind to sialic acid conjugates on host cells (see below) and therefore the degree to which CD46 and sialic acid promote virus attachment and infection remains to be determined. It is also possible that sialic acid residues on CD46 are recognized by the HAdV-D37 fiber.

3.2.3 *Desmoglein 2*

Although earlier reports suggested that certain species B adenoviruses that cause respiratory and urinary tract infections such as types B3, B7, B11, and B14 attach to CD46 (Segerman *et al.*, 2003; Sirena *et al.*, 2004), it seems that these viruses bind to this receptor with much lower affinity than other virus types such as HAdV-B16 and HAdV-B35 (Marttila *et al.*, 2005; Trinh *et al.* 2012; Tuve *et al.* 2006). These observations raised the possibility that certain B type adenoviruses use a different receptor for cell attachment. An early clue to an alternative receptor suggested that HAdV-B3 binds to a ~130 kDa protein that was tentatively referred to as X (Di Guilmi *et al.*, 1995). Subsequently, HAdV-B3-affinity capture and tandem mass spectrometry of cell membrane extracts identified desmoglein 2 (DSG2) as receptor X. Further studies showed that HAdV types 11, 3, 7, and 14 bind to DSG2 with relatively high affinity (Wang *et al.*, 2011b). DSG2 is a calcium-binding member of the cadherin family of proteins and in epithelial cells it mediates cell-cell adhesion (Chitaev & Troyanovsky, 1997). The penton dodecahedron (PtDd) of HAdV-B3, which contains the HAdV-B3 fiber in association with the dodecahedral penton base, was shown to bind to DSG2 with a K_D of 2.5 nM (Wang *et al.*, 2011a).

More recently, random mutagenesis of the HAdV-B3 fiber knob domain helped pinpoint the residues in the fiber knob domain that mediate DSG2 association (H. Wang *et al.*, 2013). Several clusters of amino acid residues located in the grooves on the extreme distal end of the fiber knob were found to be critical as their alterations reduced or ablated DSG2 binding.

An interesting consequence of adenovirus binding to DSG2 is that this interaction triggers the epithelial to mescenchymal cell transition, which is accompanied by the transient opening of intercellular junctions

(Wang *et al.*, 2011a). Such an opening process promotes increased access to other receptors such as the CAR or CD46 that are initially hidden in tight junctions. The PtDd of HAdV-B3 mimics the same cell junction opening as intact virus particles and has therefore been referred to as junction opener (JO-1). An excess of PtDd is normally produced during HAdV-B3 infection and likely aids cell to cell spread of progeny particles. In support of this concept, a mutant HAdV-B3 that is unable to produce the PtDd had reduced cell to cell spread compared to wild-type virus (Lu *et al.*, 2013). Disruption of tight junctions by virus binding to DSG2 may also have utility in the development of cancer therapeutics. For example, the JO-1 complex has been shown to increase intratumoral penetration of the anti-Her2/neu monoclonal antibody (Herceptin) (Beyer *et al.*, 2011), as well as the penetration of several chemotherapeutic drugs in mouse models of different cancers (Beyer *et al.*, 2012). Interestingly, mutations introduced into the HAdV-B3 fiber knob within or near the extended EF loop appear to make this structure more flexible as well as increase affinity for DSG2 (Wang *et al.*, 2013a). As a consequence, a JO-1 complex with these EF loop-enhancing mutations significantly improves the intratumoral penetration and efficacy of chemotherapeutic agents.

To facilitate analysis of type B virus infections *in vivo*, a human DSG2 (hDSG2) receptor transgenic mouse model has been established by Wang and colleagues (2012). The tissue pattern of hDSG2 expression in this murine model is similar to that in humans and nonhuman primates. Moreover, hDSG2 expression allowed virus transduction of bronchial and alveolar epithelial cells following intranasal administration of a HAdV-B3 vector encoding GFP (green fluorescent protein) in these transgenic mice. Thus, the hDSG2-expressing small animal model may provide the means to test various species B adenoviral vectors for gene delivery *in vivo*.

3.2.4 *Sialic acid containing glycoconjugates*

Early studies of the ability of neuraminidase treatment to destroy virus binding suggested that certain species D adenoviruses associated with severe ocular infections, such as HAdV-D37, -D8, and -D19a, use sialic acid as the primary receptor on epithelial cells (Arnberg *et al.*, 2000a; Arnberg *et al.*, 2000b; Arnberg *et al.*, 2002b). These findings are consistent with the observation that fiber knobs of pathogenic species D adenovirus types 19a and 37 have a prominent cluster of basic amino acids unlike fiber knob domains from other viruses that do not bind to sialic acid (Arnberg *et al.*, 1997). The overall predicted isoelectric point of the HAdV-D37 and HAdV-D8 fiber knob domains is also uncharacteristically high (~9.0), a property that is consistent with their predicted electrostatic interactions with the negatively charged sialic acid residues (pKa = 2.6) (Arnberg *et al.*, 2002a). In contrast, the attachment receptors for the non-pathogenic species D types D9 and D19p (type 19 prototype) are α_v integrins rather than sialic acid as, as indicated by inhibition of binding of these viruses to cells by function-blocking α_v integrin monoclonal antibodies but not by neuraminidase treatment (Arnberg *et al.*, 2000b). A current mystery is that, despite their roles in attachment, α_v integrins do not appear to be required for HAdV-D19p infection whereas they are needed for infection by HAdV-D9. The 19p virus has a rather high particle to PFU ratio (1,600:1), indicating its lower competency for replication in tissue culture, a property that complicates evaluation of receptor recognition. As discussed below, α_v integrins also serve prominent roles in internalization of many adenovirus types. As mentioned earlier, HAdV-D37 has also been reported to use CD46 as a receptor (Wu *et al.*, 2004), although it has not yet been determined whether binding to this receptor occurs via sialic acid moieties associated with one or more of the four isoforms of this glycoprotein. HAdV type 52, a species G virus associated with gastroenteritis, contains two different fiber proteins, long and short. Interestingly these two fibers bind two distinct receptors: CAR (long fiber) and sialic acid (short fiber), but most of the cell attachment is contributed by the association of the short fiber with sialic acid (Lenman *et al.*, 2015). Moreover, structural analyses of the short fiber bound to an analog, 2-O-methylated sialic acid, by Lenman and coworkers showed that the attachment sites on the fiber knob domain for sialic acid are distinct from those of the species D HAdVs and are located on shallow grooves on the side of the fiber knob.

The crystal structure of a species D HAdV fiber knob and the sialic acid binding site (Burmeister *et al.*, 2004) has enabled the rational design of inhibitors for these pathogenic viruses. In particular, a multivalent conjugate of a sialic acid derivative, 3'-sialyllactose, has been shown to inhibit binding of HAdV-D37 to human corneal epithelial cells and infection (Johansson *et al.*, 2005). Recently, a triazole linker-based trivalent sialic acid inhibitor for HAdV-D37 was reported to have low nanomolar (IC_{50} = 1.4 nM) capacity to restrict virus attachment and infection (IC_{50} = 2.9 nM) (Caraballo *et al.* 2015). Although this promising antiviral compound did not appear to have ocular toxicity in an animal model of HAdV-D37 disease, its safety and efficacy in humans has not yet been reported.

The growing knowledge of adenovirus attachment receptors underscores the diversity of cell binding sites for different adenovirus types, and has provided increased opportunities to develop antiviral agents (Aplander *et al.* 2011; Arnberg, 2012), as well as gene therapy vectors with selected tissue tropism.

3.3 Integrins as Receptors for Adenovirus Internalization

In some of the earliest studies of adenovirus host cell interactions, a virus-encoded soluble factor was reported to trigger host cell detachment from glass or plastic surfaces *in vitro* (Everett & Ginsberg, 1958; Pereira, 1958; see also Fig. 3.2). This so-called toxic factor was later shown to be the

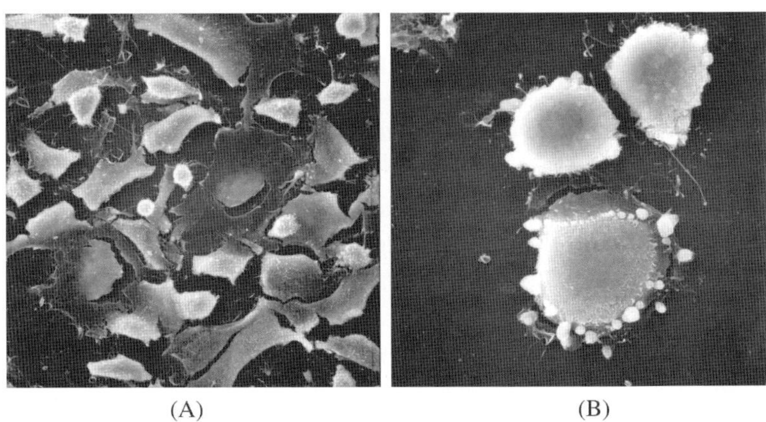

Figure 3.2. Scanning electron micrographs of cells undergoing detachment mediated by adenovirus penton base. Human A549 lung epithelial cells were incubated for 5 minutes (panel **A**) or 45 minutes (panel **B**) following the addition of purified HAdV-C2 penton base. Note that cells become rounded and eventually detach without undergoing membrane damage, and will re-attach and spread if the penton base is removed from the culture media.

penton base (Boudin *et al.*, 1979) that is normally produced in excess over that incorporated into viral capsids during HAdV-C2 infection. It was not until 1993 that Wickham and colleagues reported that the penton base is not actually toxic, but does cause cell detachment and can mediate cell adhesion. These properties are due to the presence in the penton base of HAdV-C2, as well as many other adenovirus types, of an arginine–glycine–aspartic acid (RGD) sequence (Neumann *et al.*, 1988) that associates with the vitronectin-binding integrins $\alpha_v\beta_3$ and $\alpha_v\beta_5$ on a variety of different human epithelial cells. Integrins are heterodimeric cell membrane glycoproteins that recognize the RGD motif in different extracellular matrix proteins such as vitronectin allowing cell adhesion and cell motility (Cheresh & Harper, 1987; Dong *et al.*, 2012; Hynes, 1992). Importantly, function-blocking antibodies to α_v integrins prevent adenovirus internalization into cells (Wickham *et al.*, 1993), but not initial virus binding (with the exception of HAdV-19). Moreover, transfection of human M21-L4 melanoma cells with cDNA encoding α_v integrin restores both cell adhesion to vitronectin and enhances adenovirus infection. Mutations that alter the RGD sequence in the HAdV-C2 penton base also decrease the kinetics of infection *in vitro* (Bai *et al.*, 1993). Certain HAdV types, such as HAdV-F41, lack an RGD sequence in the penton base and are less efficient in cell entry and infection (Albinsson & Kidd, 1999). In some cell types, such as human B lymphocytes, the absence of α_v integrins profoundly reduces HAdV-C5-mediated gene delivery, and this restriction can be reversed by transformation of B cells by Epstein–Barr virus, which induces increased α_v integrin synthesis (Huang *et al.*, 1997). The integrin-binding RGD motif is conserved in most but not all penton base proteins of different HAdV types and promotes virus entry and infection (Albinsson & Kidd, 1999; Cuzange *et al.* 1994; Mathias *et al.*, 1994). Thus, most adenovirus types use two different host cell receptors for attachment and internalization. Integrins are now well recognized for their capacity to promote cell entry of and infection by diverse bacterial and viral pathogens (Nemerow & Cheresh, 2002; Cattaneo, 2004).

The role integrins in adenovirus infection *in vivo* appears to be more complicated to assess at this point in time, in part because of the diverse cell types and receptor repertoires found in various tissues. For example, β_5 integrin knock out mice were reported to be as susceptible to adenovirus infection as normal mice (Huang *et al.*, 2000), suggesting that this particular integrin is not absolutely required for infection. Other integrins such as $\alpha_v\beta_1$ and β_3 might fulfill this role (Li *et al.*, 2001). Alternatively, association of virus particles with soluble blood proteins can circumvent normal receptor interactions, as discussed below.

Integrin-mediated adenovirus entry into non-polarized epithelial cells occurs primarily via clathrin-mediated endocytosis as demonstrated by ultrastructure and immunofluorescence microscopy (Chardonnet & Dale, 1970; Greber *et al.*, 1993; Patterson & Russell, 1983). Efficient adenovirus entry also requires the large GTPase dynamin (Wang *et al.*, 1998) as well as the early endosome marker Rab5 GTPase (Rauma *et al.*, 1999). Integrin clustering on the cell plasma membrane is thought to occur via associations with the five RGD sites, located on extended loops on each penton base (Chiu *et al.*, 1999; Stewart *et al.*, 1997) and this triggers uptake of virus particles into cells. Consistent with this notion, multimeric penton base, but not monomeric RGD peptides, was shown to stimulate integrin-mediated cell signaling and cell adhesion (Stupack *et al.*, 1999). The spacing of each of the five RGD loops on the penton base is approximately 60 Å, similar to the optimal spacing (~50 Å) needed to activate integrin-mediated cell motility (Maheshwari *et al.*, 2000).

Adenovirus type 2 internalization via α_v integrins requires activation of a signal transduction pathway composed of phosphatidylinositol 3-OH kinase (Li *et al.*, 1998b; Li *et al.*, 2000b), Rho family GTPases (Li *et al.*, 1998a) and the large adapter molecule, p130/CAS (Li *et al.*, 2000a). The primary result of activation of this signaling cascade is the reorganization of actin filaments. Polymerized actin can serve as a scaffold to prolong the half-life of signaling complexes and provides the mechanical force needed for the formation of endocytic vesicles (Fujimoto *et al.*, 2000; Qualmann *et al.*, 2000). As noted above, association of species B adenoviruses with integrins on non-polarized epithelial cells or hematopoetic cells also leads to activation of Rho family GTPases and actin remodeling, events that may promote cell entry via macropinocytosis (Meier *et al.*, 2002; Wolfrum & Greber, 2013).

Although the precise role of cell signaling in adenovirus internalization has yet to be fully delineated, particularly *in vivo*, activation of common cell signaling pathways has been used to enhance adenovirus-mediated gene delivery to cells that lack α_v integrins. In one scenario, bifunctional antibodies that simultaneously ligated the penton base and a cell surface growth factor receptor (e.g. EGF), triggered a similar signal transduction pathway as that engendered by adenovirus-integrin ligation. These bifunctional antibodies increased adenovirus-mediated gene delivery to cells lacking α_v integrins by 10–50 fold (Li *et al.*, 2000a). Other investigators have re-engineered adenovirus vectors to display the integrin-binding RGD sequence in the virus fiber (Laborda *et al.*, 2014; Ueyama *et al.*, 2014) or the hexon or protein IX (Kurachi *et al.*, 2007a) to enhance virus-mediated gene transfer to cells lacking primary receptors (CAR or CD46).

Certain non-HAdVs such as canine adenovirus type 2 (CAV-2) and murine adenovirus type 1 (MAV-1) lack an RGD sequence in their penton base proteins, raising the possibility that these viruses do not need integrins for infection. As discussed earlier, CAV-2 attaches to and enters certain host cells exclusively via CAR, thus circumventing the need for integrins. In the case of MAV-1 however, the RGD motif is displayed in its fiber protein, rather than on the penton base. Based in part on competition experiments with soluble RGD peptides, it appears that MAV-1 also enters cells by association with α_v integrins (Raman *et al.*, 2009). MAV-1 may also use heparin sulfate glycosaminoglycans (HSPGs) as receptors for infection, although it is not clear whether integrins and HSPGs work additively (two-step model) or synergistically.

3.4 Scavenger Receptors and Blood-Derived Proteins

Despite the accumulated knowledge of receptors for adenovirus in tissue culture systems, adenovirus tissue tropism *in vivo* is much more varied (Baker *et al.*, 2013), as was first brought to light in studies of gene delivery to the liver following intravenous administration of a HAdV-C5 vector. When HAdV-C5 vectors are delivered systemically into mice, Kupffer cells (KCs) of the liver rapidly remove the virus from the blood stream (Alemany *et al.*, 2000; Tao *et al.*, 2001; Wolff *et al.*, 1997). However, none of the major adenovirus receptors (i.e. CAR and integrins) appear to play a role in this process (Smith *et al.*, 2008a). A clue to the actual blood removal mechanism was obtained from studies that showed that poly-inosine (poly I), a known ligand for scavenger receptors (SRs), reduced uptake by Kupffer cells and increased the circulating half-life of adenovirus (Haisma *et al.*, 2008; Smith *et al.*, 2008b). Moreover, poly I inhibits the uptake of HAdV-C5 into these cells both in culture and *in vivo* (Xu *et al.*, 2008). Recently, manipulation of recombinant adenoviruses, and of the hexon protein in particular, has shown that the hexon hypervariable loop regions 1, 2, 5, and 7 likely contribute to SR interactions on Kupffer cells based on the ability of targeted addition of polyethylene glycol (PEG) to these regions to block association of virus particles with Kupffer cells (Khare *et al.*, 2012).

Other cell types in the liver also exhibit alternative modes of adenovirus binding and infection. For example, modifications of the capsid that eliminated CAR or integrin association in tissue culture had virtually no effect on virus transduction of liver hepatocytes *in vivo* (Nicklin *et al.*, 2005), suggesting that other host factors mediated entry. Shayakhmetov and coworkers (2005) demonstrated that soluble blood proteins such as coagulation factor IX and serum complement proteins (e.g. C4BP) bind to adenovirus and act as a bridge to link up with HSPGs on liver hepatocytes. Studies in cells in culture also showed that several other vitamin K-dependent coagulation factors such as FX, protein C, and FVII can enhance HAdV-C5 gene delivery to hepatocytes (Parker *et al.*, 2006). Moreover, warfarin treatment, used to deplete plasma of coagulation factors *in vivo,* also significantly reduced HAdV-C5 transduction, lending additional support for the notion that certain blood proteins promote adenovirus liver tropism *in vivo*. Subsequently, multiple investigators showed that FX selectively binds to the top of the hexon trimer on the HAdV-C5 capsid and hence allows the bridging of virus particles to hepatocytes via FX interactions with HSPGs (Kalyuzhniy *et al.*, 2008; Vigant *et al.*, 2008; Waddington *et al.*, 2008). This hepatocyte transduction pathway is operative in mice as well as in non-human primates (Alba *et al.*, 2012). Modification of the hexon protein of HAdV-C5 to abolish the FX bridging pathway has no effect on transduction of liver Kupffer cells, indicating the virus follows a separate virus entry pathway in these cells (Waddington *et al.*, 2007).

Although not all adenovirus types bind to FX (Waddington *et al.*, 2008), HAdV-C5 association with FX is capable of activating the host proinflammatory response (Doronin *et al.*, 2012), a response that could lead to deleterious consequences for various organs transduced with this viral vector. More recently, Xu *et al.* (2013) found that FX association with HAdV-C5 can also protect the virus from IgM antibody and complement-mediated neutralization, suggesting that *in vivo* FX serves as a shield against natural immunity to the virus.

Despite the growing awareness that vitamin K-dependent coagulation factors such as FX play a clear role in liver tropism by adenovirus *in vivo*, it remains to be determined whether other soluble factors can facilitate binding or entry of the virus in other tissues. Previous studies hint that this is indeed a possibility. For example, Johansson and colleagues (2007) reported that lactoferrin, a protein found in human tears, can also act as a bridging molecule to enhance adenovirus infection of epithelial cells in a CAR-independent manner.

3.5 What Next?

The identification of different attachment receptors (CAR, CD46, DSG2, and sialic acid) recognized by distinct HAdVs and non-HAdVs has established a better understanding of how these viruses recognize host cells. In most, but not all cases, the intrinsic affinity of these receptors for the fiber protein is strong and in the nanomolar range. The multivalency of the fiber protein, which displays three independent receptor-binding sites (36 per virion), provides a large boost in receptor avidity, thereby supporting even low intrinsic affinity interactions (micromolar range), such as those of HAdV-D37 association with sialic acid.

The broad tissue distribution of adenovirus receptors also allows virus entry into many cell types *in vivo*. Nonetheless, certain receptors such as CAR are sequestered on the basal lateral surface of polarized epithelial cells, a property that prevents virus attachment and infection until the receptor becomes exposed when tight junctions are breached. Thus, adenovirus use of a receptor on various host cells *in vivo* can be complicated by receptor accessibility. Moreover, in some cancer cell types, such as those derived from malignant mesothelioma, CAR expression can vary greatly and dictate therapeutic efficacy of oncolytic HAdV-C5 vectors (Takagi-Kimura *et al.*, 2013). Consequently, current and future gene therapy endeavors have explored and will continue to seek ways to increase adenovirus vector binding to cancer cells by modifying the fiber or substituting the CAR-binding HAdV-C5 fiber with a HAdV-B35 fiber that recognizes an alternative cell receptor such as CD46.

Knowledge of adenovirus receptors has also opened the door for the development of antiviral approaches to block the earliest steps in adenovirus binding to host cells. For example, synthetic analogs of multivalent sialic acids have shown promise as antiviral agents against EKC-associated species D HAdVs. Whether these receptor analogs will be safe and effective in humans, particularly as topical agents to prevent or treat ocular infections, remains to be determined. Antiviral agents used in systemic delivery might be even more difficult to bring to the clinic, as they require good bioavailability and appropriate pharmacokinetic profiles.

In addition to serving as attachment receptors, the α_v integrin family of receptors promote cell entry of many adenovirus types. Integrin ligation by the viral penton base RGD sequences can activate signaling pathways that aid virus uptake into the cell. However, this effect is a double-edged sword, in that integrin-mediated signaling also enhances proinflammatory responses that can be potentially harmful for the host. Efforts to re-engineer adenovirus vectors with integrin-binding RGD motifs will therefore need to balance increased virus infectivity with the possibility of stimulating host immunity.

Despite the ongoing efforts to understand adenovirus receptor interactions based on tissue culture cell systems, HAdV-C5 tissue tropism *in vivo* is regulated in systemic situations by association of the virus capsid with soluble blood proteins such as FX. These molecules act as a bridge to bring the virus to the surfaces of hepatocytes in the liver. This process underscores the need to investigate host cell interactions *in vivo* as well as in cells in culture. Remaining questions for future studies include: what are the precise structural details underlying HAdV-C5–FX association and can we take advantage of this information to re-engineer the HAdV-C5 hexon to avoid FX association? Finally, it will be important to understand fully the consequences of FX association with adenovirus particles, particularly with regard to the immune responses that are engendered.

Intracellular Trafficking   4

4.1 Introduction

Once adenoviruses undergo stable receptor-mediated binding to host cells, there are additional major challenges awaiting the virion to accomplish cell entry (Fig. 4.1). Foremost among these is the requirement to escape the confines of the early endosome and to deliver the genome into the nucleus, where production of new virions occurs. To begin the cell trafficking journey, the adenovirus capsid must undergo partial disassembly. This step is primarily limited to the dissociation of the fiber, penton base, and peripentonal hexons. Loss of these proteins from the vertex region allows the exposure and then release of protein VI from the interior of the capsid, leading to the disruption of the cytoplasmic vesicle in which the virus would otherwise be trapped. Importantly, membrane damage may serve as a signal of invasion by a foreign entity in certain host cells to set off signaling events that result in immunologic or membrane repair responses. Once released into the cytoplasm, the partially uncoated virus particle hitches a ride on microtubules allowing rapid transit to the nuclear pore complex (NPC; Fig. 4.1). At this location, the particle undergoes further disassembly enabling genome import into the nucleus. Although we have a fairly good understanding of the major cell trafficking steps that precede virus replication, many details remain to be discerned. Recent investigations have revealed a few surprises that are described in this

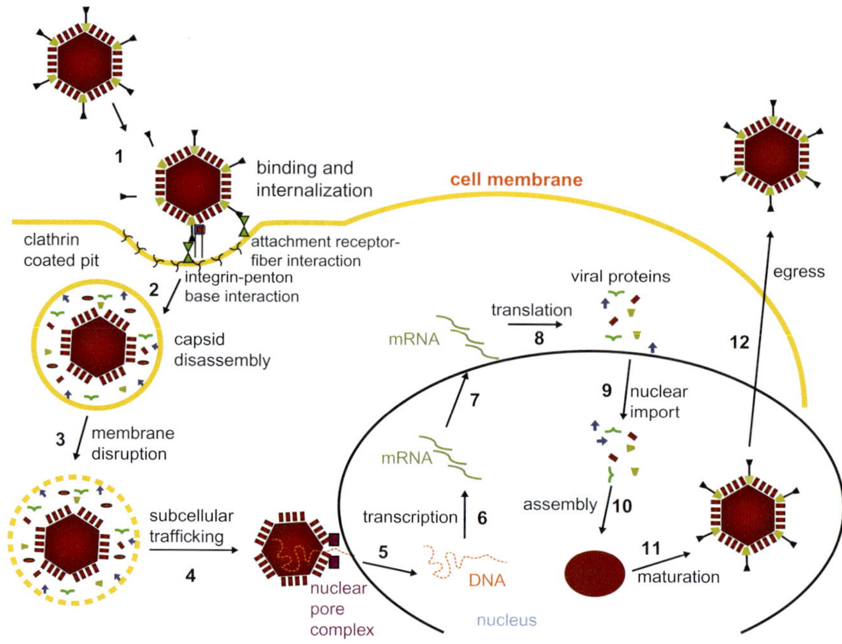

Figure 4.1. The infection cycle of human adenovirus. The different steps of the adenovirus infection cycle including attachment (**1**), internalization (**2**), endosome disruption (**3**), cytoplasmic trafficking (**4**), nuclear import (**5**), transcription of viral mRNA (**6**), cytoplasmic export of mRNA (**7**), translation (**8**) and nuclear import of viral proteins (**9**), assembly of virus particles (**10**), maturation (**11**), and egress (**12**) are depicted in this cartoon. Note that the details of several of these processes are still unknown and require further investigation. Specific events related to viral gene transcription and DNA replication are discussed in Chapters 5 and 6, respectively.

chapter. The knowledge gained should help complete our understanding of virus cell entry and perhaps open up new avenues for gene transfer as well as identify potential new targets for antiviral applications.

4.2 Early Events in Capsid Disassembly

In some of the earliest studies of adenovirus disassembly, it was reported that removal of the capsid vertex region containing the penton base and fiber is temperature dependent, and likely to occur at the cell surface based on sensitivity of the internal viral genome to DNase treatment

(Boulanger & Hennache, 1973; Lawrence & Ginsberg, 1967; Lonberg-Holm & Philipson, 1969). Greber and colleagues (1993) subsequently showed that the virus particle undergoes a series of protein dissociations as it penetrates the cell and progresses through the cytoplasm. Using a biochemical approach with radiolabeled virus particles, these investigators noted loss of the fiber with a $t_{1/2}$ of 12 minutes after infection at 37°C, coinciding with endocytosis. These studies did not determine whether it was only the fiber that was released or whether it was the entire penton. As we now know that the association of the penton base and fiber is virtually irreversible (Zubieta *et al.*, 2005), it seems more likely that these two proteins are released together as a unit. However, fluorescence resonance energy transfer (FRET) analyses of human adenovirus (HAdV)-C5 labeled with fluorophores indicated that the fiber protein is released with a half-time of 3 minutes coincident with endocytosis while the penton base dissociates much later (Martin-Fernandez *et al.*, 2004). Moreover, other studies have also suggested that the loss of the fiber protein alone could be the signal for capsid disassembly. Anti-fiber antibodies, but not antibodies to other capsid proteins, render virus particles sensitive to DNase treatment (Everitt *et al.*, 1992; Wohlfart *et al.*, 1985). Recombinant adenovirus particles that lack only the fiber protein are also less stable and their genomes are released prematurely compared to normal fiber-containing virus particles (Von Seggern *et al.*, 1999).

Greber *et al.* (1997) showed that an inhibitor of the low pH environment in the early endosome, ammonium chloride, inhibited endosomal release of adenovirus. The investigators also reported that the process of endosome rupture by the virus occurs with a halftime of ~15 minutes post-infection, and estimated that the loss of the fiber protein from the virus capsid occurred prior to endosome rupture by the virus. Following endosomal escape, particles required 35–45 minutes to reach the NPC where a majority of the viral nucleic acid was released from disassembled particles.

More recently, investigators have used live-cell immunofluorescence analyses to show that fiber–CAR (Coxsackie virus and adenovirus receptor) and penton base–integrin associations play significant roles in triggering initial capsid disassembly at the cell plasma membrane (Burckhardt *et al.*, 2011). These studies showed that the binding of the capsid proteins to the different receptors results in distinct motions at the cell surface, which lead to loss of the fiber proteins and the exposure of protein VI from the interior of the particle. Specifically, CAR–fiber interactions generate CAR clustering, leading to diffusive motion on the cell membrane, while penton base–integrin interactions produce much less movement. Disruption of the underlying cytoskeletal network by treatment with cytochalasin D prevents CAR-fiber mediated drifting motions as well as subsequent exposure of protein VI, consistent with earlier studies defining a role for actin and integrin ligation to mediate fiber release at the cell surface (Nakano *et al.*, 2000). These findings suggest that via different membrane interactions the virus capsid receives dual environmental cues that somehow promote penton disassembly. Although the precise mechanism has yet to be elucidated, there are a number of structural and biophysical studies that support this model of capsid disassembly. First, binding of soluble $\alpha_v\beta_5$ integrin to adenovirus causes a visible conformational change in the penton base that is consistent with the initial phase of disassembly (Lindert *et al.*, 2009). Second, atomic force microscopy (AFM) nanoindentation studies revealed that integrin $\alpha_v\beta_5$ binding to adenovirus selectively loosens the fivefold axes (vertex region) (Snijder *et al.*, 2013). Moreover, binding of a human alpha defensin, HD5, to the HAdV-C5 penton stiffens the fivefold axes and restricts virus capsid disassembly as well as impedes endosome rupture (Smith & Nemerow, 2008; Smith *et al.*, 2010).

In addition to receptor ligation, which clearly promotes capsid disassembly, other environmental cues may be needed for cell entry. As noted earlier, the low pH environment in the early endosome has been proposed to contribute to capsid disassembly and/or membrane penetration by the virus (Greber *et al.*, 1993). However, the role of endosome acidification remains controversial, in part due to the use of inhibitors of acidification such as weak bases (ammonium chloride, chloroquine, methylamine) or of the vacuolar H^+-ATPase (bafilomycin A1) to study virus uncoating and cell entry/infection. The challenge with the use of these agents is that they can have pleiotropic effects on cells depending on their concentrations and use in different cell types, and their potential toxicity requires careful evaluation. Earlier studies of adenovirus endosome penetration using a co-entry assay with a *Pseudmonas* toxin (Seth *et al.*, 1984a; Seth *et al.*, 1984b), showed that 10 mM ammonium chloride, or 5 mM methylamine, amongst other agents, caused half maximal inhibition of toxin delivery by adenovirus. However, none of these agents completely inhibited virus-mediated toxin delivery at their non-toxic concentrations. A requirement for low pH in membrane penetration reported in these earlier studies is supported by more recent findings showing that adenovirus-mediated endosome penetration is restricted by 0.3 μM bafilomycin A1 (Wiethoff *et al.*, 2005) or by a synthetic cyclic D,L-α-peptide that eliminates the low pH in endosomes (Horne *et al.*, 2005). The dissociation of the vertex region of the capsid has been reported to occur at low pH but this reaction appears also to be dependent on ionic strength (Everitt *et al.*, 1992; Prage *et al.*, 1970). The thermostability of adenovirus particles can also be reduced by exposure to low pH (Wiethoff *et al.*, 2005). Nonetheless, there are a number of other studies that raise some doubt about the need for low pH in adenovirus disassembly or escape from the endosome. Svensson and Persson (1984) reported that 40 mM methylamine or 40 mM ammonium chloride had no effect on virus internalization or disassembly. In a particularly careful evaluation, several different lysosomotropic inhibitors, including 10 mM methylamine and ammonium chloride, did not impair adenovirus disassembly or endosome penetration or even a later event in infection, transcription of viral genes (Meunier-Durmort *et al.*, 1997; Rodriguez & Everitt, 1996). One possibility to explain the discrepancies for an acid pH requirement in adenovirus disassembly or membrane penetration is that the proton gradient itself is actually more crucial than a particular pH value, as was proposed by Carrasco (1994). Another is that acid-dependent proteases such as cathepsins, commonly found in endolysosomes, might facilitate capsid disassembly and ultimately endosome rupture, although this possibility has not yet been investigated.

The viral cysteine protease (AVP), an enzyme that cleaves the preprotein forms of IIIa, VI, VII, VIII, μ, as well as the 52/55K protein as described in Chapter 7, may also facilitate adenovirus cell entry. The supposition is based on biochemical studies in which reduction and alkylation reduced infectivity without disrupting virus particles (Greber *et al.*, 1996). The fact that AVP was one of the components of the virus capsid most strongly labeled by the reducing agent suggested that it was the target of virus inactivation. Copper chloride, another AVP inhibitor, also blocked infection as measured by gene transfer (Cotten & Weber, 1995). Based on cell entry analyses, Greber and colleagues proposed that integrin ligation by the penton base causes exposure of protein VI that is then cleaved by capsid-associated AVP, which is activated by the reducing environment of the endosome. As protein VI functions as a cement protein, its cleavage by AVP could in principle, weaken the virus particle and thereby facilitate release of the genome upon nuclear pore binding. It is also plausible that proteolytic cleavage of protein VI is carried out by cellular proteases rather than by the action of AVP.

4.2.1 *Identification and analyses of the membrane lytic activity of protein VI*

A temperature-sensitive version of HAdV-C2, known as *ts1*, contains uncleaved precursor forms of IIIa, VI, VII, VIII, μ, and TP (terminal protein) as a result of its inability to incorporate the approximately seven molecules of AVP into the capsid (Begin & Weber, 1975; Weber, 1976). Interestingly, these immature-like *ts1* particles are still able to bind to receptors and enter cells normally. However, they do not disrupt cell endosomes and thus are either degraded or recycled out of the cell (Greber *et al.*, 1996). The *ts1* mutant HAdV-C2 provided the means to identify the endosome-disrupting protein of adenovirus by Wiethoff and colleagues (2005). These investigators found that *ts1* particles, unlike normal Ad2 particles, were hyperstable and failed to undergo conformational changes leading to the loss of the peripentonal hexons, penton base, and fiber proteins upon exposure to elevated temperatures or to low pH. In addition, *ts1* particles were unable to mediate co-delivery of a ribotoxin (α-sarcin) indicating that they lacked the ability to disrupt the endosome. Importantly, the vertex proteins derived from partially disassembled wild-type virus particles were capable of disrupting artificial lipid bilayers, while immunodepletion of protein VI from the collection of dissociated capsid proteins, selectively reduced membrane disruption (Wiethoff *et al.*, 2005). Further studies showed that recombinant protein VI possessed pH-independent membrane lytic properties similar to those of intact virions. Furthermore, the predicted amphipathic α-helix (AAH) domain in protein VI (residues 34–54) was essential for full membrane lytic activity. In the virus particle, protein VI is located on the inner capsid shell (Reddy *et al.*, 2014b; Wiethoff & Nemerow, 2015) and is anchored inside the base of the peripentonal hexons at the fivefold axis via its N-terminal 33 amino acids (Snijder *et al.*, 2014). Subsequent cleavage of pVI at glycine 33 by AVP during virus maturation separates the membrane lytic domain (amino acids 34–79) from the preprotein N-terminal moiety. Cleavage at this position in preprotein VI appears to have an additional role in the virus replication cycle as revealed by mutagenesis studies. A HAdV-C5 mutant virus containing a single, conservative, amino acid substitution — glycine 33 to alanine — in preprotein VI exhibits substantially reduced AVP cleavage of this protein, virus infectivity, and most importantly, maturation of virus particles after assembly (Moyer *et al.*, 2015). However, the precise mechanism underlying this latter defect has yet to be defined.

Although the studies outlined above indicate that protein VI is responsible for adenovirus-mediated endosome disruption, it is possible that other capsid proteins participate in this event. To address this possibility, random mutagenesis studies, directed to the AAH domain of VI, were performed to evaluate the functions of this capsid protein (Moyer *et al.*, 2011). One particular mutant VI-adenovirus containing a single amino acid substitution (L40Q) in the predicted AAH domain reduced infectivity as well as endosome rupture by ~10-fold compared to these activities of wild type virus particles. Additional biophysical studies showed that the lack of membrane penetration by the L40Q-VI mutant virus was due to impaired membrane insertion of the mutant protein. As protein VI is also a true cement protein that helps to stabilize the inner capsid surface, the L40Q-VI mutation may also cause premature release of the virus vertex region, thereby compromising cell entry and infection (Martinez *et al.*, 2015). In separate analyses, antibodies directed against VI inhibited adenovirus-mediated membrane disruption as well as delivery of the viral genome to the nucleus (Maier *et al.*, 2010). These studies provide additional evidence that protein VI is the primary mediator of endosome rupture during cell entry by adenovirus.

Although it is now clear that protein VI plays a major role in endosome rupture during adenovirus cell entry, the precise mechanism is still under investigation. Fluorescence imaging (Maier & Wiethoff, 2010) of artificial lipid vesicles (liposomes) showed that recombinant protein VI rearranged these spherical compartments into elongated tubules, akin to how some detergents or antimicrobial peptides remodel lipid bilayers (Domanov & Kinnunen, 2006). Based on incorporation of different phospholipids into artificial lipid bilayers, it was determined that protein VI causes extensive positive membrane curvature, thereby imparting increasing stress to the lipid bilayer, and eventually leading to membrane fragmentation (Maier *et al.*, 2010). Using tryptophan fluorescence quenching analyses with brominated phospholipids, Maier and colleagues (2010) determined that the N-terminal 80-residue domain of protein VI containing the AAH inserts into membranes in an oblique orientation, consistent with its propensity to cause positive membrane curvature (Epand & Epand, 2000). The precise number of protein VI molecules needed to rupture a single endosome has not yet been determined. However, the potential release of hundreds of VI molecules from the capsid into the lumen of the endosome would produce a high local concentration of this membrane lytic factor.

4.2.2 *Role of protein VI in cell signaling and immunity*

The ~90 nm diameter adenovirus particle causes significant membrane damage when protein VI is released from the interior of the capsid. Membrane disruption can have consequences beyond allowing the partially disassembled capsid to enter the cytoplasm, such as producing signals that the cell is being invaded by a microbial pathogen. There are at least two major consequences that can result from pathogen-mediated endosome rupture: membrane repair (Gurcel *et al.*, 2006; Luisoni *et al.*, 2015; Tam *et al.*, 2010) and immune activation (Hara *et al.*, 2007; Thurston *et al.*, 2012). A recent study by Greber (2016) suggested that some exposure of protein VI occurs earlier than anticipated and begins at the cell surface upon receptor engagement of the penton. Exposure of VI was suggested to generate small lesions in the plasma membrane that create a conduit for calcium ion flux into the cell as well as to stimulate the process of lysosomal exocytosis (Luisoni *et al.*, 2015). This membrane repair response results in the release of acid sphinogmyelinase with subsequent degradation of sphingomyelin to generate ceramide lipids in the plasma membrane. The production of ceramide was also reported to enhance protein VI-mediated membrane rupture. Thus, this signaling pathway appears to establish a link between a specific membrane repair response and enhancement of adenovirus-mediated endosome rupture during cell entry and infection.

Membrane rupture by adenovirus can also elicit a danger signal that leads to innate immune responses. For example, HAdV-C5, but not the *ts1* mutant HAdV-C2 virus that fails to disrupt endosomes, induces a proinflammatory response in mice associated with an accelerated activation of p38 and ERK mitogen activated kinases (Smith *et al.*, 2011). Capsid disassembly and endosome escape also causes activation of other MAP kinases such as JNK 1/2, and leads to interferon production in macrophages (Fejer *et al.*, 2008). Species B adenoviruses such as HAdV-B35 can enter plasmacytoid dendritic cells (pDCs) via the CD46 receptor and activate type 1 interferon production (Iacobelli-Martinez & Nemerow, 2007). Interferon production in pDCs is initiated by recognition of viral dsDNA by Toll-like receptor 9 (TLR9) in endo-lysosomes. Several investigators have reported that adenovirus-induced interferon production in macrophages occurs via multiple signaling pathways and is likely initiated by different cytosolic DNA sensors (Muruve *et al.*, 2008; Nociari *et al.*, 2009; Zhu *et al.*, 2007b; see also Chapter 9). Di Paolo and colleagues (2009) also provided evidence that adenovirus-mediated endosome disruption in macrophages located in the marginal zone of the spleen produces a signal for transcription of the gene encoding IL-1α, a proinflammatory cytokine. However, the exact signaling molecules that act downstream of endosome rupture are still not fully characterized. One such molecule that has been identified is cathepsin B, a resident of endolysosomes that is released into the cytosol of macrophages upon vesicle rupture by HAdV-C5, leading to mitochondrial membrane damage and the generation of reactive oxygen species (Barlan *et al.*, 2011b; McGuire *et al.*, 2011). HAdV-C5 mediated endolysosome disruption in macrophages also leads to activation of NALP3 and ASC, two components of the cytosolic inflammasome pathway (Muruve *et al.*, 2008). As proinflammatory molecules can enhance elimination of virus-infected cells, adenoviruses may possess mechanisms to evade immune recognition. In contrast to species C adenoviruses, certain members of species A, B, and D appear to traffic preferentially to lysosomes rather than the early endosome (Miyazawa *et al.*, 1999; Shayakhmetov *et al.*, 2003; Teigler *et al.*, 2014). This difference in trafficking is most likely linked to the engagement of different receptors by the types A, B, and D virus fibers (see Chapter 3). An important consequence of trafficking to the lysosome is that it leads to membrane rupture of these compartments with the release of cathepsin B, which activates the NLRP3 inflammasome (Barlan *et al.*, 2011a). Thus, by exiting from the early endosome instead of the later lysosomal compartments, HAdV-C5 may produce a less potent innate immune response.

Another damage response stemming from the invasion of cells by microbial pathogens is triggered by the exposure of certain sugar molecules, known as galectins, which are originally located on the inner membranes of cell vesicles but become re-oriented to the cytoplasm (Dupont *et al.*, 2009; Maier *et al.*, 2012; Paz *et al.*, 2010). Maier and colleagues reported that galectin 8 is recruited to membranes ruptured by adenovirus. Galectin 8 in particular is known to recruit components of the autophagy pathway (Thurston *et al.*, 2012) and protect cells from invading organisms. Whether the autophagy pathway influences adenovirus host cell entry or minimizes immune recognition of the virus has not yet been fully investigated. One recent report indicated that adenovirus-mediated cell autophagy might actually enhance adenovirus infection and gene delivery, but exactly how this occurs remains uncertain (Zeng & Carlin, 2013).

4.3 Transport of Adenovirus to the Nucleus via Microtubules

Following escape from the endosome, the disassembled adenovirus particle travels to the nucleus where many reactions essential for virus reproduction occur. Given the very elaborate and crowded milieu of the cytoplasm, it is highly unlikely that such particles could arrive at the nucleus in a reasonable period of time by simple passive diffusion. Early negative stain-electron microscopy studies suggested that after endosome escape, adenovirus achieves a more rapid cytoplasmic journey via association with microtubules (Dales & Chardonnet, 1973; Miles *et al.*, 1980). More recently, live-cell fluorescence imaging has provided strong evidence for the participation of microtubules and microtubule motors in adenovirus cytoplasmic transport (Bremner *et al.*, 2009; Leopold *et al.*, 2000; Suomalainen *et al.*, 1999). These live-cell imaging analyses also indicted that adenovirus particles travel on microtubules in both the minus and plus end directions (Gazzola *et al.*, 2009), which are likely mediated by kinesin and dynein motors, respectively. However, microinjection of cells with anti-kinesin 1 antibody did not alter virus transport (Leopold *et al.*, 2000), raising the question of whether adenovirus trafficking depend on a distinct kinesin motor protein, such as kinesin 2 that mediates trafficking of some human herpesviruses (Radtke *et al.* 2010; Sathish *et al.* 2009).

Imaging analyses of adenovirus particle movements in the cytoplasm have been bolstered by the use of microtubule inhibitors. For example, treatment of infected cells with nocodazole to depolymerize microtubules substantially reduces virus movement as well as infection (Mabit *et al.*, 2002). Most of our knowledge about microtubule-mediated adenovirus transport comes from studies on the role of dynein motors. Inhibitors of dynein, including specific RNAi and dynamitin, severely restrict virus movement on these cytoskeletal elements (Bremner *et al.*, 2009; Leopold *et al.*, 2000; Suomalainen *et al.*, 1999). There is also evidence that dynein binds directly to capsids, and that the hexon is the primary binding site for this motor protein along with dynactin-associated proteins NUDE/NUDEL (Bremner *et al.*, 2009). Interactions of dynein with the hexon occur via association of the dynein intermediate and light chains. Phosphorylation of the light chain of dynein by protein kinase A (PKA), and perhaps by other cell kinases, is required for hexon binding (Scherer *et al.*, 2014). Interestingly, exposure of the HAdV-C5 hexon to low pH (pH 4.4–5.4) is also important for efficient dynein motor binding (Scherer & Vallee, 2015). This finding suggests that a conformational change must occur in the hexon to allow dynein light chain association. This notion is supported by the observation of increased sensitivity of the acid-treated hexon to dissociation by SDS and proteolysis by dispase. However, the structural details of the conformational changes of the hexon that support dynein binding have not yet been delineated. Further, molecular genetic and biochemical analyses suggested that it is hypervariable region one (HVR1) loop of the hexon that plays the primary role in dynein associa-tion (Scherer *et al.*, 2015). It is also possible that other adenovirus capsid proteins contribute to trafficking to the nucleus along microtubules. In that regard, a PPXY motif in protein VI appears to facilitate microtubule motor-dependent trafficking to the nucleus (Wodrich *et al.*, 2010). However, as protein VI interpolates into cellular membranes during endosome disruption, it is not evident how much of this protein would be available for nuclear targeting of the partially disassembled capsid following endosome escape.

4.4 Nuclear Import of the dsDNA Genome

One final major challenge for the partially disassembled adenovirus particle, containing most of the outer capsid proteins minus the vertex region, after it reaches the NPC via microtubule transport, is that the viral genome nucleic acid has to pass through this nuclear gate. Generally, molecular complexes of more than 40 nm cannot travel through the NPC via passive diffusion (Pante & Kann, 2002), thus creating an additional encumbrance for the virus. Transit of larger cellular or viral cargoes through the NPC occurs in a directed way and is accomplished by associations with the host cells nuclear import machinery, including soluble karyopherins of the importin-beta family and Ran-GTP (Flatt & Greber, 2015). The outer (cytoplasmic facing) surface of the NPC is composed of many different proteins called nucleoporins (NUPs) and these molecules help to stabilize the NPC interaction with the nuclear membrane and also regulate the passive and directed diffusion of cargoes into the nucleus. Several models have been proposed to explain the gating of cargoes by NPC (Lim *et al.*, 2006; Ribbeck & Gorlich, 2001; Rout *et al.*, 2003), but this process is still incompletely understood.

The binding of the partially uncoated adenovirus particle to the NPC remains poorly characterized, but the interaction appears to be mediated by a low-affinity association with the unstructured phenylalanine–glycine (FG) repeats in NUP214 (Trotman *et al.*, 2001). Presumably, a high-avidity association with NUP214 stems from multiple contacts of the virus with this NPC protein. Binding to the NPC requires association of the hexon protein with the amino terminal domain of NUP214 (Cassany *et al.*, 2015).

After NPC binding, the genome must somehow be delivered into the nucleus, an event that is generally thought to occur without the majority of the virus capsid entering. Strunze and colleagues (2011) have proposed an interesting molecular mechanism that may explain how adenoviral DNA is imported into the nucleus. These authors suggested that partially disassembled capsids are able to recruit kinesin-1, the plus-end MT motor protein, upon arrival at the NPC. The association of kinesin-1 and its light chain Klc 1/2 in particular, is mediated by binding of protein IX located on the outer capsid surface. Simultaneously, the kinesin-1 heavy chain, KIF5C, mediates attachment to NUP358 and other associated NPC proteins. The action of the kinesin-1 motor is then proposed to generate mechanical force that helps to dissociate the capsid further, as well as to dislodge NUP358, NUP214, and NUP62 from the central channel of the NPC. These processes facilitate entry of the viral DNA into the nucleus via a widened NPC. The dsDNA genome of adenovirus enters the nucleus in association with several viral proteins, including the covalently attached TP, protein VII, and protein μ that help condense the nucleic acid (Chatterjee *et al.*, 1986a; Greber *et al.*, 1997; Puntener *et al.*, 2011).

Although the complicated mechanism of nuclear import of the adenoviral genome that is facilitated by kinesin-1 mediated capsid disassembly is still under investigation, there is experimental evidence that supports this model. Live-cell fluorescence imaging and other analyses demonstrated that capsid fragments along with different NUPs and kinesin-1 motors are displaced from NPCs following adenovirus infection (Strunze *et al.*, 2011). In addition, genetic deletion of individual FG domains in NUPs has been reported to increase the permeability barrier of the NPC (Patel *et al.*, 2007).

4.5 What Next?

During the earliest stages of infection, an adenovirus particle is subjected to a series of disassembly steps initiated by receptor engagement at the cell surface. Receptor engagement not only mediates high affinity virus binding and entry into cells, but also prepares the virus particle for membrane disruption in an early endosome or macropinosome. The binding to distinct cell receptors (i.e. CAR and α_v integrins) that have different types of mobility in the plasma membrane may generate the mechanical forces necessary to dislodge the fiber and subsequently, the penton base from the outer capsid. Early exposure of protein VI on the cell surface may also help potentiate subsequent membrane penetration events in the endosome. Thus, in the right time and place (early endosome), the virus particle has undergone conformational changes that allow release of protein VI for membrane disruption. Although the role of acidification in membrane penetration is unclear at present, exposure to low pH in the endosome may cause conformational changes in the hexon protein that facilitate subsequent transit of particles on microtubule motors (dynein).

The importance of capsid disassembly for successful adenovirus cell entry and trafficking is underscored by the defects in a temperature-sensitive mutant adenovirus, *ts1*. This mutant virus cannot undergo the conformational transitions needed to release protein VI and therefore remains trapped in endocytic vesicles or gets recycled out of the cell. Several HAdV species also interact with human alpha defensins and these innate immune proteins restrict virus disassembly and prevent subsequent membrane penetration. These findings not only add further insight into the virus trafficking cycle but also point to potential targets for antiviral therapy.

Molecular genetic, biochemical, and structural studies have established the role of protein VI in membrane penetration, a critical event in adenovirus cell entry. However, the precise mechanism of membrane disruption by VI remains to be determined. Currently, the evidence favors a model in which multiple copies of protein VI insert in an oblique orientation into membranes causing positive membrane curvature that leads to stress on the lipid bilayer and ultimately its rupture. Although further studies are needed to validate this model fully, recombinant forms of protein VI have already been explored as a way to enhance gene delivery. One example of this approach is the engineering of cellular vault nanoparticles (Poderycki *et al.*, 2006) to display multiple copies of protein VI fused to the major vault protein (Han *et al.*, 2011; Lai *et al.*, 2009). These virus-like particles enhanced transfer of a ribotoxin or a DNA plasmid into cells. As vault nanoparticles are derived from naturally occurring cellular elements (Kickhoefer *et al.*, 2005), they have the advantages of being less immunogenic than viral vectors and can also be targeted to specific cell receptors (Kickhoefer *et al.*, 2009). Further studies will be required to determine whether these or other *in vitro* applications of protein VI nanoparticles can be successfully extended to human therapeutic applications.

Adenovirus has also proven to be a highly successful vehicle that has furthered our understanding of how macromolecular assemblies traffic to the nucleus via microtubules and their associated motor proteins. Ongoing studies have revealed the underlying molecular mechanisms in this process, including the location of the dynein motor binding site on the virus as well as the components of the dynein motor that participate in binding. Acidification may also be needed for the conformational changes in the hexon that expose dynein binding sites although the exact structural details of dynein–hexon association remain to be elucidated.

Import of the viral nucleic acid into the nucleus also appears to be a complicated process that requires capsid associations with NUP proteins on the cytoplasmic face of the NPC as well as with the kinesin-1 motor protein that mediates final capsid dissociation. As with import of other large cellular cargoes, the details of these events require additional investigations. While much remains to be accomplished in understanding the full repertoire of cellular trafficking events necessary for adenovirus cell entry, the information uncovered thus far has not only provided potential targets for intervention in the early stages of virus infection but also enhanced our understanding of normal cellular processes.

Gene Expression   **5**

Investigation of the mechanisms that effect efficient expression of adeno-viral genes has provided important insights into both the molecular biology of mammalian cells and how viral gene products redirect, or subvert, cellular processes to achieve virus-specific ends. These processes have been subjected to intense scrutiny and many mechanisms critical for efficient viral gene expression in permissive human cells can be described in some detail.

5.1 The Viral Genome

The double-stranded DNA genomes of most, if not all, known human adenoviruses (HAdVs) have been sequenced, as have those of many non-human adenoviruses, including representatives of each of the five genera (see Chapter 1). These genomes range in size from just over 26 to 48.4 kbp, with *Mastadenovirus* genomes spanning 30–38 kbp (Harrach *et al.*, 2012), and include a set of common coding sequences (see below). However, the studies that identified and mapped transcription units and associated *cis*-acting elements relied almost exclusively on HAdV-C2 and HAdV-C5, as did investigation of regulation of gene expression and the functions of viral proteins. Consequently, the organization, expression, and replication of the viral genome are described for the HAdV species C archetype, with major differences of other adenoviruses noted.

In addition to the coding and control sequences described below, the viral genome carries two other sequences critical for virus reproduction, the inverted terminal repetitions (ITRs) that include the origins of viral DNA synthesis (Chapter 6) and the packaging sequence. By the convention established in early experiments to map HAdV-C2 coding sequences (Flint *et al.*, 1975), the sequences that specify encapsidation of viral DNA molecules lie near the left end of the genome, between the ITR and the E1A transcription unit. This packaging signal (described in Chapter 7) overlaps an enhancer required for efficient transcription of the E1A gene (Fig. 5.1).

An unanticipated property of the adenoviral genome is covalent linkage of the 5′ ends to a protein, named not surprisingly the terminal protein (TP) (Rekosh *et al.*, 1977; Robinson *et al.*, 1973). The TP precursor (preTP) provides a primer for initiation of viral DNA synthesis from each terminal origin (Chapter 6), and its proteolytic processing during

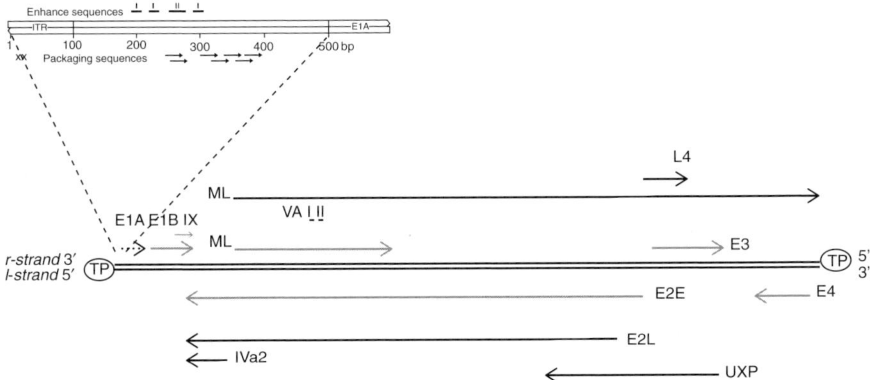

Figure 5.1.   Organization and expression of the HAdV-C2/5 genome. RNA polymerase II transcription units are drawn to scale with the arrows indicating the direction of transcription, and immediate early, early, and late transcription units shown by dotted, gray, and black lines, respectively. The promoter of the UPX gene has been mapped (Ying *et al.*, 2010), but its termination site has not, and transcription may therefore extend beyond the site indicated. The VA RNA genes transcribed by RNA polymerase III are shown as lines without arrowheads. In the expansion above, the positions of the *cis*-acting sequences present at the left-end of the genome are indicated. The inverted terminal repetition (ITR), which includes the terminal origin of viral DNA synthesis, is repeated at the right end. TP, terminal protein.

maturation of newly assembled virus particles liberates TP. When attached to the genome, the fully processed TP facilitates efficient viral DNA synthesis in *in vitro* systems, and greatly increases the infectivity of viral DNA (Sharp *et al.*, 1976). Alterations in both the TP- and precursor-specific segments of the pre-TP induce changes in the intranuclear localization of the genome, and impair viral gene expression (Kato *et al.*, 2012; Schaack *et al.*, 1990; Webster *et al.*, 1997). It is therefore thought that interactions of the products of processing of preTP with cellular components establish, or direct the genome to, an intracellular environment optimal for expression and replication of the genome.

5.2 Viral Gene Products and Their Coding Sequences

5.2.1 *RNA polymerase II transcription units and protein coding sequences*

Initial mapping of the viral mRNAs produced during different periods of the infectious cycle delineated eight transcription units (Berk, 2013; see also Fig. 5.1), which are transcribed by cellular RNA polymerase II (Price & Penman, 1972). These units were originally designated as early (E) and late (L), depending on whether they appeared in infected cells before or only following the onset of viral DNA synthesis, and distinguished by Arabic numerals, i.e. E1, E2, etc. (Flint *et al.*, 1975). However, this distinction was subsequently found to be an overly simplistic representation of the temporal progression of gene expression: some viral genes were named for the proteins they encode, and different systems were adopted for naming the multiple proteins encoded within the majority of transcription units. Unfortunately, these differences have never been rationalized into a coherent nomenclature. Here, we use the gene and protein names in most common use. Two additional HAdV-C5 promoters and transcription units were identified only quite recently. One, the U exon protein (UXP) gene lies in the leftward-transcribed ('l') strand near the E2 transcription unit (Tollefson *et al.*, 2007; Ying *et al.*, 2010), and the second, controlled by the L4 promoter, is entirely included within the major late (ML) transcription unit in the 'r'-stand (Morris *et al.*, 2010). As can be seen in Fig. 5.1, most sequences of both strands of the viral genome serve as templates for transcription by RNA polymerase II.

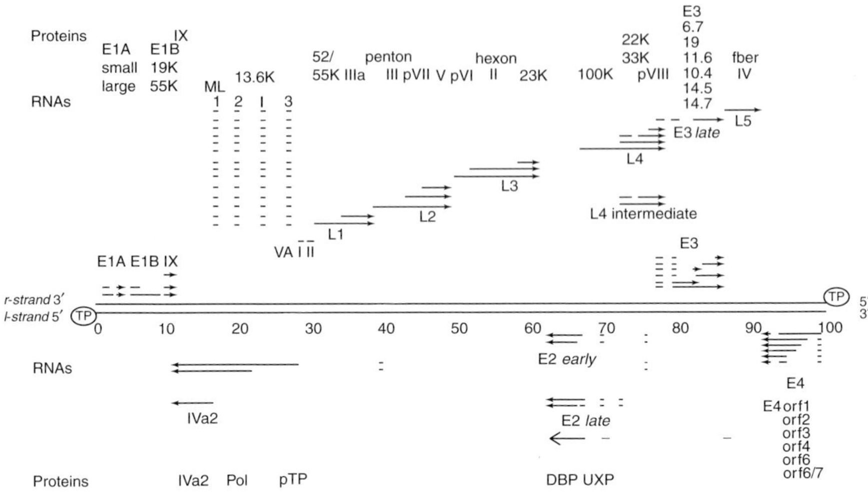

Figure 5.2. HAdV-C2/5 mRNAs and protein coding sequences. The exons of the various mRNAs are drawn as lines in which the gaps represent introns removed during pre-mRNA processing, and arrowheads sites of poly(A) addition. Only the major E1A and E1B mRNAs are shown. The proteins synthesized from mRNAs encoded in the r- and l- strands are listed above and below, respectively, in the order in which their mRNAs are depicted. The multimeric structural proteins assembled from protein III, II and IV are also indicated. The L1 52/55-kDa protein is a single protein, but is so named because phosphorylation leads to its migration as a doublet in SDS-polyacrylamide gels. The coding sequences for the UXP mRNA were determined by sequencing of cDNA copies of the mRNA (Tollefson *et al.*, 2007). Adapted from Berk (2013), with permission.

Most of the transcription units, regardless of when they are expressed, contain coding sequences for two or more proteins (at least 14 in the case of the ML transcription unit), the exceptions being the IX, IVa2 and UXP genes (Fig. 5.2). This arrangement necessitates production of multiple mRNAs via alternative processing of primary transcripts. Mapping of the genomic sequences that are copied into mRNAs by such techniques as visualization of RNA–DNA hybrids in the electron microscope (Berget *et al.*, 1977; Chow *et al.*, 1977; Chow *et al.*, 1979), nuclease S1 mapping (Berk & Sharp, 1977), and sequencing of cDNAs (Freyer *et al.*, 1984) established that primary transcripts may be alternatively spliced (e.g. E1A pre-mRNA) or subjected to both this modification and alternative polyadenylation (e.g. E2, E3 pre-mRNAs). The most extreme example of the latter phenomenon among any known viral pre-mRNAs is processing of ML pre-mRNAs:

these RNAs can be polyadenylated at any one of five different sites to give rise to five families of mRNAs, designated L1 to L5, and all but the L5 family includes multiple, alternatively spliced mRNAs linked to a common 5′ tripartite leader sequence built from three small exons (11, 12, and 13) (Fig. 5.2). Furthermore, additional 5′ exons that are included in some, but not spliced ML mRNAs have been identified, notably the so-called i-leader (Chow *et al.*, 1979; see also Fig. 5.2).

The compression of the open reading frames for more than 40 viral proteins into only 10 transcription units reduces the need to devote limited genetic information to control sequences such as promoters. Sizeable segments of the viral genome also contribute to more than a single transcription unit. Such "multiple usage" of genetic information is, again, epitomized by the ML transcription unit, which includes the sequences of two additional genes transcribed by RNA polymerase II (the E3 gene and that expressed from the L4 promoter), as well as the two VA-RNA genes transcribed by RNA polymerase III (Fig. 5.1). Similarly, the protein IX gene lies entirely within, and shares a poly(A)-addition site with, the E1B transcription unit, and the IVa2 gene in the l-strand overlaps the 3′ end of the E2 transcription unit, and lies within the coding sequence for the viral DNA polymerase (Figs. 5.1 and 5.2). Although coding and control sequences on one strand are largely interspersed with one another and among those on the other strand, there are several examples of sequences that function in both orientations. For example, the r-strand sequence that forms the control and 5′ transcribed sequence of the E3 transcription unit also encodes the C-terminal, 96 amino acids of protein VIII and the 3′ untranslated region of the L4 mRNAs. Similarly, the ML promoter, the first ML three exons and the introns between them in the r-strand encode the viral DNA polymerase or pre-TP in the l-strand (Figs. 5.1 and 5.2).

Proteins that perform related functions are often encoded in the same transcription unit; for example, all but one of the structural protein in the ML and the replication protein in the E2 transcription units (Fig. 5.2). Beyond such grouping, no obvious guiding principle is evident from inspection of the organization of mastadenoviral genomes. However, when the arrangements of coding sequences are compared among representatives of all genera, it is clear that genes for proteins that perform conserved core functions, such as the structural and replication proteins, are flanked on either side by non-conserved sequences that vary considerably in length, and in the number and nature of the proteins specified (Fig. 5.3).

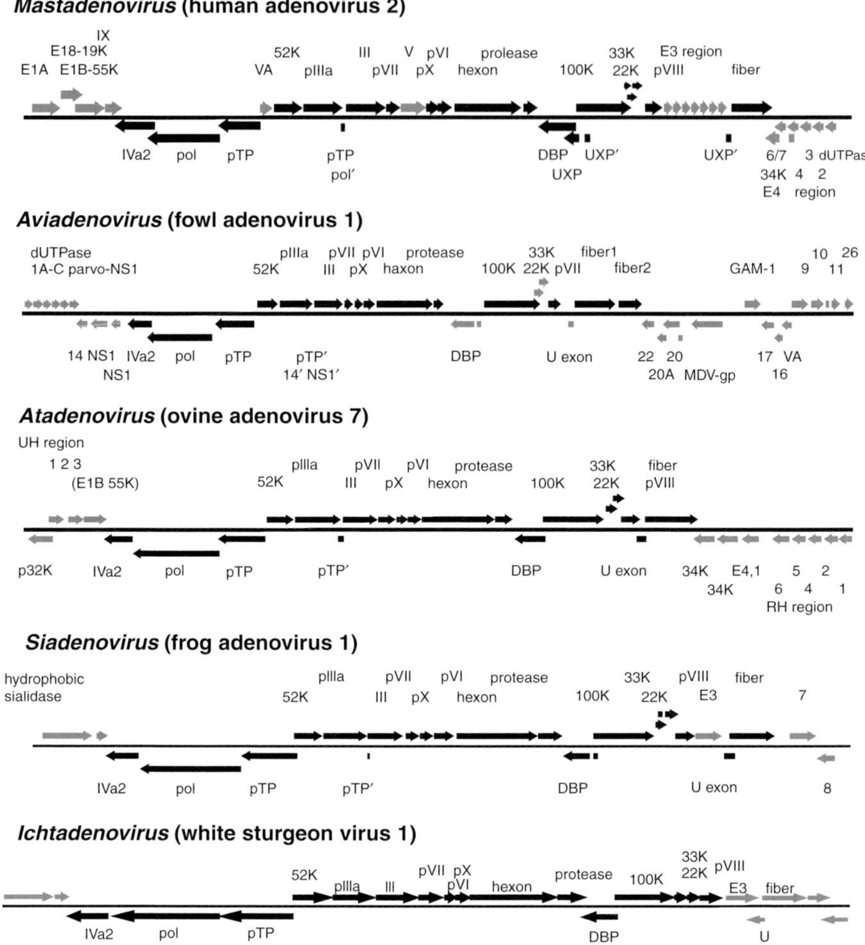

Figure 5.3. Adenovirus coding sequences. The organization of the coding sequences of representative genomes of the five adenovirus genera are shown schematically. Coding sequences conserved in all genera are indicated in black and those represented in more than one genus or unique to a genus in gray. pX is also known as pμ. Adapted from Harrach *et al.* (2012) and Kovacs *et al.* (2003), with permission.

The little information available suggests that major aspects of transcriptional organization (and hence strategy) are also conserved, including a ML transcription unit with a leader sequence common to the mature mRNAs (Payet *et al.*, 1998), and the kinetic patterns of expression of most genes (Wu *et al.*, 2013). On the other hand, not all structural proteins are:

protein IX and core protein V are unique to mastadenoviruses (Fig. 5.2). Furthermore, the major transcriptional regulators of these adenoviruses, the E1A proteins (see below), are not encoded in the genomes of members of the other genera (Fig. 5.3). Related functions may be performed by proteins unique to a particular genus: the FAdV-1 GAM-1 protein has been reported both to block apoptosis (Chiocca *et al.*, 1997) and, in conjunction with the viral Orf 22 protein, to interact with the retinoblastoma (RB) protein to activate E2F (Lehrmann & Cotton, 1999), as do HAdVC E1A proteins (see below). In this context, it is noteworthy that the coding sequence for the viral proteins that ensure efficient expression of the ML transcription unit, the IVa2, L4 33K, and L4 22K proteins, are conserved in all genera, as is the gene for the major translational regulator, the L4-100K protein (Fig. 5.2), which also acts as a hexon chaperone (Chapter 7).

5.2.2 *RNA polymerase III transcription units*

The ~160 nucleotide viral RNA now known as virus-associated (VA) RNA I was the first non-coding viral RNA to be described (Ohe & Weissman, 1970), and also the first identified viral RNA polymerase III transcript (Price & Penman, 1972; Weinmann *et al.*, 1974). Genes specifying a related RNA, VA RNA II, are present in the genomes of other species C HAdVs (Mathews, 1975; Pettersson & Philipson, 1975), located close to the VA RNA I gene in the ML transcription unit (Fig. 5.1). Transcription of these genes is directed by internal sequences typical of many RNA polymerase III promoters, A and B boxes that are binding sites for the essential initiation protein TFIIIC, and necessary for synthesis of VA RNAs in infected cells (Bhat & Thimmappaya, 1984; Fowlkes & Shenk, 1980; Lassar *et al.*, 1983; Pruzan & Flint, 1995). Both VA RNA genes are transcribed during the early phase of infection, but the rate of VA RNA I synthesis accelerates continuously following entry into the late phase, whereas that of VA RNA II does not (Söderlund *et al.*, 1976). This difference has been attributed, in part, to genome replication-dependent relief of repression of VA RNA I transcription by a limiting inhibitory protein that binds to the intragenic (van Dyke & Roeder, 1987). However, efficient VA RNA I transcription during the late phase also reduces VA RNA II transcription; mutations in the VA RNA I promoter that impair transcription of this gene lead to increased synthesis of VA RNA II with kinetics resembling those of VA RNA I synthesis in wild-type virus-infected cells (Bhat & Thimmappaya, 1984). Presumably as a result of this difference in the rates of transcription, VA RNA I accumulates to exceptionally high concentrations during the late phase, about 10^8 copies per cell, similar to the concentration of ribosomes, whereas the concentration of VA RNA II is some 10-fold lower (Mathews & Pettersson, 1978). Whether this difference is related to, or necessary for, the unique functions fulfilled by VA RNA I is not clear.

Genetic and biochemical analyses indicated that VA RNA I allows efficient protein synthesis during the late phase of infection (Reichel *et al.*, 1985; Schneider *et al.*, 1985; Thimmappaya *et al.*, 1982), a function that facilitates virus reproduction in the face of the antiviral actions of interferons (Chapter 9). Subsequently, this small viral RNA was reported to antagonize the host RNA-editing enzyme adenosine deaminase (Lei *et al.*, 1998), and to serve both as a pathogen-associated molecular pattern (PAMP) to activate innate antiviral defenses (Chapter 9), and as a precursor for viral miRNAs (see next section). Deletion of the VA RNA II gene has no effect on the reproduction of HAdV-C5, at least in established lines of human cells in culture, but further decreases the yield of a mutant lacking the VA RNA I gene (Thimmappaya *et al.*, 1982). No function was ascribed to VA RNA II until more recent studies of the origin of viral miRNAs (see next section).

The genomes of all mastadenoviruses that infect primates contain at least one VA RNA gene, always of the VA RNA I type, but no analogous RNA polymerase III transcription units are present in the genomes of other members of this, or any of the other genera (King & Krebsbach, 2013). The FAdV-1 (a.k.a. chicken embryo lethal orphan virus) genome does include a gene transcribed by this enzyme (Fig. 5.2), but the ~90-nucleotide RNA has no sequence homology with VA RNAs (Larsson *et al.*, 1986), and its function has not been established.

The HAdV-C5 genome also contains a third RNA polymerase III transcription unit, with a promoter and transcribed sequences superimposed on the E2E RNA polymerase II promoter and 5′ exon, that directs synthesis of two short RNAs (45 and 90 nucleotides) (Ellsworth *et al.*, 2001; Huang *et al.*, 1994; Pruzan *et al.*, 1992). In contrast to the VA RNAs, these small E2E RNAs accumulate to only very low concentrations in infected cells and are unstable (Huang & Flint, 2003). These properties, the superposition of the RNA polymerase II and III E2E promoters between positions −30 to +20 (Ellsworth *et al.*, 2001), and similar efficiencies of E2E transcription by the two enzymes during the early phase of infection (Huang & Flint, 2003) suggest that RNA polymerase III transcription serves to reduce production of E2 mRNAs, rather than to produce functional RNAs. However, the superimposition of the RNA polymerase II and III E2E promoters with another in the l-strand and with the coding sequence for the L4 33K protein in the r-strand have made it difficult to investigate the contribution of the RNA polymerase III E2E transcription unit to viral reproduction.

5.2.3 *miRNAs*

Mammalian microRNAs (miRNAs) are cleaved from their initial precursors in the nucleus to form pre-miRNAs that are then exported to the cytoplasm via the export receptor exportin-5 (XPO5). Within the cytoplasm, ds miRNAs of 22 bp are generated via cleavage by the RNase III enzyme dicer and loaded into RNA-induced silencing complexes (RISCs) that contain an argonaute (AGO) protein and other proteins. During this process, the RNA is unwound and one strand (the guide) retained while the second is discarded. Base-pairing of sequences near the 5′ end of the miRNA, called the seed sequences, with complementary sequences in an mRNA leads to cleavage of the mRNA when the RISC contains AGO2, or inhibition of translation and subsequent mRNA degradation when an alternative AGO protein (the human genome encodes four) is present (Cullen, 2004; Bartel, 2009; Fabian & Sonenberg, 2012; Peters & Meister, 2007). VA RNAs can disrupt this process, concurrent with synthesis of viral miRNAs, by two mechanisms.

The VA RNAs and pre-miRNAs have some secondary structure features in common, and, like the host RNAs, VA RNAs are exported from the nucleus by XPO5 (Gwizdek *et al.*, 2003). Synthesis of VA RNA I or II in the absence of other viral gene products to concentrations less than those attained during the late phase of infection blocks miRNA-mediated repression of reporter genes (Andersson *et al.*, 2005; Lu & Cullen, 2004), in part by efficient competition for binding to XPO5 (Lu & Cullen, 2004).

In addition, VA RNA I binds specifically to dicer to interfere with processing of cellular pre-miRNAs (Andersson *et al.*, 2005; Lu & Cullen, 2004). As these observations predict, silencing of reporter genes by RNA interference is impaired in HAdV-C5 infected cells. Such inhibition correlates temporally with competitive inhibition of cleavage of their substrates by dicer and RISC (Andersson *et al.*, 2005). Examination of cells infected by mutants defective for synthesis of either VA RNA I or II established that both small viral RNAs inhibit dicer, but VA RNA I to a greater degree (Andersson *et al.*, 2005). This difference may simply be the result of the greater quantities of VA RNA I described above. In fact, both VA RNAs can serve as substrates of dicer *in vitro* and in infected cells, and are cleaved to produce viral miRNAs that are incorporated into RISCs and can silence reporter mRNAs that contain complementary sequences (Andersson *et al.*, 2005; Aparicio *et al.*, 2006; Sano *et al.*, 2006; Xu *et al.*, 2007). Such viral miRNAs are also produced in persistently infected lymphoid cells (Furuse *et al.*, 2013).

More recently, the miRNA repertoire of permissive cells infected by HAdV-C5 has been catalogued in depth by high-throughput sequencing of the pools of 20–40-nucleotide RNAs after increasing periods of infection (Zhao *et al.*, 2013) or of the small RNAs associated with endogenous RISCs, isolated by immunoprecipitation with antibodies that bind to the four human AGO proteins (Bellutti *et al.*, 2015). In both cases, the vast majority of the viral miRNAs contains sequences of VA RNA I or VA RNA II, indicating that other regions of the viral genome do not encode miRNAs. However, the representation of viral miRNAs derived from different regions of the VA RNAs was different in the two populations. Notably, small RNAs that map to the central regions of the VA RNAs were detected in the total small RNA pool (Zhao *et al.*, 2013), but were not present to a significant degree in the viral miRNAs incorporated into RISCs (Bellutti *et al.*, 2015). These RNAs therefore appear to represent side products of the dicer and RISC-processing reactions that liberate viral miRNAs from their VA RNA precursors. Most of the VA RNA I-derived miRNAs are generated from the 3′ end by cleavage at various positions in the terminal stem, whereas a single VA RNA II-derived miRNA, also from the 3′ end, predominates in RISCs isolated from infected cells (Bellutti *et al.*, 2015; see also Fig. 5.4). Overall, miRNAs from VA RNA II are

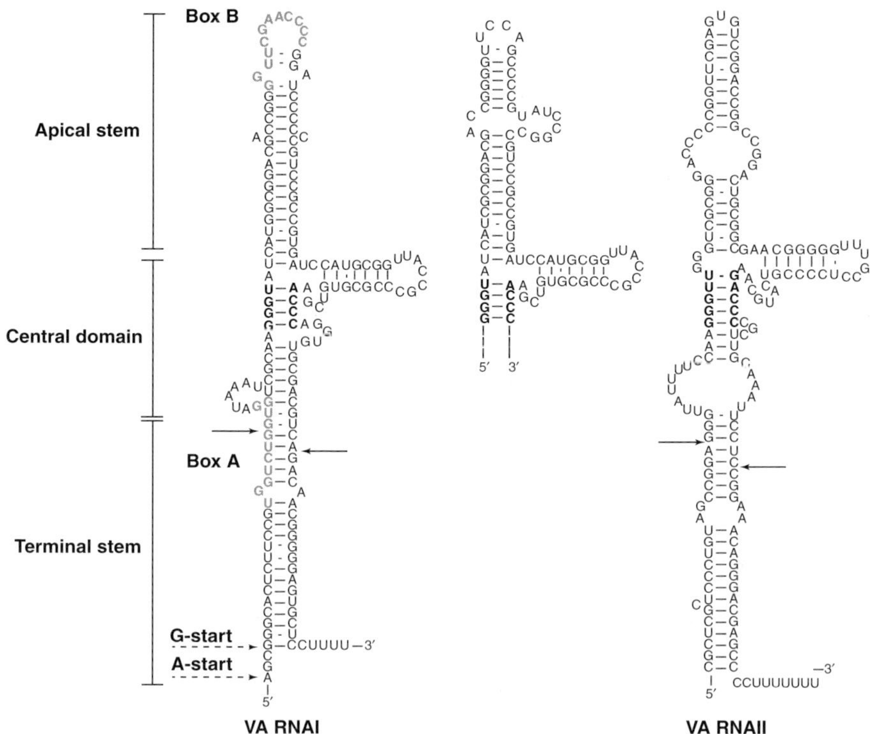

Figure 5.4. The species C HAdV VA RNAs. The secondary structures of HAdV-C2 VA-RNA I and VA-RNA II shown were deduced from the results of a combination of RNAse protection assays, analysis of the sensitivity of residues to chemical probes and site-directed mutagenesis (Ma & Mathews, 1993; Coventry & Conn, 2008). As indicated, the apical stem of VA RNA I can adopt alternative structures (Wahid *et al.*, 2009). The sequences in gray indicate the A and B box motifs of the RNA polymerase III promoters, and nucleotides in bold face a conserved, tetranucleotide, base-paired sequence critical for the structure and function of VA RNA I (Clarke *et al.*, 1994; Ma *et al.*, 1993). The alternative A and G initiation sites for synthesis of VA RNA I, which contribute to the heterogeneity of miRNAs processed from this RNA, are indicated by the dashed arrows, and sites of dicer cleavage shown by the solid arrows. Adapted from Punga & Akusjarvi (2013), with permission.

more abundant, as judged by the number of their sequences detected in both this population (Bellutti *et al.*, 2015) and in RISCs isolated from cells overproducing AGO2 (Xu *et al.*, 2007). The VA RNAs of species B, D, and E also serve as precursors for production of RISC-associated miRNAs (Kamel *et al.*, 2013).

Putative cellular mRNA targets of various VA RNA-derived miRNAs have been sought based solely on searches for seed-complementary sequences (Zhao *et al.*, 2013), and by cataloguing of mRNAs either decreased in concentration in cells synthesizing VA RNAs or mimics of individual viral mRNAs (Aparicio *et al.*, 2010; Bellutti *et al.*, 2015) or in the mRNA populations associated with RISCs under such conditions (Bellutti *et al.*, 2015). These methods have identified mRNAs that appear to be regulated by the two major miRNAs generated from the 3′ end of VA RNA I. The mRNAs encode proteins that participate in such processes as signal transduction, regulation of apoptosis, cell growth, and DNA repair. These observations indicate that the viral miRNAs can modulate the expression of specific cellular genes, but the impact of such changes on virus reproduction is not clear; mutations that eliminate the seed sequences of the VAI-derived miRNAs did not impair synthesis of viral late proteins, or viral cytotoxicity in mutant virus-infected 293 or HeLa cells (Kamel *et al.*, 2013). It is possible that the many genetic and epigenetic changes characteristic of these transformed human cells prevented detection of beneficial consequences of inhibition of cellular gene expression by the adenoviral miRNAs, or that such changes influence intrinsic or adaptive responses to infection that were not examined in these experiments.

5.3 Temporal Regulation of Viral Gene Expression

Adenoviral protein-coding genes are expressed in a stereotyped temporal sequence by mechanisms that directly or indirectly activate transcription of all but the E1A gene, the first to be expressed in infected cells. The parameters that govern the transitions between one phase of expression and the next are reasonably well understood. Immediate early (IE) E1A gene products induce efficient transcription of viral E genes and hence synthesis of a variety of viral non-structural, regulatory proteins. Such proteins include those necessary for viral DNA synthesis, which commences as the E2 replication proteins accumulate, typically around 6–8 hours p.i. in established cultured human cell lines. The increase in concentration of viral DNA templates allows transcription from additional viral promoters, including the IVa2 promoter. By convention, these are late genes, because their expression requires viral DNA synthesis. Consequently, a fitting name for this class is 'immediate late,' although they are often termed intermediate (or misleadingly 'delayed early'). Proteins encoded within the IVa2 and L4 transcription units substantially increase transcription from the ML promoter, which controls production of most the structural and several important non-structural proteins (Fig. 5.2).

Although transcriptional regulation drives progression through the infectious cycle, viral proteins that operate post-transcriptionally are essential to achieve synthesis of the complete panoply of viral mRNAs, particularly of the ML mRNAs. Such proteins also induce selective expression of viral genetic information during the late phase of infection.

5.3.1 *Viral templates for transcription*

At the beginning of infection viral genomes enter nuclei associated with a major core protein of virus particles, protein VII (Chapter 4). Characterization of the products of micrococcal nuclease digestion of intranuclear viral DNA led to the suggestion that genomes are repackaged by cellular nucleosomes (Daniell *et al.*, 1981; Sergeant *et al.*, 1979; Tate & Philipson, 1979). Nevertheless, direct examination of proteins associated with viral DNA by UV light-induced crosslinking (Chatterjee *et al.*, 1984) or chromatin immunoprecipitation (ChIP) (Haruki *et al.*, 2003; Johnson *et al.*, 2004; Xue *et al.*, 2005) led to the conclusion that protein VII remains bound to viral DNA during the early phase of infection, and then dissociates. During this period, the core protein forms discrete foci, which then become diffusely localized (Xue *et al.*, 2005). This morphological change is transcription dependent and appears to be the result of alteration in the intranuclear organization of protein VII, rather than in its binding to viral DNA (Chen *et al.*, 2007; Komatsu *et al.*, 2011).

Histone chaperones that can remodel chromatin, such as template acting factor I (TAFI, a.k.a. nucleosome assembly protein 1) and TAFII stimulate replication and transcription of HAdV-C5 DNA–protein VII templates *in vitro*, and TAFI colocalizes with protein VII during the early phase of infection (Xue *et al.*, 2005). The four core histones as well as acetylated histone H3 were also found (by ChIP) to become associated with viral DNA as the early phase of infection progressed (Komatsu *et al.*, 2011). Furthermore, viral DNA molecules that were immunoprecipitated with antibodies against acetylated H3 were also associated with RNA polymerase II and protein VII (Komatsu *et al.*, 2011). Although TAF1 was not required for recruitment of acetylated H3 to viral DNA, inhibition of its production by RNAi decreased the efficiency of the IE and E gene expression by a factor of two. Comparison of the ability to support viral gene expression in transfected cells of purified HAdV-C5 DNA linked to the TP and the complete viral core indicated that protein VII enhances both recruitment of histones and early mRNA synthesis, independently of any effects of entry of genomes into the nucleus. These observations are consistent with a model in which recruitment of histones (including acetylated H3) to viral DNA associated with protein VII facilitates transcription during the early phase of infection. Whether the two classes of proteins are differentially located on the genome (e.g. at promoters or internal transcribed regions), and organized in the same manner on individual viral DNA molecules is not known. In this context, it would be of considerable interest to compare transcriptionally active templates with viral DNA molecules that reach the nucleus but do not function as templates.

The nature of the template(s) for transcription of late genes following the onset of viral DNA synthesis has received relatively little attention. However, loss of protein VII is accompanied by an increase in the binding of core histones in the ratio of which they are present in cellular chromatin, suggesting that nucleosomes are deposited on viral genomes (Komatsu & Nagata, 2012). Only a small fraction of viral DNA molecules present during the late phase is transcribed (Beyer *et al.*, 1981), and whether this subset, transcriptionally inert molecules, or both, are nucleosome-associated has not been addressed. However, core histones have also been reported to be associated with mutant viral DNA genomes that are transcribed and replicated to only very limited degrees (Ferrari *et al.*, 2012; Ross *et al.*, 2011), perhaps an indication of a silencing role for these cellular proteins. Viral DNA templates for late transcription are not distributed throughout the nucleus, but rather localized to the surfaces of the viral replication centers described in Chapter 6. Synthesis and initial processing of viral late pre-mRNAs takes place at these sites (Aspegren *et al.*, 1998; Bridge & Pettersson, 1996; Bridge *et al.*, 1995; Pombo *et al.*,

1994; Puvion-Dutilleul *et al.*, 1992), to which initiation competent RNA polymerase II and essential initiation proteins like the TATA binding protein are recruited (Fuchsova *et al.*, 2015).

5.3.2 *The immediate early EIA transcriptional activators*

5.3.2.1 *The E1A gene and its products*

The E1A transcription unit (see Fig. 5.1) is expressed within an hour or two after infection of established human cell lines by species C HAdVs (Nevins *et al.*, 1979). It includes a typical RNA polymerase II promoter with a TATA sequence required for transcription (Hearing & Shenk, 1983a; Osborne *et al.*, 1982) preceded by two enhancer elements (termed enhancers I and II) that are necessary for efficient E1A gene expression in infected cells (Hearing & Shenk, 1983b; Hearing & Shenk, 1986). Deletion of the entire enhancer region reduced E1A mRNA concentrations in infected cells some 15-fold (Hearing & Shenk, 1983b). In contrast to enhancer I, enhancer II is required in *cis* for efficient expression of all viral early genes (Hearing & Shenk, 1986). Cellular proteins that bind to repeated core element of enhancer I have been identified, including GA binding proteins (Bruder & Hearing, 1989; Bruder & Hearing, 1991; Yoshida *et al.*, 1989), one of which proved to be a member of the Ets family (Higashino *et al.*, 1993). Whether the distribution, concentration or activity of these proteins contributes to reported cell differentiation- or proliferation-dependent differences in E1A expression (Dieckmann & Krippl, 1994; Gonzalez *et al.*, 2006; Herbst *et al.*, 1990) is not known.

Primary E1A transcripts are polyadenylated at a single site, but alternative splicing yields five mRNAs that encode distinct proteins (Fig. 5.5). The two larger mRNAs are the most abundant during the early phase of infection, with the others accumulating during the late phase (Chow *et al.*, 1979; Perricaudet *et al.*, 1979; Stephens & Harlow, 1987). The proteins specified by these mRNA share a short N-terminal sequence, and all sequences of the 243R, 217R and 171R of HAdV-C5 are present in the largest 289R E1A protein (Fig. 5.5). A comparison of the E1A sequences of seven human and one simian adenovirus identified three conserved regions, CR1 and CR2 in the 5′ exon common to the 243R and 289R proteins, and CR3 unique to the latter (Kimelman *et al.*, 1985). Subsequent

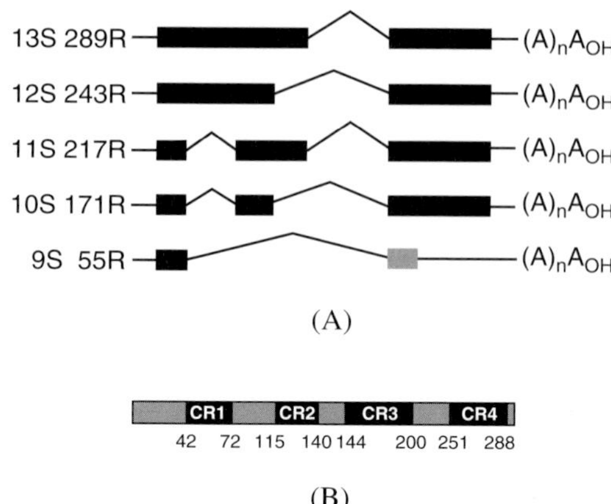

Figure 5.5. HAdV-C5 E1A proteins. (**A**) The alternatively spliced E1A mRNAs, named for their apparent sizes, are drawn in the direction of transcription, with introns and poly(A) indicated by gaps and (A)nA$_{OH}$, respectively. Coding sequences are indicated by the boxes. The lengths of the proteins, listed for HAdV-C5 at the right, vary with *Mastadenovirus* type. The alternative reading frame translated to produce the smallest E1A protein is shown in light gray. (**B**) Conserved regions 1–4 (CR1–4) are shown in the 289R E1A protein, based on sequence comparisons of the E1A proteins of 15 HAdVs, representing species A–F, and three SiAdV types (Avvakumov *et al.*, 2002).

analyses confirmed the conservation of these E1A regions among representatives of species A–F HAdVs, identified CR4, which is located near the C-terminus (Fig. 5.5), and established that CR3 exhibits the highest degree of sequence identity or similarity (Ablack *et al.*, 2010; Avvakumov *et al.*, 2002; Schaeper *et al.*, 1998). As would be anticipated, CR1–CR4, which serve as binding sites for numerous cellular proteins, are important for the regulation of transcription by the major E1A proteins (see below).

The 217R and 181R proteins share sequences with the larger E1A proteins (Fig. 5.5), but little is known about their functions or roles in infection: they are not required for reproduction of HAdV-C5 in transformed human cells and cannot cooperate with activated Ras or other oncogenes in transformation of rodent cells (Stephens & Harlow, 1987; Ulfendahl *et al.*, 1987). The 55R E1A protein has been shown to stimulate expression of viral early and late genes in quiescent human fibroblasts, but to a much more modest degree from the major E1A activators (Miller *et al.*, 2012). This E1A protein alone also supports limited virus reproduction, an activity that depends on its association with the proteasome subunit S8 (Miller *et al.*, 2012). It contains an activation domain and interacts with unliganded thyroid hormone receptor when overproduced in mammalian cells (Arulsundaram *et al.*, 2014), but the contributions of these properties to the effects of the 55R proteins on viral gene expression and reproduction have not been investigated.

5.3.2.2 *Activation of viral gene expression by CR3*

Mutants of HAdV-C5 with alterations that prevent synthesis of the 289R E1A protein are severely impaired for reproduction in non-complementing cells, as a result of substantial defects in transcription from viral early promoters (Berk *et al.*, 1979; Jones *et al.*, 1979; Nevins, 1981). This protein is also auto-regulatory, stimulating transcription from the E1A promoter in infected cells (Berk *et al.*, 1979; Hearing & Shenk, 1986; Nevins, 1981). The importance of CR3-dependent stimulation of transcription for successful virus reproduction and the then mysterious ability of this short E1A segment to promote transcription from a wide variety of promoters without sequence-specificity provoked intense investigation of its mechanism of transcriptional activation.

A 65-residue CR3 sequence (amino acids 139–204 in the HAdV-C5 289 protein) is sufficient for strong activation of transcription when recruited to promoters by fusion with the DNA-binding domain of the yeast transcriptional regulator Gal4 (Lillie & Green, 1989; Martin *et al.*, 1990), and essential for stimulation of transcription by the full-length 289R protein (Webster *et al.*, 1991). This segment includes an essential N-terminal zinc finger (Fig. 5.6) (Culp *et al.*, 1988; Webster *et al.*, 1991). The adjacent C-terminal sequence is also absolutely required, as it directs recruitment of the E1A protein to promoters via interaction with various cellular sequence-specific DNA-binding regulators such as members of the ATF family (Chatton *et al.*, 1993; Lee *et al.*, 1987; Liu & Green, 1994). The following segment of 10–12 residues (depending on the HAdV type), which is rich in negatively charged amino acids, is required for maximal stimulation of transcription from viral early promoters in transient expression assays (Strom *et al.*, 1998), but its molecular function is still unknown.

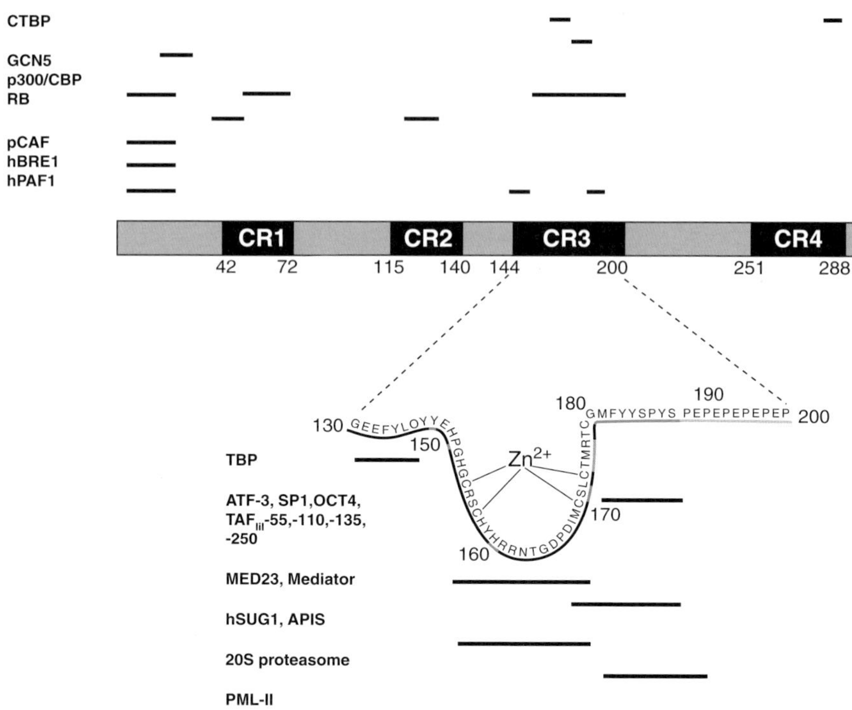

Figure 5.6. E1A-interacting proteins implicated in regulation of transcription of viral genes. The 289R E1A sequence and conserved regions are depicted as in Fig 5.5B, with the CR3 region expanded below. Although the structure of this region has not been determined, the central region has been shown to bind Zn^{2+} via the residues indicated and formation of the Zn finger is essential for CR3-dependent activation of transcription (see text). E1A sequences required for binding to cellular proteins that interact with the N-terminal region CR1, CR2, and CR4, and in some cases also with CR3 are indicated by the lines above, with the interacting proteins listed at the left. Regions required for binding of proteins that interact with CR3 are summarized below. Most of these proteins are described in the text. CBP, CREB-binding protein; OCT 4, POU domain transcription factor OCT 4; TAFII 55 (etc.), transcription initiation factor TFIID 55-kDa subunit (etc.).

The breakthrough in elucidation of the mechanisms by which CR3 stimulates transcription came from what seemed an unlikely experimental approach, analysis of proteins that bound under stringent *in vitro* conditions (0.7M KCl) to the zinc finger (Boyer & Ketner, 2000). The single protein identified in this way, now designated MED23, proved to be a subunit of human mediator, a large, multi-subunit (26 in humans) co-activator that is essential for the regulation of transcription of most, if not all, genes by RNA polymerase II, and conserved from yeast to humans (Allen & Taatjes, 2015; Conaway & Conaway, 2011; Taatjes, 2010). Subsequent biochemical and proteomic studies established that CR3 binds to the complete mediator (Cantin *et al.*, 2003; Vijayalingam & Chinnadurai, 2013). This interaction stimulates assembly of RNA polymerase II preinitiation complexes (Cantin *et al.*, 2003; Wang *et al.*, 2001), and also appears to facilitate transcriptional elongation: the set of CR3-associated proteins include subunits of the super-elongation complex (SEC), and RNAi-mediated knockdown or inhibition of CDK9, the catalytic SEC subunit, impaired synthesis of viral early proteins and virus reproduction (Vijayalingam & Chinnadurai, 2013). CR3-dependent expression of reporter genes and production of viral E2 mRNA following infection are severely defective in murine MED23 -/- cells (Stevens *et al.*, 2002), as is reproduction of murine MAdV-1 and synthesis of its early mRNAs (Fang *et al.*, 2004). These observations argue that the mechanism of CR3-depdendent stimulation of transcription is conserved among mastadenoviruses. Consistent with this view, the CR3 regions of HAdVs representative of species A-F fused to the Gal4 DNA-binding domain all stimulate reporter gene expression in both normal and transformed human cells, albeit to different degrees, the full-length E1A proteins exhibit CR3-dependent binding to MED23, and MED23 is required for activation of transcription by the CR3 segments (Ablack *et al.*, 2010).

A number of other cellular proteins that associate with CR3 (Fig. 5.6) have also been implicated in regulation of transcription. These proteins include the hSUG1 subunit of the 19S ATPase independent of 20S (APIS) and, independently, the 20S proteasome, both of which can be detected at viral early promoters by ChIP (Rasti *et al.*, 2006). Interaction of the former protein is conserved among species A–F large E1A proteins, and inhibition of its production by RNAi delayed synthesis of viral early proteins in HAdV-C5-infected cells (Rasti *et al.*, 2006). Comparison of the half-lives of the larger E1A protein and of derivatives with amino acid substitutions that impair activation of transcription, in conjunction with examination of the effects of proteasome inhibition, indicated that E1A protein turnover is necessary for efficient transcriptional activation (Rasti *et al.*, 2006). It has been proposed that this property, which has been reported for other viral activators, could facilitate multiple cycles of initiation of transcription, but exactly how such re-initiation might occur is not known.

Several lysine acetyltransferases initially shown to bind to more N-terminal E1A sequences (Fig. 5.6) also interact with CR3, including GCN5 (a.k.a. KAT2A), PCAF (a.k.a. KAT2B), p300 (a.k.a. KAT3B), and CBP (a.k.a. KAT3A) (Ablack *et al.*, 2012; Pelka *et al.*, 2009a; Pelka *et al.*, 2009b). Binding to these enzymes is conserved among species A–F large E1A proteins (Ablack *et al.*, 2010; Ablack *et al.*, 2012). Both p300 and GCN5 are associated with the E4 promoter, dependent on synthesis of the 289R E1A protein, in infected cells (Ablack *et al.*, 2012; Pelka *et al.*, 2009b). However depletion of GCN5 by RNAi or specific inhibition of its acetyltransferase activity increased expression from this viral promoter in wild-type mouse embryonic fibroblasts, but not in *GCN5* knockout cells. Such inhibition was associated with reduced recruitment of active RNA polymerase II to the E4 promoter and decreased virus yield (Ablack *et al.*, 2012), indicating that GCN5 negatively regulates CR3-dependent transcription. This finding suggests that at least for some promoters, such as that controlling production of E4 proteins, which can be cytotoxic (Lavoie *et al.*, 1998), E1A-dependent transcription must be curtailed to support optimal virus reproduction. The C-terminal binding protein (CTBP), first identified by virtue of its association with E1A CR4 (Fig. 5.6) and subsequently shown to be a corepressor (see below) may also limit CR3-dependent transcription. This protein also binds, albeit less strongly, to CR3 and reduces stimulation of transcription by this E1A segment in a reporter assay (Bruton *et al.*, 2008).

In contrast to GCN5, the other E1A-interacting lysine acetyl transferases appear to be necessary for maximal stimulation of transcription by CR3 (Ablack *et al.*, 2012; Pelka *et al.*, 2009a). The participation of these enzymes in regulation of viral transcription by the 289R E1A protein is generally interpreted in terms of modification of histones (the first KAT substrates to be identified) associated with viral DNA (Ablack *et al.*, 2012; Pelka *et al.*, 2009a). Histones acetylated at particular lysine residues are present on viral DNA molecules in infected cells. However, as discussed previously, whether the presence of these cellular proteins is necessary for expression of viral early genes has not been established. Nor has participation in E1A-dependent transcription of other KAT substrates, which include a variety of transcriptional regulators (Appella & Anderson, 2001; Glozak *et al.*, 2005), been eliminated.

Association of CR3 with 1 of the 6 human isoforms of the promyelocytic leukemia (PML) protein (isoform II) has also been reported to stimulate synthesis of E2 mRNA and increase virus yield modestly in HAdV-C5 infected human hepatocytes (Berscheminski *et al.*, 2013). The substantial number of cellular proteins that can interact with CR3 (Fig 5.6) to promote stimulation of transcription by the large E1A protein raises the question of the relative importance of each. However, no direct comparisons have been reported, and the effects of inhibition of the synthesis or activity of the individual proteins were examined under different conditions, for example in different cell lines. It would therefore be of considerable interest to compare CR3-dependent transcription of viral early genes in normal human cells, particularly cells resembling those infected in nature, when they do and do not produce the various KATs or other CR3-interacting proteins.

5.3.2.3 *Regulation by CR1 and CR2*

In contrast to CR3, the preceding E1A segment common to this and the 243R protein appears to be intrinsically disordered (Pelka *et al.*, 2008). The highly conserved CR1 and CR2 sequences have received the most attention, in large part because they are required for E1A-dependent transformation and the mitogenic activity of the protein (Braithwaite *et al.*, 1983; Howe *et al.*, 1990; Spindler *et al.*, 1985; Stein *et al.*, 1990; Zerler *et al.*, 1987). Each of these regions, as well as the N-terminal segment, binds alone, or in conjunction with one another, to a considerable number and variety of cellular proteins (Fig. 5.6). The association of relatively short segments with so many different proteins, in many cases via overlapping sequences, has been attributed to the intrinsic plasticity of the disordered protein and hence the ability to adopt multiple folded structures upon contact with other proteins (Pelka *et al.*, 2008). The molecular consequences and physiological importance of all these interactions are not known, but some serve to prime the host cell to support viral DNA synthesis (Chapter 6) or to counter intrinsic antiviral defenses (Chapter 9).

Furthermore, the association of these E1A regions with several cellular proteins is important for optimal expression of viral early genes, a function epitomized by the first E1A-interacting protein to be identified, RB (Harlow *et al.*, 1986).

During the early phase of infection, the E2 transcription unit that encodes viral replication proteins is transcribed from the upstream E2E promoter (Fig. 5.1). This control sequence includes a pair of inverted binding sites for transcriptional regulators of the E2F family, which are required for stimulation of E2E promoter activity by the 243R E1A protein (Adam *et al.*, 1986; Bagchi *et al.*, 1990; Kovesdi *et al.*, 1986; Phelps *et al.*, 1991). The E2F proteins are heterodimers of a DP1 or DP2 subunit with one of the six E2Fs that regulate cell cycle progression, as well as quiescence and senescence (Berk, 2005; Frolov & Dyson, 2004; La Thangue, 2003). Binding of RB to an E2F activator represses transcription from E2F-depdendent promoters, as RB recruits histone deacetylases (HDACs) and methylases (Brehm *et al.*, 1999; Frolov & Dyson, 2004; Ross *et al.*, 1999). The 243R E1A protein binds to RB primarily via CR1 and CR2 (Berk, 2005). The latter E1A region contains a high affinity RB-binding site with the motif LXCXE subsequently identified in other viral and cellular proteins that bind to RB, such as SV40 large T antigen and HDACs (Berk, 2005; DeCaprio, 2009; Helt & Galloway, 2003). Displacement of RB from E2F, and hence activation of E2-dependent transcription, also depends on CR1 sequences, which compete with E2Fs (Fattaey *et al.*, 1993; Ikeda & Nevins, 1993) for binding to overlapping sites on RB (Liu *et al.*, 2007).

A second viral early protein, the E4 Orf6/7 protein, augments 243R E1A-dependent transcription from the E2E promoter, where this protein interacts with E2F to promote its cooperative binding to the pair of binding sites in the E2E promoter (Cress *et al.*, 1994; Huang & Hearing, 1989; Marton *et al.*, 1990; Obert *et al.*, 1994). This activity of the E4 Orf6/7 proteins leads to increased synthesis of E2E mRNAs and proteins in infected cells (Marton *et al.*, 1990; Swaminathan & Thimmapaya, 1995). Production of the protein depends on stimulation of transcription from the E4 promoter by the 289R E1A protein. Consequently, maximal stimulation of E2E transcription must be delayed relative to that of other early genes, perhaps to prevent premature entry into the late phase of infection.

The closely related lysine acetyl transferases p300 and CBP, first identified via their association with E1A proteins (Egan *et al.*, 1988; Stein *et al.*, 1990; Whyte *et al.*, 1989), are necessary for maximal E1A protein-dependent stimulation of transcription. Although these proteins can bind to CR3, N-terminal and CR1 sequences form a higher-affinity binding site (Barbeau *et al.*, 1994; Stein *et al.*, 1990; Wang *et al.*, 1993; Whyte *et al.*, 1989). More recently, the E1A N-terminal region has been reported to be associated with hBRE1 (human brefeldin A-sensitivity 1, aka RNF20, zing finger protein 20) (Fonseca *et al.*, 2012), which functions in mono-ubiquitinylation of histone H2B (Osley, 2004). Deletion of the E1A sequences required for hBRE1 binding or RNAi-mediated knockdown of the cellular protein impaired synthesis of E3 and E4 mRNAs in infected cells. hBRE1 recruits the PAF1 subunit of RNA polymerase II-associated factor homolog (PAF) (Fonseca *et al.*, 2013), and depletion of PAF1 decreased production of E3 and E4 mRNAs (as well as E2 mRNAs) considerably, as a result of reduced elongation of transcription by RNA polymerase II (Fonseca *et al.*, 2014). This effect, as well as the decreased association of histone H3 trimethylated at K36 with the 3′ ends of the viral transcription units, are consistent with previously described activities of PAF1 (Kim & Roeder, 2009; Tomson & Arndt, 2013).

5.3.2.4 *Regulation via CR4*

The C-terminal segment of the major E1A protein contains a nuclear localization signal (Lyons *et al.*, 1987) and the binding site for yet another cellular protein first identified by its association with this viral protein, C-terminal binding protein (CTBP1). This and the related CTBP2 are corepressors (Chinnadurai, 2002) that impair transformation of rodent cells by the 243 E1A protein in cooperation with the cellular oncogene H-Ras (Boyd *et al.*, 1993). Other CR4-interacting proteins identified more recently also modulate E1A transforming activity (Cohen *et al.*, 2013), but the relevant molecular mechanisms remain poorly understood (Berk, 2005; Chinnadurai, 2002; Chinnadurai, 2004; Cohen *et al.*, 2013). E1A substitutions that impair binding to CTBP result in a 3–5-fold reduction in virus yield in normal human epithelial and transformed A549 cells, and modest decreases in synthesis of early and late proteins (Subramanian *et al.*, 2013). The molecular function(s) of this interaction in productively infected cells, and hence whether CTB modulates viral gene expression directly or indirectly, remain unknown. The same caveat applies to the DNA replication element-related protein, more recently identified as a direct CR4-binding protein that modestly stimulates synthesis of viral early and late mRNAs (Radko *et al.*, 2014).

5.3.3 *The switch to the late transcriptional program*

Transcription of late genes depends on viral genome replication, but two classes of these genes can be defined. One set is transcribed for the first time following the onset of viral DNA synthesis, the IX, IVa2, UXP, and L4 genes (Berk, 2013). In addition, E2 transcription is taken over by a late phase-specific promoter, E2L. In contrast, the ML promoter is active during the early phase, but is transcribed to its end and at its maximal rate only during the late phase. This pattern is the result of a two-step transcriptional switch, in which the products of some of the first late genes to be expressed stimulate ML transcription and regulate processing of ML transcripts.

5.3.3.1 *Repressor titration*

The IVa2 promoter, which lies in the l strand of the genome some 200 nucleotides upstream of the ML promoter, lacks a TATA sequence but includes an initiator element and an intragenic sequence essential for efficient transcription (Carcamo *et al.*, 1990; Chen & Flint, 1992; Kasai *et al.*, 1992). This intragenic region is also recognized by a repressor; mutations that specifically impair repressor binding (without altering the reading frame for the DNA polymerase in which the promoter lies) increased IVa2 transcription *in vitro*, in reporter assays (Chen *et al.*, 1994; Lin & Flint, 2000; Huang *et al.*, 2003), and in infected cells (Iftode *et al.*, 2004). Furthermore, they led to the early onset of IVa2 transcription in infected cells (Iftode *et al.*, 2004). These observations indicate that initial repression of the IVa2 promoter during the early phase is relieved when viral DNA synthesis increases the concentration of the promoter to above the value at which the repressor, which is not altered in concentration or activity as infection progresses (Lin & Flint, 2000), can occupy all copies. The identity of this repressor has not been established.

Only one sequence related to the IVa2 internal sequence contacted by the repressor, with identity at 13/17 positions, is present in the HAdV-C5 genome (Lin & Flint, 2000). It lies some 300 bp upstream of the initiation site of the E2L promoter and within the more recently defined promoter that controls the L4 transcription unit (Morris & Leppard, 2009; Wright & Leppard, 2013). It is therefore possible that a single repressor, or family of repressors, contributes to restriction of transcription of multiple viral genes to the late phase. Substitutions within this sequence increased the basal activity of the L4 promoter in a transient expression assay (Wright *et al.*, 2015), but whether such mutations stimulate L4 transcription in infected cells has not been determined. Similarly, little is known about the mechanisms that govern the activity of the E2L promoter, apart from the report that it is not sensitive to E1A proteins (Dery *et al.*, 1987), or of the presumed benefit of the switch from the E2E to the E2L promoters. The mechanisms responsible for temporal regulation of UXP transcription are also unknown, although synthesis of UXP depends on viral DNA synthesis (Tollefson *et al.*, 2007; Ying *et al.*, 2010).

Transcription from the pIX promoter is also controlled, at least in part, by relief from repression by a cellular protein. The cellular recombination signal sequence-binding protein 2n binds specifically to a site immediately upstream of the pIX promoter TATA sequence, and mutations that impair such binding stimulate pIX mRNA synthesis specifically during the early phase of infection (Dou *et al.*, 1994). It has also been reported that viral DNA synthesis allows relief of occlusion of the pIX promoter by transcription from the upstream E1B promoter that is active during the early phase (Vales & Darnell, 1989).

5.3.3.2 *Activation and repression of the L4 promoter*

Early studies of late mRNAs established that L4 mRNAs appeared before L2, L3, and L5 ML mRNAs (Larsson *et al.*, 1992). Subsequently, it was demonstrated that the L4 33K and 22K proteins are necessary for synthesis of the full set of ML mRNAs (Farley *et al.*, 2004; see also Section 5.4). These observations strongly suggested that initial production of the L4

33K and 22K proteins could not be the result of transcription from the ML promoter. In fact, a previously unrecognized L4 promoter was identified, for example, by the ability of sequences upstream of the initiation codon common to the 33K and 22K L4 reading frames to direct expression of reporter genes (Morris *et al.*, 2010). Expression of reporter genes from the L4 promoter is stimulated by E1A proteins, and increased substantially when cells were transfected with the HAdV-C5 genome (Morris *et al.*, 2010). The latter property led to the identification of additional viral activators, the IVa2 and E4 Orf3 proteins. As described previously, production of the IVa2 protein requires viral DNA synthesis. Whether transcription from the L4 promoter also depends on genome replication directly, or indirectly because of the need for the IVa2 protein, is not yet clear. However, the L4 promoter includes a sequence of 8 bp that is an exact match to a sequence in the ML element recognized by IVa2 protein alone (see below).

The L4 promoter includes a match to the consensus initiator sequence recognized by the cellular protein TFII-I, and is associated with this general initiation protein in infected cells. Mutation of the initiator sequence increased L4 promoter-dependent reporter gene expression (by some threefold) (Wright *et al.*, 2015). The E4 Orf3 protein has been reported to induce sumoylation of TFII-I (Sohn *et al.*, 2015), suggesting that the viral protein might overcome repression of L4 promoter activity by the cellular protein. Consistent with this model, the modest stimulation of L4 promoter activity by the E4 Orf3 protein was eliminated by mutation of the initiator sequence, and by an E4 Orf3 substitution that ablates all known function of the viral protein (Wright *et al.*, 2015). Furthermore, in HAdV-C5-infected cells the E4 Orf3 protein induced sumoylation, ubiquitinylation and proteasomal degradation of TFII-I, with an increase in L4 22K mRNA synthesis similar to that observed in reporter assays (Bridges *et al.*, 2016).

In reporter assays, the activity of the L4 promoter is also stimulated by overproduction, and impaired by RNAi-mediated knockdown, of p53, a crucial tumor suppressor and transcriptional regulator, even in the presence of the viral activators of L4 transcription (Wright & Leppard, 2013). Furthermore, p53 was specifically associated with the L4 promoter during a short window (some 4 hours from 10 hours p.i.) preceding synthesis of

ML gene products in infected HeLa cells (Wright & Leppard, 2013). The participation of p53 in the regulatory circuits that govern temporal regulation of HAdV-C5 transcription may seem surprising, as it is well established that several viral gene products counter the accumulation or activity of this tumor suppressor. The E1A proteins induce increased production and stability of p53 (Braithwaite *et al.*, 1990; Lowe & Ruley, 1993). However, binding of the E1B 55-kDa protein to p53 (Sarnow *et al.*, 1982b) blocks p53-dependent transcription (Yew & Berk, 1992; Yew *et al.*, 1994). This viral protein is also an E3 sumo ligase, which sumoylates a small fraction of the p53 population present in infected cells to induce partition of most p53 with PML bodies and nuclear export (Kindsmuller *et al.*, 2007; Pennella *et al.*, 2010). Furthermore, modification of p53 by the virus-specific E1B-55 kDa and E4 Orf6 protein-containing E3 ubiquitin ligase targets the cellular protein for proteasomal degradation (Harada *et al.*, 2002; Querido *et al.* 2001a; Steegenga *et al.*, 1998). The E4 Orf3 protein also prevents p53-dependent transcription in infected cells, at least to some degree (DeHart *et al.*, 2015; Soria *et al.*, 2010), by inducing repressive methylation of histone H3 at p53-responsive promoters (Soria *et al.*, 2010). As efficient transcription of early genes requires E1A proteins, the viral proteins that negatively regulate p53 or remove it from infected cells are made later than the E1A proteins. Consequently, there is a relatively short window during the infectious cycle where p53 increases in concentration (Chahal & Flint, 2012; Grand *et al.*, 1994; O'Shea *et al.*, 2004; Querido *et al.*, 1997a), and presumably retains the ability to stimulate transcription from the L4 promoter.

5.3.4 *Regulation of ML transcription*

5.3.4.1 *RNA synthesis to the end of the transcription unit*

The termination site for ML transcription lies downstream of the L5 poly(A)-addition site near the right end of the genome (Fraser *et al.*, 1979). However, examination of RNA labeled during very short periods in infected cells indicated that ML transcription does not proceed beyond the middle of the transcription unit during the early phase, and terminates gradually downstream of the L1 poly (A)-addition site (Iwamoto *et al.*, 1986). Co-infection experiments indicated that only replicated genomes can support the late pattern of ML transcription (Thomas & Mathews, 1980). This property suggested that some feature(s) of the viral DNA molecules that enter nuclei at the beginning of the infectious cycle slows and blocks RNA polymerase II transcription across the middle of the genome, until genome replication produces templates in which the ML transcription unit is fully accessible. Such a feature is presumed to be determined by the way in which the viral genome is associated with viral and/or cellular proteins, but our understanding of the viral nucleoproteins present in infected cells is quite limited (see above). Consequently, the mechanism underlying this very unusual regulatory switch remains a complete enigma, despite the fact that it is crucial for synthesis of most structural proteins and successful virus reproduction.

5.3.4.2 *Stimulation of transcription from the ML promoter*

The final stage in the adenoviral transcription program is characterized by a large increase in the efficiency of ML transcription. The ML promoter, which is responsive to 289R E1A protein during the early phase, exhibits quite typical architecture, including an initiator element, TATA sequence, and upstream binding sites for the cellular activators USF and CBF (Carthew *et al.*, 1985; Concino *et al.*, 1984; Reach *et al.*, 1990; Reach *et al.*, 1991; Sawadogo & Roeder, 1985). However, late phase-specific stimulation of transcription from this promoter is mediated via intragenic sequences located in the first ML intron, termed DE1 and DE2, which are recognized by proteins made only in infected cells during the late phase (Jansen-Durr *et al.*, 1989; Leong *et al.*, 1990). The DE2 sequence is bound specifically by a dimer of the IVa2 protein (Lutz & Kedinger, 1996; Tribouley *et al.*, 1994), which also recognizes related sequence in the packaging signal at the left end of the viral genome (Chapter 7). Mutations that prevent IVa2 protein production or alter the sequences of DE1 and DE2 reduce ML transcription to a modest degree, but the latter were severely inhibitory when combined with mutations that impair USF binding (Pardo-Mateos & Young, 2004; Zhang & Imperiale, 2003). DEI is also recognized by the IVa2 protein, but in association with (an) additional protein(s) (Lutz & Kedinger, 1996) that has proved difficult to identify definitively.

Purification and characterization of infected cell-specific DE1 binding activity indicated that IVa2 interacted with a hypophosphorylated form of the L4 33K protein, which was sufficient to stimulate ML transcription from a reporter gene (Ali *et al.*, 2007). However, the L4 22K protein can also bind specifically to ML DNA that includes the DE1 and DE2 sequences (Backstrom *et al.*, 2010; Ostapchuk *et al.*, 2006), and has been reported to stimulate ML transcription *in vitro* and in reporter assays (Backstrom *et al.*, 2010). Both of these L4 proteins also contribute to regulation of ML pre-mRNA processing (see next section) and are required for packaging of the genome during assembly (Chapter 7); mutations that prevent synthesis of either protein in infected cells, or alter them, exhibit defects in production of progeny virus particles, encapsidation of the genome, and expression of viral late genes (Fessler & Young, 1999; Finnen *et al.*, 2001; Guimet & Hearing, 2013; Morris & Leppard, 2009; Morris *et al.* 2010; Wu *et al.*, 2012; Wu *et al.*, 2013). However, whether either or both of these L4 proteins directly stimulate transcription from the ML promoter in infected cells cannot be discerned from the data available. One complication is that in most experiments, the steady state levels of ML mRNAs were examined (Guimet & Hearing, 2013; Morris & Leppard, 2009; Wu *et al.*, 2013), a parameter that is determined by the rate of not only transcription, but also of pre-mRNA processing and mRNA turnover. Nevertheless, comparison of ML RNA that was not processed at the L3 poly(A)-addition site in the presence and absence of the L4 22K protein indicated this protein did not increase pre-mRNA concentration, but acted post-transcriptionally (Morris & Leppard, 2009). Further compounding the difficulty of mechanistic interpretation of mutant virus phenotypes is the modulation of the production or properties of the proteins implicated in stimulation of ML transcription by one another. The L4 22K protein is required for efficient production of the L4 33K mRNA (Guimet & Hearing, 2013) and also appears to stabilize the IVa2 protein, resulting in a small increase in ML promoter activity (Morris & Leppard, 2009). It should also be noted that the stimulation of ML transcription by the IVa2 protein in combination with the L4 33K or 22K proteins achieved in reporter assays (Ali *et al.*, 2007; Backstrom *et al.*, 2010; Morris & Leppard, 2009) is much less than the increase of some 100-fold in ML transcription observed upon entry into the late phase of the infectious cycle (Shaw & Ziff, 1980). This difference might reflect less than optimal concentrations of the viral proteins in such assays, the absence of infected cell-specific posttranslational modifications that govern their activity, or the absence of other proteins; both protein IX and the E2 DBP have been reported to stimulate ML transcription in transient expression assays (Chang & Shenk, 1990; Lutz & Kedinger, 1996).

5.4 Regulation of Pre-mRNA Processing

A second mechanism important for setting relative mRNA concentrations or the time of synthesis of individual mRNAs during the infectious cycle is regulation of the recognition of alternative poly(A)-addition or alternative splice sites in viral pre-mRNAs. Indeed, viral proteins that regulate these processes are essential for synthesis of the complete panoply of ML mRNAs.

5.4.1 *Alternative polyadenylation*

Production of the E2 and E3 pre-mRNAs depends on recognition of two polyadenylation sites in the pre-mRNAs (Fig. 5.2) at the appropriate frequency. For example, the poly(A)-addition site proximal to the E2E promoter is used some 10- to 20-fold more efficiently than is the distal site, as judged by the relative concentrations of the E2A and E2B mRNAs (Binger & Flint, 1984; Chow *et al.*, 1979). This ratio is important for production of the E2 DBP, which functions stoichiometrically (E2A mRNA; see Chapter 6 for more details), in much larger quantities than the DNA polymerase or preTP primer (E2B mRNAs), and hence efficient virus reproduction. Mutations that alter the E2A polyadenylation site reduce the concentrations of this mRNA, as expected, and decrease virus yield, particularly when combined with alterations at the beginning of the 5′ terminal exon (Caravokyri & Leppard, 1996). Features of the promoter proximal E3 poly(A)-addition site that render it suboptimal, as well as *cis*-acting signals that govern the efficiency of splicing at specific sites, ensure production of the complete set of E3 mRNAs (Brady *et al.*, 1992).

ML transcripts are subject to more extreme alternative polyadenylation, at one of five possible sites (L1–L5) (Fig 5.2), and utilization of these sites is temporally regulated. Early in infection, ML transcription continues beyond the L2 and L3 poly(A)-addition sites (Iwamoto *et al.*, 1986), but only the L1 site is utilized efficiently. In contrast, during the late phase, the five polyadenylation sites present in full-length ML pre-mRNA are recognized with similar frequencies (Chow *et al.*, 1979; Nevins & Darnell, 1978b; Nevins & Wilson, 1981). The decreased use of the L1 site late in infection stems in part, from the reduced activity during this period of the cellular processing protein cleavage factor 1, which recognizes GU-rich sequences downstream of pre-mRNA cleavage sites (Mann *et al.*, 1993); in combination with its relative low affinity for the L1 site, such reduced activity would decrease the frequency of L1 polyadenylation. The L1 site is, in fact, a relatively poor substrate for 3′ end determination *in vitro* and in cells (Prescott & Falck-Pedersen, 1994). The great preponderance of L1 mRNA over L3 mRNAs in the early mRNA population has been attributed to sequences upstream of the L1 poly(A)-addition site that prevent accumulation of mRNA processed at the downstream L3 site (Prescott *et al.*, 1997). It has been proposed that this signal is disrupted upon L1 polyadenylation, allowing L1 mRNA to accumulate. How such a signal might be overridden later in the infectious cycle, when all ML mRNAs are synthesized, is not known.

5.4.2 *Alternative splicing*

Early studies established that the majority of viral pre-mRNAs are alternatively spliced to mRNAs that encode different proteins, a process that greatly expands the coding capacity of the compact viral genome. Recently, the repertoire of adenoviral splicing has been defined with much greater sensitivity by high throughput sequencing of RNA from HAdV-C2-infected cells (Zhao *et al.*, 2014). This approach identified several previously unrecognized splice sites, and confirmed (or refined) those mapped originally, as well as the temporally regulated splicing reactions.

The utilization of splice sites in several viral pre-mRNAs changes with entry into the late phase of infection. The E1A mRNAs for the minor proteins are rare during the late phase, but become the predominant species late in infection (see Chow *et al.*, 1979; Stephens & Harlow, 1987). Similarly, the late phase of infection is marked by a large (some 20-fold) increase in production of the E1B mRNA for the 19-kDa protein (13S), which is produced in similar quantities to the 55-kDa mRNA (22S) early in infection. By far the most dramatic, and best understood, example of regulation of alternative splicing is the early-to-late switch in ML pre-mRNA splicing.

Polyadenylated L1 transcripts made during the early phase carry two 3′ splice sites for ligation to the spliced tripartite leader (tpl) sequence to yield mRNAs for the L1 52/55-kDa and IIIa proteins (Fig. 5.2). However, only that for the 52/55-kDa protein is recognized efficiently during the early phase, and production of the alternatively spliced IIIa mRNA depends on entry into the late phase of infection. This pattern is the result of both negative and positive regulation of splicing at the IIIa 3′ splice site by a *cis*-acting splicing suppressor and enhancer, respectively, both located just upstream of the splice site. The IIIa repressor element blocks assembly of splicesomes on the 3′ splice site as a result of its binding by hyper-phosphorylated forms of cellular SR (serine–arginine-rich) splicing proteins (Kanopka *et al.*, 1998). Hypophosphorylation of such proteins, primarily SPF1, induced by the viral E4 Orf4 protein relieves such repression, and renders accessible the IIIa branch point at which the lariat intermediate is formed during splicing (EstmerNilsson *et al.*, 2001; Kanopka *et al.*, 1998). This E4 protein binds to and modulates the activity of protein phosphatase 2A (Kleinberger & Shenk, 1993) and also interacts with specific phosphorylated SR proteins (Estmer Nilsson *et al.*, 2001), suggesting that it alters the substrate specificity of the cellular enzyme. Although the E4 gene is transcribed during the early phase, production of some spliced E4 mRNAs, including that encoding the Orf4 protein, has been reported to depend on viral DNA synthesis (Dix & Leppard, 1993), a pattern consistent with the timing of the switch in L1 pre-mRNA splicing. Dephosphorylation of cellular SR proteins also appears to contribute to temporal regulation of alternative splicing of E1A transcripts; overproduction of the SR proteins prevents the early-to-late switch in splicing of both L1 and E1A pre-mRNAs (Molin & Akusjarvi, 2000).

L1 pre-mRNA splicing is also governed by a splicing enhancer (3′ UDE) that spans the IIIa branchpoint and 3′ splice site. This sequence both inhibits recognition of this splice site in extracts of uninfected cells and activates splicing at the site in infected cell extracts (Muhlemann *et al.*, 2000). The L4 33K protein stimulates spliceosome assembly on IIIa pre-mRNA *in vitro*, and production of spliced IIIa mRNA in a 3′ UDE-depdendent manner *in vitro* and in transient expression assays (Tormanen *et al.*, 2006), establishing that this viral protein is sufficient to induce the late pattern of splicing. Consistent with these observations, very little IIIa mRNA is synthesized in non-complementing cells infected by a mutant virus defective for production of the L4 33K protein (Wu *et al.*, 2013). The ability of this protein to enhance IIIa splicing depends on its unique C-terminal sequence, which is well conserved in species A–F HAdVs (Biasiotto & Akusjarvi, 2015), specifically one of the Scr residues in a short sequence that includes three arginine–serine repeats (Östberg *et al.*, 2012; Tormanen *et al.*, 2006).

Early studies identified L4 33K as one of three viral proteins that are highly phosphorylated in infected cells (the others being the E2 DBP and the L4 100K protein) (Axelrod, 1978; Gambke & Deppert, 1981). Subsequently, the L4 33K protein was shown to be a substrate for both DNA-dependent protein kinase (DNA-PK) and protein kinase A (PKA), which phosphorylate one or more serine residues in the C-terminal domain (Tormanen Persson *et al.*, 2012). Although the form of the L4 33K protein that acts as a splicing enhancer has not been identified, phosphorylation by DNA-PK is inhibitory; depletion of this kinase stimulated synthesis of IIIa mRNA in HAdV-C5-infected cells (Tormanen Persson *et al.*, 2012). The timing of induction of alternative L1 pre-mRNA splicing may therefore be controlled, in part, by differential phosphorylation of the L4 33K protein, as DNA-PK is inhibited by the E4 Orf3 and Orf6 proteins (J. Boyer *et al.*, 1999). The contribution of PKA to regulation of alternative splicing in infected cells, and the mechanism(s) by which phosphorylation of the L4 33K protein regulates its activity, remain to be determined.

Comparison of synthesis of ML mRNAs from a plasmid carrying the full-length transcription unit or one truncated at the end of L3 indicated that only the L1 52/55K mRNA was synthesized from latter, unless the L4 33K and L4 100K coding sequences were expressed *in trans* (Farley *et al.*, 2004). This finding suggested a broader role for the L4 33K protein in the regulation of ML pre-mRNA processing. The L4 33K region included in the plasmid used in these experiments was subsequently shown to give rise to a second (unspliced) mRNA that specifies the L4 22K protein (Davison *et al.*, 2003; Ostapchuk *et al.*, 2006), raising the question of the relative contributions of the 33K and 22K proteins to the regulation of ML pre-mRNA splicing. The phenotypes of mutant viruses defective for production of the L4 33K or 22K proteins, or with mutations that introduce substitutions into the latter (Guimet & Hearing, 2013; Wu *et al.*, 2012; Wu *et al.*, 2013), as well as synthesis of ML RNAs and proteins in cells transfected with mutant genomes (Morris & Leppard, 2009), have been examined to address this question. In the absence of the L4 33K protein, the steady state concentrations of ML mRNAs produced by recognition of more downstream 3′ splice site were reduced to a greater (L1 IIIa, L2 V, and L3 pVI mRNAs) or lesser (e.g. L2 pVIII and L5 fiber mRNAs) degree. The former set of mRNAs correspond to those that depend to the greatest degree on the L4 33K protein for synthesis in transient expression assays, and are produced via weak 3′ splice sites (Tormanen *et al.*, 2006), consistent with the conclusion that the L4 33K protein functions as a major activator of alternative splicing in infected cells.

Cells infected with an L4 22K-null mutant virus (or transfected with the mutant genome), as well as with a virus with point mutations that introduce H166Q and H177Q substitutions into the unique C-terminal portion of this protein, display defects in synthesis of many ML proteins and overproduction of early proteins (Guimet & Hearing, 2013; Wu *et al.*, 2012). Such decreases correlated with reduced steady-state concentrations of the corresponding ML mRNAs, notably L1 IIIa, L2 V and pVII mRNAs, and L4 pVIII mRNAs (Guimet & Hearing, 2013; Morris & Leppard, 2009). The stabilization of the IVa2 protein by the L4 22K protein discussed previously would make a modest contribution to such impairment ML gene expression. However, the absence of corresponding decreases in the concentration of pre-mRNAs indicated that L4 22K protein primarily stimulates pre-mRNA processing (Morris & Leppard, 2009).

The concentration of the L4 33K mRNA were also reduced below the level of detection by the L4 22K mutations, and the unspliced L4 22K mRNA accumulated to higher than normal concentrations, consistent with regulation of splicing of L4 transcripts by the 22K protein (Guimet & Hearing, 2013). Such dependence of synthesis of the L4 33K regulator of alternative ML splicing on the L4 22K protein could account for defects in production of other ML mRNAs induced by L4 22K mutations. However, synthesis of the L4 33K protein from a cDNA coding sequence from which the 22K mRNA cannot be produced failed to complement defects in ML protein production in cells transfected with an L4 22K-null mutant genome (Morris & Leppard, 2009), indicating the L22K protein fulfills a distinct function in ML gene expression. It has been suggested this protein also stimulates alternative splicing, but with a specificity different from that of the 33K protein (Guimet & Hearing, 2013). This hypothesis has not yet been tested directly, for example using *in vitro* splicing systems to compare the effects of these two L4 proteins on splicing of different substrates.

5.5 Post-Transcriptional Inhibition of Cellular Gene Expression During the Late Phase of Infection

Early experiments to monitor protein synthesis during the infectious cycle established that production of cellular proteins is progressively inhibited during the late phase until eventually becoming undetectable by 24 hr. p.i. under typical conditions of infection in HeLa cells (Anderson *et al.*, 1973; Beltz & Flint, 1979). In fact, both export of cellular mRNAs from the nucleus and their translation in the cytoplasm are inhibited.

5.5.1 *Selective export of viral late mRNAs*

The vast majority of mRNAs that enter the cytoplasm of HAdV-C5 infected cells from the beginning of the late phase are viral in origin (Beltz & Flint, 1979). The intranuclear processing (capping, polyadenylation, and splicing) of cellular pre-mRNAs is not perturbed, indicating that export of cellular mRNA is inhibited (Dobner & zhyshkowska, 2001; Flint & Gonzalez, 2003). In contrast, export of other classes of cellular RNAs is not blocked, suggesting that a pathway specific for mRNA export is targeted specifically. Indeed, export of viral late mRNAs is decreased by RNAi-mediated inhibition of synthesis of NXF1 the major export receptor for cellular mRNAs (see Erkmann& Kutay, 2004; Kohler & Hurt, 2007) or by the overproduction of a dominant negative derivative of this protein during the late phase of infection (Yatherajam *et al.*, 2011). Selective export of viral late mRNAs requires the E1B 55-kDa and E4 Orf6 proteins (Babich *et al.*, 1983; Bridge & Ketner, 1990; Cutt *et al.*, 1987; Pilder *et al.*, 1986; Williams *et al.*, 1986) and the virus-specific E3 ubiquitin ligase they form (Blanchette *et al.*, 2008; Woo and Berk, 2007). These observations suggest that ubiquitin-dependent destruction, or modification, of one or more components of the NXF1 export pathway is necessary for induction of selective export of viral late mRNAs. However, the relevant substrates of the virus-specific E3 ubiquitin ligase have not been identified, but it is not likely to be NXF1 itself; this protein is not degraded or altered in mobility during the late phase of infection (Yatherajam *et al.*, 2011).

A major unsolved mystery is how viral and cellular mRNAs are distinguished, when the viral late mRNAs that are exported carry the same features and are made by the same pathway as cellular and viral early mRNAs (U.-C. Yang *et al.*, 1996) that do not leave the nucleus. One possible basis for such discrimination among mRNAs is suggested by the correlation of the efficient export of viral late mRNAs with the localization of the E1B 55-kDa and E4 Orf6 proteins to the peripheral zones of viral replication centers (Gonzalez & Flint, 2002; Ornelles & Shenk, 1991), which are the site of synthesis and initial processing of viral late pre-mRNAs (Aspegren *et al.*, 1998; Bridge & Pettersson, 1996; Moyne *et al.*, 1978; Pombo *et al.*, 1994; Puvion-Dutilleul & Puvion, 1991; Puvion-Dutilleul *et al.*, 1992). It has been proposed that recruitment of a limiting protein necessary for mRNA export to such sites would favor export of viral late mRNAs (Ornelles & Shenk, 1991), a process that might be facilitated by increased targeting for degradation of the critical protein(s) by the virus-specific E3 ubiquitin ligase. In fact, the distinction between mRNAs that are or are not exported during the late phase of infection appears to be based on the time of activation of transcription of the corresponding genes; when transcription of cellular genes is induced during the late phase as a result of HAdV-infection or exogenous stimuli, the mRNAs made are not subjected to export inhibition (Moore *et al.*, 1986; U.-C. Yang *et al.*, 1996). The now well-established dependence of all steps in production of a eukaryotic mRNA, from transcription of the genes to export from the nucleus, on one another (Erkmann & Kutay, 2004; Reed & Cheng, 2005) seems likely to account for this unexpected finding. Furthermore, this property suggests that a protein(s) that coordinates these intranuclear processes, such as a component(s) of the TREX complex (Katahira, 2015; Muller-McNicoll & Neugebauerm, 2013; Reed & Cheng, 2005), might be regulated by the E1B-55 kDa and E4 Orf6 protein-containing E3 ubiquitin ligase.

5.5.2 *Selective translation of viral late mRNAs*

As a result of inhibition of entry of cellular mRNAs into the cytoplasm, synthesis of proteins specified by unstable mRNAs with short half-lives must cease during the late phase of infection. It has also been proposed that export may govern translation (Palazzo & Akef, 2012). Be that as it may, the late phase of infection is also characterized by a poor translational environment and selective translation of viral late mRNAs.

During this period, phosphorylation of the mRNA 5′ cap-binding protein elF4E is severely reduced (Huang & Schneider, 1991; Zhang & Schneider, 1994). Such reduction is the result of binding of the viral L4 100K protein to the C-terminus of elF4G, the adaptor for assembly of the cap binding complex (Hentze, 1997), with concomitant dissociation of the MAP kinase-activated kinases MNK1 and MMK2 (Cuesta *et al.*, 2001; Cuesta *et al.*, 2004), which phosphorylate and activate elF4E bound to elF4G (Pyronnet, 2000). Mutational analyses established that the N-terminal 66 amino acids of the L4 100K protein are sufficient to bind to elF4G, displace MNK1, block elF4E phosphorylation, and inhibit cap-dependent translation. This segment of the viral protein and MNK1 contain a related elF4G-binding motif, but the L4 100K proteins binds more strongly and hence outcompetes MNK1 (Cuesta *et al.*, 2004).

Despite such inhibition of cap-dependent translation, the capped viral late mRNAs are translated efficiently. Such selective translation also depends on the L4 100K protein, which allows initiation of translation of all viral late mRNAs examined (Hayes *et al.*, 1990). The 5′ tripartite leader sequence common to ML mRNA also stimulates translation during the late, but not the early, phase of infection (Logan & Shenk, 1984) as well as late mRNA export (Huang & Flint, 1998). Structural and mutational studies indicated that this ML mRNA 5′ untranslated region contains a group of stable stem-loop hairpins preceded by a largely unstructured segment at the 5′ end (Dolph *et al.*, 1990; Zhang *et al.*, 1989). During the late phase of infection, the tripartite leader sequences direct initiation of translation by a non-canonical mechanism known as ribosome shunting (Yueh & Schneider, 1996; Yueh & Schneider, 2000). In this process, 40S ribosomal subunits associated with elF4G bind to the cap, but do not scan to the downstream initiation AUG codon, but rather bypass the intervening RNA sequences. The L4 100K protein induces such shunting by binding to both the 5′ tripartite leader sequence of ML mRNAs and elF4G; substitutions that impair tripartite leader sequence recognition or interaction with elF4G inhibit ribosome shunting-dependent translation (Xi *et al.*, 2004).

This activity of the L4 100K protein is regulated by tyrosine phosphorylation, a modification that enhances preferential binding to mRNAs that carry the 5′ tripartite leader sequence (Xi *et al.*, 2005). The ability of the interaction of this viral protein with elF4G to disrupt the normal mechanism of cap-dependent initiation of translation while promoting ribosome shunting on ML mRNAs is an efficient way to ensure selective synthesis of viral proteins. It should be noted, however, that translation of late mRNAs that do not carry the 5′ tripartite leader sequence, such pIX and IVa2 mRNAs, was also observed to depend on the L4 100K protein (Hayes *et al.*, 1990). Whether the 5′ untranslated regions of these and other late mRNAs also facilitate ribosome shunting has not been investigated.

5.6 What Next?

Sequential expression of specific sets of viral genes is a common feature of the infectious cycle of viruses with DNA genomes, regardless of the type of host organism or the intracellular site of virus reproduction. The temporal program of adenoviral gene expression is, at present, the most elaborate of any known; it includes four distinct periods or phases, and is elicited by the actions of viral proteins that regulate transcription as well as by viral DNA synthesis-dependent relief of transcriptional repression and the intricate regulation of alternative processing of viral pre-mRNAs that is essential for synthesis of the complete repertoire of viral proteins. The mechanisms that govern the transitions from one phase of viral gene expression to the next and the roles played by viral regulatory proteins have been elucidated, in some cases (for example, the E1A transcriptional activators) in considerable molecular detail. Nevertheless, several processes crucial for successful completion of the infectious cycle remain poorly or incompletely understood. Prominent examples are the mechanisms that allow RNA polymerase II to synthesize RNA to the end of the ML transcription unit on replicated DNA templates, that achieve a large increase in the efficiency of ML transcription during the late phase, and that lead to production of all ML mRNAs by regulation of alternative polyadenylation and splicing. Furthermore, the mechanisms that underlie the orderly and efficient expression of viral genes have been studied almost exclusively in transformed cell lines in culture. Consequently, it may be that in normal primary human cells these regulatory circuits include additional processes and players.

The complement(s) of proteins associated with viral templates for transcription and the architecture of such nucleoproteins also remain enigmatic, in part because only a few of the entering or replicated viral DNA molecules serve as transcriptional templates. Comparisons of the properties of these templates, isolated by virtue of their association with RNA polymerase II, with viral DNA molecules that are not transcribed would be anticipated to clarify the consequences of association of cellular histones and nucleosomes with viral DNA, and might shed further light on the unusual mechanisms that govern the early-to-late transition.

The more recent discovery of adenoviral miRNAs, derived exclusively form the VA RNAs made by RNA polymerase III, has expanded the functional repertoire of viral gene products. Nevertheless, although these viral miRNAs have been shown to have the ability to suppress cellular gene expression, how they facilitate adenovirus reproduction is as yet an unsolved mystery.

Replication of the Genome　　6

6.1 Introduction

In rapidly proliferating transformed human cells, viral DNA synthesis typically commences 6–8 hours after infection, once the E2 replication proteins accumulate. The major proteins encoded in the IE EIA gene activate transcription from the E2E promoter by the multiple mechanisms described in the previous chapter. They can also induce large scale reprogramming of cellular gene expression to induce cells that are not cycling to enter S phase, and hence establish an intracellular environment favorable for viral DNA synthesis. Continued viral DNA synthesis requires additional viral products that counter antiviral host responses induced by accumulation of viral genomes. The concerted actions of these viral early proteins allow synthesis of tens of thousands of copies of the viral genome within a period of about 12 hours. Both the simple and elegant replication mechanism, in which both strands of the template DNA are copied continuously from a protein primer and the establishment of discrete intranuclear sites of adenoviral DNA replication (replication centers or compartments) seem likely to promote efficient viral DNA synthesis.

In this chapter, we describe the mechanisms of viral DNA synthesis and the viral proteins that support this process directly or indirectly, focusing once again on the well-characterized human adenovirus (HAdV)-C2 and -C5.

6.2 Priming the Host Cell: Induction of Entry into S Phase

The core machinery for adenoviral DNA synthesis is viral in origin, but several cellular proteins are necessary for efficient genome replication (see next section). Perhaps just as importantly, the host cell provides the large quantities of dNTP precursors that are needed for synthesis of tens of thousands of copies of the viral genome. However, dNTPs and the enzymes that produce them are present at relatively low concentration in cells that are withdrawn from the cell cycle in G0, or proliferating only slowly, and hence likely to be limiting for viral DNA synthesis. The mitogenic activity of E1A proteins, originally characterized in the context of transformation (Chapter 1), was proposed to allow HAdVs to replicate successfully in cells in these states, such as the differentiated epithelial cells of the upper respiratory tract naturally infected by species C HAdVs. Consistent with this view, the HAdV-C5 243R E1A protein is necessary for efficient genome replication and virus reproduction in G0-arrested human lung fibroblasts, but not in HeLa cells (Spindler *et al.*, 1985). Furthermore, HAdV-C5 infection results in changes in expression of only a few hundred genes in HeLa cells (Granberg *et al.*, 2005; Granberg *et al.*, 2006), but of thousands in infected normal human cells (Miller, 2006; Zhao *et al.*, 2012).

Progression through the G1 restriction point in mammalian cells is controlled by the sequential activation of cyclin-dependent kinases (CDKs), largely via tightly controlled accumulation of their regulatory cyclins (Reed, 1997; Sherr, 1994). The genes that encode these proteins as well as others that support DNA synthesis are controlled by promoters that contain binding sites for E2F transcriptional regulators (Dyson, 1994; Johnson & Schneider-Broussard, 1998; La Thangue, 1996). The interactions of E1A CR1 and CR2 sequences with RB proteins to release associated E2F family members and with the cellular lysine acetyl transferases KAT2A and KAT3B described in the previous chapter (Fig. 5.6) are necessary for the mitogenic activity of E1A and induction of cellular DNA synthesis (Berk, 2005; Berk, 2013). Synthesis of the 243R E1A protein in quiescent murine cells leads to reorganization of chromatin at the promoters of E2F-regulated genes, such as *Cd6* and *cyclin A*, as well as loss of E2F4 and the histone deacetylases HDAC1/2-mSIN3B (Ghosh and Harter, 2003; Sha *et al.*, 2010). These observations suggested that the 243R E1A protein induces extensive chromatin remodeling by removal of repressors and repressive chromatin modifications to allow subsequent recruitment of transcriptional activators and active histone modifications. It is now clear that such alterations are wrought on a global scale by this viral protein.

The large impact of the 243R E1A protein on the organization and expression of cellular genes has been examined by ChIP (chromatin immunoprecipitation) and microarray hybridization in G1-arrested human fibroblasts infected by a mutant of HAdV-C5 that directs synthesis of only the 243R E1A protein (Ferrari *et al.*, 2008; Ferrari *et al.*, 2012). The viral protein interacted sequentially with three distinct sets of cellular genes, initially genes associated with response to pathogenic organisms, then those associated with cell cycle progression and macromolecular synthesis, and, late after infection, genes that encode differentiated cell functions. Expression of the second group of cellular genes was increased from 17 hours p.i., concomitant with loss of RB and related proteins from their promoters, association of KAT3A and KAT3B, and acetylation of histone H3K19 at and downstream of these promoters (Fig. 6.1). In conjunction with the dependence of viral genome replication in arrested human cells on the 243R protein (Spindler *et al.*, 1985), these observations reinforce the conclusion that extensive reprogramming of cellular gene expression supports efficient viral DNA synthesis. The E1A-dependent removal of

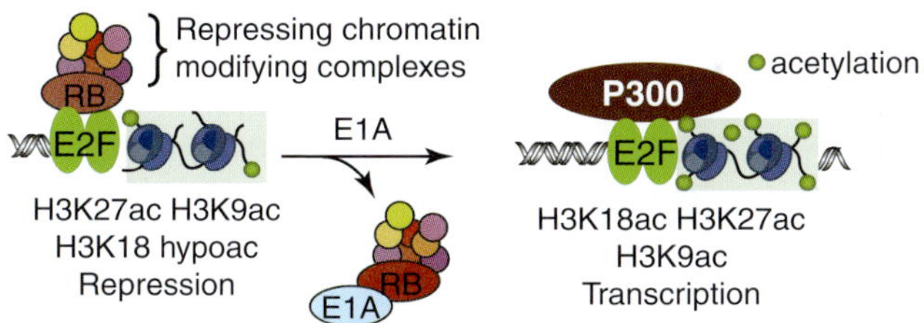

Figure 6.1. Model for activation of E2F-regulated gene by viral E1A proteins. Many promoters of the genes encoding proteins that allow progression through the G1 restriction point into S phase of the cell cycle contain binding sites for E2F family members, which are required for efficient transcription. When RB or a related protein is bound to E2F at such a promoter sequence, it recruits histone deacetylases (HDACs) and other chromatin-modifying enzymes that repress transcription. The interaction of the viral E1A proteins with RB results in disruption of the RB–E2F association and removal of the repressive complex from E2F bound to the promoter. Subsequently, the lysine acetyl transferase KAT3B (a.k.a. p300) is recruited, nucleosomes are modified at the histone residues indicated and transcription is activated. Adapted from Ferrari *et al.* (2014), with permission.

RB from both these cellular promoters and the viral E2E promoter that controls production of viral replication proteins ensures a propitious intracellular environment for viral DNA synthesis.

6.3 Replication of the Genome

Viral replication proteins accumulate from 4–6 hours after infection, and an increase in the number of viral genomes can be detected a few hours thereafter. The lag is presumed to reflect the need for the E2 replication proteins to attain sufficient concentrations to associate with one another and with the viral DNA template in nuclei of infected cells. Viral DNA synthesis continues for some 12 hours as 10^4 or more viral DNA genomes are made (Pina & Green, 1969). Under optimal conditions *in vitro*, the HAdV-C2 DNA polymerase and human DNA polymerase α exhibit similar rates of polymerization (Field *et al.*, 1984), suggesting that viral DNA is not synthesized at an unusually high rate in infected cells. In fact, the number of viral base pairs made, $\sim 7 \times 10^8$ bp for 2×10^4 copies of the 3.5×10^4 bp genome, in 10–12 hours, is roughly equivalent in terms of biosynthetic demand to complete replication of the cellular genome, $\sim 3 \times 10^9$ bp in some 8 hours, in proliferating cells in culture. In such cells, cellular DNA synthesis is inhibited as the infection cycle progress (Ledinko & Fong, 1969; Pina & Green, 1969). The mechanism has not been elucidated, but such inhibition presumably maximizes the supply of dNTPS available for synthesis of viral genomes.

6.3.1 *An all-continuous DNA synthesis mechanism via protein priming*

The mechanism of species C HAdV DNA synthesis was one of the first steps in the infectious cycle to be elucidated, by a combination of microscopic, genetic and biochemical approaches. Characterization of replication intermediates by electron microscopy revealed a very unusual mechanism of DNA synthesis, in which one strand of the double-stranded (ds) DNA template is displaced, as the other is copied (Lechner & Kelly, 1977). The observation of single strands with short duplex stems formed by base pairing of the inverted terminal repetitions suggested a mechanism in which both template strands are replicated continuously from terminal origins (Rawlins *et al.*, 1984), but not at the

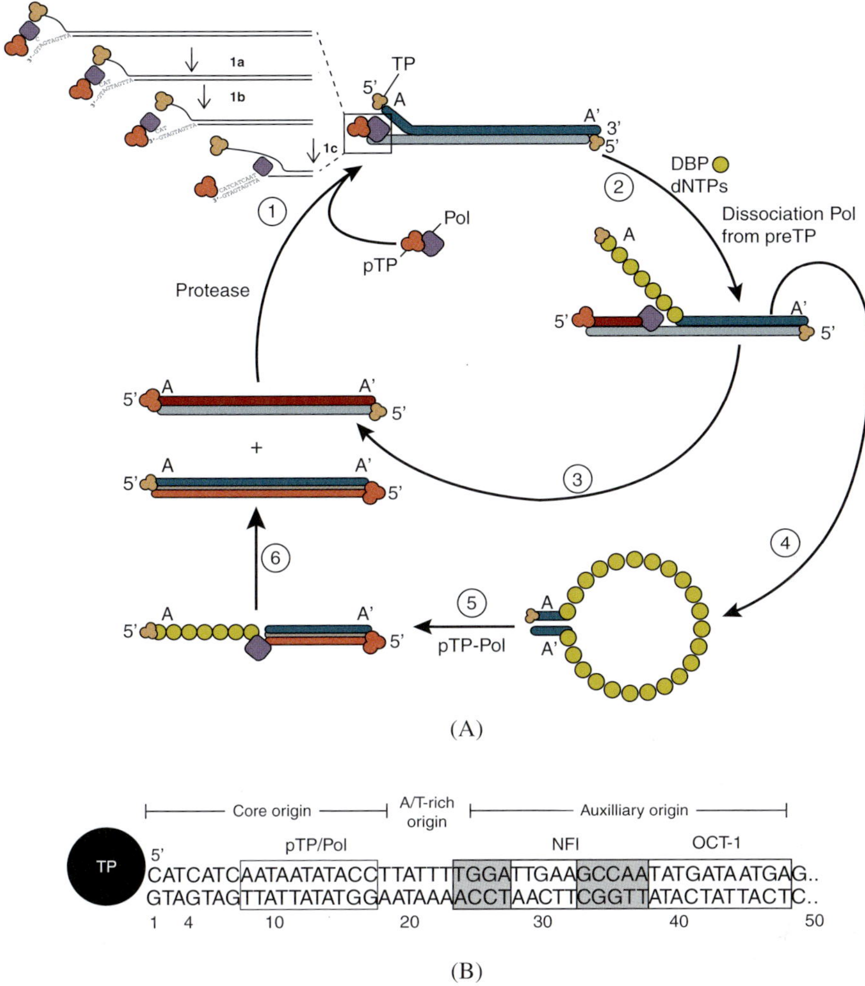

Figure 6.2. The HAdV-C5 origin and replication of the genome. (**A**) Replication of the viral genome by protein priming and all-continuous DNA synthesis. The pTP–Pol complex assembles on the terminal origins, and Pol catalyzes the addition of a priming dCMP to a specific serine residue (Ser) in pTP (step 1). As shown in the expansion, this dCMP residue is templated by a G at position 4 in the origin, and two additional dNMPs are then added (step 1a). The pTP–Pol–CAT complex is then displaced to pair with nucleotides 1–3 in the template strand (step 1b). Elongation then continues with disruption of the pTP–Pol interaction (step 1c) shortly after pairing of the CAT initially made with the end of the template strand. One strand of the genome is then copied, with displacement of the other, which becomes coated with the E2 DBP (steps 2 and 3). The displaced strands can be

←

Figure 6.2. (*Continued*)   replicated after annealing of either complementary strands, or of the inverted terminal repetitions of a single strand to reform ds origins (steps 4 and 5). Each strand of the parental DNA template is therefore replicated continuously. Adapted from Flint *et al.* (2015), with permission and the insert adapted from Hoeben & Uli (2013), with permission. (**B**) The sequence of the HAdV-C5 origin at the left end of the viral genome is shown, with core and auxiliary origins and the sequences required for binding of viral and cellular proteins indicated. Adapted from Liu *et al.* (2003), with permission.

same time (Fig. 6.2A). Sequence and mutational analyses delineated a core (or minimal) origin of replication, comprising the terminal 18 bp at each end of the genome and including a highly conserved sequence (positions 9–18) that supports a low level of replication (Bailey & Mautner, 1994; Hay, 1985; Rawlins *et al.*, 1984; Stillman *et al.*, 1982; van Bergen & van der Vliet, 1983). The core origin is followed by binding sites for the cellular transcriptional activators NF1 and Oct1 (Fig. 6.2B), which increase the efficiency of viral DNA synthesis considerably (Liu *et al.*, 2003; van der Vliet, 1995; van der Vliet *et al.* 2006).

The development of *in vitro* systems (initially extracts of infected cells) in which the species C HAdV genome was replicated (Challberg *et al.*, 1979) was of the greatest importance, allowing elucidation of the mechanism of viral DNA synthesis and the molecular functions of viral replication proteins. Characterization of an activity necessary for viral DNA synthesis in these extracts identified a heterodimer that included a 140-kDa DNA polymerase with distinctive properties (Enomoto *et al.*, 1981; Ikeda *et al.*, 1982; Lichy *et al.*, 1982). This enzyme was shown to be viral in origin by biochemical complementation of extracts of cells infected by a HAdV-C5 mutant temperature-sensitive for initiation and elongation of viral DNA synthesis (Stillman *et al.*, 1982; van Bergen & van der Vliet, 1983). The second protein proved to be the precursor to the TP, pTP (Challberg *et al.*, 1980; Enomoto *et al.*, 1981). These two proteins, and the third viral protein required for genome replication, the single-stranded (ss) DNA-binding protein (DBP), are all encoded in the E2 transcription unit (Chapter 5).

The pTP serves as the protein primer for the initiation of DNA synthesis from the terminal origins; a specific Ser residue (Ser 580 in the HAdV-C5 pTP) is esterified to dCMP by the viral DNA polymerase (Pol) (Desiderio & Kelly, 1981; Smart *et al.*, 1982). The first reaction in initiation is assembly of a preinitiation complex, in which binding of the Pol/pTP heterodimer to the conserved origin sequence is stabilized by interactions of Pol and pre-TP with the cellular NF1 and OCT1 proteins, respectively, bound to their adjacent recognition sites (Fig. 6.2B) (Armentero *et al.*, 1994; Chen *et al.*, 1990; Coenjaerts *et al.*, 1994; de Jong *et al.*, 2002; Mul *et al.*, 1990). Interaction of pTP with the template-bound terminal protein may also stabilize the pre-initiation complex (Pronk & van der Vliet, 1993). Binding of OCT1 and NF1 to the origin also bends the DNA substantially (Mysiak *et al.*, 2004a; Mysiak *et al.*, 2004b). Such bending may be essential for initiation, as these two proteins act synergistically to increase the rate some 100-fold (Mysiak *et al.* 2004b). The DBP, which binds to dsDNA and cooperatively to ssDNA, also stimulates initiation: it both increases the affinity with which NF1 binds to the origin and reduces the K_m for the Pol-catalyzed covalent linkage of dCMP to pTP (Cleat & Hay, 1989; Mul & van der Vliet, 1993; Stuiver & van der Vliet, 1990). The DBP may also facilitate the local unwinding of the origin that follows preinitiation complex assembly.

Perhaps unexpectedly, initiation takes place at bp 4 in the template strand, in the sequence 3'GTAGTA (Fig. 6.2A), by covalent linkage of dCMP to pTP, and is followed by addition of two more nucleotides (King & van der Vliet, 1994). The Pol–pTP complex then slips back, such that the newly synthesized trinucleotide base pairs with residues 1–3 in the template. This mechanism, which is conserved in protein-primed replication of the dsDNA genomes of certain bacteriophages (Hoeben & Uil, 2013), accounts for the presence of short repeated sequences (2–4 bp) in the terminal sequences of all adenoviral origins that have been examined (de Jong *et al.*, 2003). Once the initiation complex is positioned at the ends of the genome, Pol dissociates from pTP and its rate of polymerization increases (King *et al.*, 1994; King *et al.*, 1997). The processivity of the viral enzyme as well as the rate of DNA synthesis are stimulated by binding to the template by DBP, which is also essential for unwinding of the template by an ATP-independent mechanism (see below). Because of this activity, a helicase is not required for viral DNA synthesis, but cellular topoisomerases I and II are necessary (Nagata *et al.*, 1983), presumably to relieve overwinding of the template DNA.

6.3.2 *The E2 replication proteins*

The replication proteins of species C HAdVs were the first viral analogs of the mammalian DNA synthesis machinery to be identified, and consequently were characterized in some detail. In view of the serious consequences of HAdV infections in immunocompromised patients (Chapter 9), these proteins are of increasing interest as potential targets of antiviral drugs (Chapter 10).

6.3.2.1 *The adenoviral DNA polymerase (Pol)*

The structure of the large (1,198 residues) HAdV Pol has not been determined, but its organization has been established from sequence comparisons, mutational analysis, and biochemical studies of its activities. This enzyme is a member of the alpha family of DNA polymerases, which includes other DNA polymerases that operate with a protein primer (Hay, 1996; Knopf, 1998). It possesses the domains common to all DNA polymerases that form a hand-like shape, with palm, fingers and thumb (Joyce *et al.* 1994), with insertions specific to protein-primed DNA polymerases (Brenkman *et al.*, 2002; King *et al.*, 1997; Liu *et al.*, 2003). Conserved motifs associated with DNA polymerase activity lie within the C-terminal half of the protein, and substitution of conserved amino acids, such as two Asp residues found in all DNA polymerases (D685 and D1014 in HAdV-C5 Pol), eliminates both initiation and elongation activities (Chen & Horwitz, 1989; Joung & Engler, 1992; Joung *et al.*, 1991; Y. Liu *et al.*, 2000; Liu *et al.*, 2003). It is therefore considered (Liu *et al.*, 2003) that addition of nucleotides to a nascent viral DNA chain proceeds by the two-metal-ion mechanism common to these enzymes: one metal ion activates the 3′ hydroxyl group of the primer and the second facilitates the dissociation of the pyrophosphate product of each cycle of elongation (Joyce & Steitz, 1994; Steitz, 1998).

The N-terminal segment of Pol, which confers $3' \rightarrow 5'$ exonuclease activity, is necessary for the intrinsic proofreading activity of the enzyme (Brenkman *et al.*, 2002; King *et al.*, 1997; Liu *et al.*, 2003). More recently, it was observed that substitutions of conserved residues that form the metal ion-binding region of the nuclease active site, which were introduced in an attempt to decrease the fidelity of HAdV-C5 replication, blocked virus reproduction (Uil *et al.*, 2011). As similar alterations in the exonuclease domains of other viral DNA polymerases do not preclude viral replication, it was proposed that this domain of HAdV-C5 Pol also performs an essential function during DNA synthesis, specifically in strand displacement (Uil *et al.*, 2011). The corresponding residues of the related ø29 protein-primed Pol are necessary for this step, as well as for exonucleolytic activity. Substitutions that altered residues thought to contribute to stabilization of the end of the newly synthesized DNA strand in the exonuclease active site did decrease substantially the fidelity of viral DNA synthesis in infected cells (Uil *et al.*, 2011). Indeed, high-throughput sequencing of viral genomes synthesized by wild type and substituted Pols identified one alteration that increased the normally low error rate (<1 misincorporation/10^6 bp) to a value approaching those typical of viral RNA-dependent RNA polymerases and reverse transcriptases. These studies established the importance of the proofreading $3' \rightarrow 5'$ exonuclease of the HAdV-C5 Pol in maintaining the integrity of the genome. They also identified altered enzymes with increased misincorporation rates that facilitated isolation of additional mutants with increased cytotoxicity (Uil *et al.*, 2011), of interest in the context of the development of therapeutic oncolytic adenoviruses (Chapter 12).

The functions of the HAdV Pol domains found only in protein primed DNA polymerases (termed TPR1 and TPR2) have not been examined systematically, but one substitution that impairs binding to pTP (Y. Liu *et al.*, 2000) lies within TPR1. Indeed, TPR1 of the ø29 Pol can be seen in the X-ray crystal structure of this enzyme to form a discrete subdomain (Berman *et al.*, 2007; Kamtekar *et al.*, 2004), and is responsible for specific binding to the protein primer (Rodriguez *et al.*, 2005). Comparisons with TPR domains of the ϕ29 Pol also suggests that both may contribute to processivity (Uil *et al.*, 2011). However, both the interaction of pTP with three of the four fragments of HAdV Pol liberated by cleavage with endoproteinase C (Parker *et al.*, 1998) and the distribution of substitutions that reduce or prevent pTP binding over some 400 residues (Y. Liu *et al.*, 2000) indicate that non-contiguous regions in Pol form a large pTP-binding surface.

Substitutions of several amino acids in the C-terminal segment of Pol substantially reduce or eliminate binding to ss- and dsDNA (Y. Liu *et al.*, 2000). Two putative zinc fingers, one near the N- and the second near the C-terminus, have also been reported to be necessary for binding the viral origin of replication (Ori) (Joung *et al.*, 1992). In the absence of structural information, it is not possible to deduce how these various DNA-interacting regions cooperate with one another to position the template and displaced single strand for optimal catalysis; although a high-resolution structure of the ø29 Pol bound to a model primer-template has been reported (Berman *et al.*, 2007), this enzyme is less than half the size of the HAdV Pol.

6.3.2.2 *The pre-terminal protein (pTP)*

The 87-kDa pTP can be produced in large quantities by expression of its coding sequence in mammalian or insect cells (Bosher *et al.* 1990; Stunnenberg *et al.*, 1988), and its properties have been examined in some detail by both biochemical and genetic methods. During initiation of viral DNA synthesis, the pTP–Pol complex is recruited to the terminal origins of replication by the interaction of the two proteins with one another, and of both with viral DNA (Temperley & Hay, 1992). The protein primer binds specifically to ds Ori DNA via sequences within the N-terminal segment that is removed during proteolytic processing to produce the TP (see below) (Fredman *et al.*, 1991; Freimuth & Ginsberg, 1986; Pettit *et al.*, 1989; Roovers *et al.*, 1993; Webster, 1999; Webster *et al.*, 1994). It also binds to ssDNA, an activity that is thought to help maintain the interaction of Pol with the origin as it becomes unwound (de Jong *et al.*, 2003). The association of pTP with Pol is mediated via non-contiguous regions of pTP located over a large area of the protein surface (Parker *et al.*, 1998). In pre-initiation complexes, pTP also associates with OCT1, which binds specifically to positions 39–47 in the origin by means of its POU DNA-binding domain (van der Vliet, 1995; de Jong & van der Vliet, 1999). This interaction tethers pTP, and hence the pTP–Pol complex to the origin (van Leeuwen *et al.*, 1997) and induce a bend of 42° in Ori DNA (Mysiak *et al.*, 2004b). Such a conformational change is enhanced by induction of bending in the same direction (for a total 82° bend) by binding of the cellular protein NF1 to the adjacent sequence (Fig 6.2B) (Mysiak *et al.*, 2004b). The observation that these two cellular proteins synergistically stimulate initiation of DNA synthesis from the viral DNA origin suggests that induction of bending is necessary, presumably because it establishes a DNA conformation optimal for assembly of the pre-initiation complex. Multiple regions of pTP mediate its interaction with a site in the POU DNA-binding domain of OCT1 (Coenjaerts *et al.*, 1994; de Jong *et al.*, 2002; van Leeuwen *et al.*, 1997). This interaction induces alterations in the accessibility of protease cleavage sites in the pTP (Botting & Hay, 1999), suggesting that it alters conformation.

The 87-kDa pTP is proteolytically cleaved to mature TP by the viral L3 protease (AVP), which is produced late in infection (Ruzindana-Umunyana *et al.* 2002; Stillman *et al.*, 1981; see also Chapter 7). The action of the L3 protease liberates the C-terminal TP, 55 kDa in the case of HAdV-C5, via production of a 62-kDa intermediate termed iTP (Webster *et al.*, 1989). As the cleavage sites for production of both TP and iTP are conserved among HAdVs, it has been proposed that iTP plays a specific role in the infectious cycle (Webster *et al.*, 1989), one that has not been identified. As discussed in Chapter 5, the results of mutational analyses indicate that the N-terminal portion of pTP produced upon proteolytic maturation contributes to optimal intranuclear localization of the viral genome at the beginning of the infectious cycle and efficient viral gene expression.

6.3.2.3 *The ssDNA-binding protein*

The properties of temperature sensitive mutants with lesions in the DBP protein-coding sequence established that the E2 DBP is essential for replication of the viral genome (Friefeld *et al.*, 1983; Horwitz, 1978; Ginsberg *et al.*, 1977; van Bergen & van der Vliet, 1983). This protein accumulates to high concentrations in infected cells and binds with high affinity to ssDNA (van der Vliet & Levine, 1973), a property that facilitated its isolation and characterization. The DBP is essential during the elongation phase of the viral DNA synthesis, as it promotes unwinding of the ds template (Zijderveld

& van der Vliet, 1994) and enhances the processivity of Pol (Lindenbaum *et al.*, 1986). No direct interaction between these two viral replication proteins has been detected, and the DBP in thought to stimulate elongation by stabilizing the association of Pol with template DNA (de Jong *et al.*, 2003). DBP-dependent unwinding of duplex DNA is the result of cooperative binding of multiple DBP molecules to the ssDNA displaced during elongation (Dekker *et al.*, 1997; Stuiver & van der Vliet, 1990; Zijderveld & van der Vliet, 1994). The unwinding activity of the DBP does not require ATP, in contrast to that of helicases that function during replication of the cellular genome (Monaghan *et al.*, 1994; Pronk *et al.*, 1994; Zijderveld & van der Vliet, 1994). The structure of the C-terminal domain of HAdV-C5 DBP (residues 175–529) which is responsible for unwinding activity (Eagle & Klessig, 1992; Harfst & Leppard, 1999; Kitchingman, 1995; Vos *et al.*, 1988), has been determined by X-ray diffraction (Tucker *et al.*, 1994). The most striking features of DBP revealed by the crystal structure is its largely globular domain with a protruding arm formed by the C-terminal 17 residues. Binding of this arm in one molecule of DBP in the hydrophobic cleft in an adjacent molecule leads to the formation of DBP chains (Tucker *et al.*, 1994). This property is necessary for cooperative binding of DBP to DNA and stimulation of elongation, as it drives ATP-independent unwinding of the template (Dekker *et al.*, 1997). Flexibility of the DBP, conferred by the presence of a flexible loop in the globular domain, is necessary for optimally efficient genome replication (van Breukelen *et al.*, 2000). It has been proposed that this flexible segment allows conformational changes in DBP to accommodate the transitions from binding to parental dsDNA to binding to the displaced ssDNA (van Breukelen *et al.*, 2000).

The DBP also facilitates initiation of viral DNA synthesis, by stimulating formation of the initial intermediate (pTP–CAT in the case of HAdV-C5), most likely as a result of direct interaction with the pTP–Pol complex (Mul & van der Vliet, 1993). The DBP also acts indirectly to increase the affinity of NF1 for its binding site in the viral origin (Cleat & Hay, 1989; Lindenbaum *et al.*, 1986; Stuiver & van der Vliet, 1990), probably by inducing alterations in the conformation of Ori DNA that favor NF1 binding (Stuiver *et al.*, 1992). This effect of DBP is greatest when concentrations of pTP are low, suggesting that it serves primarily to recruit pTP–Pol (via the Pol–NF1 interaction) to the origin. The DBP has also been proposed to cooperate with the other replication proteins to distort and unwind the origin during initiation of viral DNA synthesis (Liu *et al.*, 2003). Finally, this protein has been reported to stimulate annealing of complementary single strands displaced during a cycle of viral DNA synthesis and hence formation of templates for additional cycles of replication (Zijderveld & van der Vliet, 1994).

The C-terminal domain described above includes several highly conserved motifs that are important for binding of DBP to ssDNA (Eagle & Klessig, 1992; Kitchingman, 1985; Neale & Kitchingman, 1990; Vos *et al.*, 1988), as well as the C-terminal hook essential for unwinding of duplex DNA (Dekker *et al.*, 1997). In fact, this C-terminal segment can substitute completely for the full length protein in *in vitro* viral DNA synthesis systems (Ariga *et al.*, 1980). Although the N-terminal domain is dispensable in such assays, alterations within it impair viral DNA synthesis in infected cells (Brough *et al.*, 1993; Vos *et al.*, 1988). Whether this domain directly modulates one or more reactions necessary for efficient viral DNA synthesis or functions indirectly, for example via reported effects on viral gene expression (Chang & Shenk, 1990; Rice & Klessig, 1984) or by promoting assembly of viral replication centers (see below) is not known.

6.3.3 *Formation of replication centers*

Viral DNA synthesis takes place at discrete intranuclear sites, termed replication centers (or replication compartments). These structures contain the E2 DBP, which permits their ready visualization by immunofluorescence (Sugawara *et al.*, 1977; Voelkerding & Klessig, 1986), and ss viral DNA replication intermediates (Pombo *et al.*, 1994; Puvion-Dutilleul & Puvion, 1990). When observed by means of the presence of the DBP, replication centers initially appear as small dot-like structures that enlarge and form ring-like structures once viral DNA synthesis commences (Sugawara *et al.*, 1977; Voelkerding & Klessig, 1986). Replication intermediates are concentrated in the center of the latter foci, whereas the peripheral zones contain newly synthesized viral transcripts (Bridge & Pettersson, 1996; Pombo *et al.*, 1994), viral proteins that induce selective export of viral late mRNAs (Gonzalez *et al.*, 2006; Ornelles & Shenk, 1991), the viral U exon protein, UXP (Tollefson *et al.*, 2007), and during the initial period of the late phase, cellular splicing components (Aspegren *et al.*, 1998; Bridge *et al.*, 1995). Several cellular proteins that participate in DNA repair including replication protein A, E1B associated protein 5 (E1B-AP5), ATR-interacting protein, breast cancer-associated 2, and RAD51 also become localized to viral replication centers (Blackford *et al.*, 2008; Stracker *et al.*, 2005), perhaps as a result of recognition of viral ssDNA and the consequent activation of cellular DNA repair mechanisms (see next section). These infected cell-specific structures are also associated with proteins modified by addition of SUMO2 or SUMO3, which increase in concentration in infected cells in response to synthesis of either the E1B 55-kDa or the E4 Orf3 protein (Higginbotham & O'Shea, 2015). Although not necessary for their formation, the recruitment of sumoylated proteins appears to modulate the morphology of replication centers.

The concentration of viral DNA templates and replication proteins in a limited number of intranuclear foci obviously increases the efficiency of the protein–protein and protein–DNA interactions necessary for genome replication. This arrangement is also likely to facilitate the recruitment of the products of the one cycle of replication as templates for the next. Replication centers do not form in type I interferon-treated cells infected by mutants defective for production of the E1B 55 kDa or E4 Orf3 protein (Chahal *et al.*, 2012; Ullman *et al.*, 2007), and the concentration of repli-cated viral DNA is reduced by two orders of magnitude (Chahal *et al.*, 2012). These observations provide some evidence that formation of repli-cation centers is indeed necessary for efficient viral DNA synthesis, and also suggest that assembly of these specialized structures is not simply the result of the passive association of viral DNA templates with replication proteins. We do not yet possess a complete catalog of the components of these infected cell-specific structures, and have little information about their detailed organization or mechanism of assembly. However, a proto-col for partial purification of replication centers form HAdV-C5 infected normal human cells has been described recently (Hidalgo & Gonzalez, 2015). The isolated structures contain known components of replication centers and, as judged by conventional confocal and super resolution microscopy, and exhibit similar morphology and organization to intranu-clear replication compartments. It has also been reported recently that overproduction and nuclear accumulation of actins carrying substitutions that impair polymerization reduced the number of HAdV-C5 infected cells in which replication centers formed, and inhibited the development of the late, enlarged forms of these structures (Fuchsova *et al.*, 2015), suggesting that polymerization of nuclear actin is involved.

Viral genomes entering the nucleus become associated with the preexisting intranuclear structures, termed nuclear domains (ND) 10s, because of the average number per mammalian cell nucleus or PML bodies for the presence of the promyelocytic leukemia protein (Ishov & Maul, 1996). This association is not mediated by the viral genome nor does it take place immediately upon nuclear entry, but has been proposed to depend on synthesis of the DBP; in the absence of other viral proteins or viral DNA, the DBP forms discrete foci at which PML and other PML body components localize (Komatsu *et al.*, 2015). PML bodies are dismantled by the viral E4 Orf3 protein (Carvalho *et al.*, 1995; Doucas *et al.*, 1996; Puvion-Dutilleul *et al.*, 1995). This protein forms discrete, track-like structures in infected cell nuclei, or when synthesized in the absence of other viral proteins, to which PML becomes localized. Biophysical experiments such as small angle X-ray scattering established that purified E4 Orf3 protein folded correctly *in vitro* forms large oligomers (Patsalo *et al.*, 2012). This property is necessary for the functions of E4 Orf3, as a derivative with a single amino acid substitution that inactivates the viral protein forms dimers, but no longer assembles into oligomers (Patsalo *et al.*, 2012). The crystal structure of a second substitution (N82A) with the same properties, which is a dominant negative inhibitor of E4 Orf3 track formation, was determined by X-ray diffraction, and found to be a compact dimer with tails formed by the C-termini of the protein (Ou *et al.*, 2012). This structural analysis suggested that oligomers form by associations among the C-terminal tails of dimers. Consistent with the structural model, removal of the C-terminal tails eliminated the dominant negative behavior and prevented track formation by the substituted E4 Orf3 protein, but allowed co-assembly in tracks with wild-type protein (Ou *et al.*, 2012). This mechanism of assembly accounts for the formation of the linear and branched E4 Orf3 tracks observed in HAdV-C5-infected cells by electron microscopy (Fig. 6.3). The importance of formation of the higher-order assemblies is emphasized by the observation that the single amino acid substitution N82A prevents disruption of PML bodies and relocalization

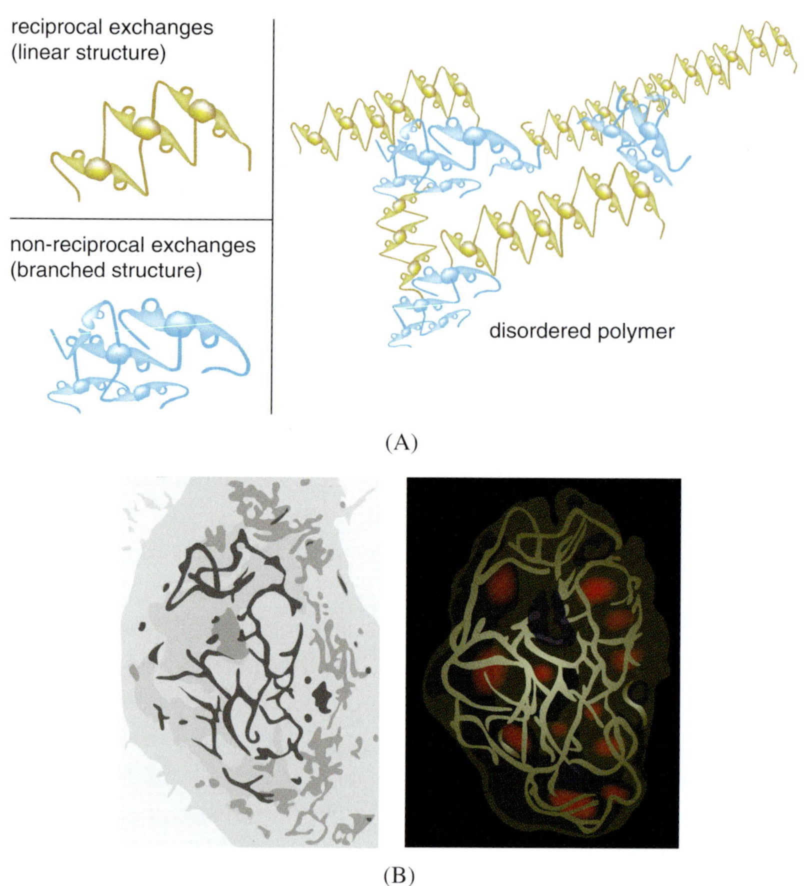

Figure 6.3. Model for assembly of HAdV-C5 E4 Orf3 protein dimers into higher-order structures. (**A**) Models for formation of linear chains or branched assemblies of E4 Orf3 protein dimers by reciprocal and non-reciprocal exchanges, respectively, of the C-terminal tails of the monomers in a dimer based on the X-ray crystal structure of an altered E4 Orf3 protein and mutational analyses. (**B**) A derivative of the E4 Orf3 protein attached to a fluorescent protein tag (minisoq), which catalyzes the formation of an electron dense polymer upon fluorescence photooxidation, was used to visualize nuclear E4 Orf3 by serial block-face scanning electron microscopy. The images show reconstructions of 150 serial section images (each 60 nm thickness) (left) and a 3D representation with the E4 Orf3 protein shown in white, nucleoli in blue, and viral replication centers in red. Scale bars, 1 μm (left) and 5 μm (right). Adapted from Ou *et al.* (2012), with permission, courtesy C. O'Shea, the Salk Institute for Biological Studies, La Jolla, CA.

of MRN (MRE11, meiotic recombination homolog 11–RAD50–NBS1, Nijmegen breakage syndrome protein 1) proteins (see next section) (Ou *et al.*, 2012).

PML reorganization depends on interaction of a particular isoform of human PML (isoform II of I–VI) with the E4 Orf3 protein (Hoppe *et al.*, 2006; Leppard *et al.*, 2009). Other components of PML bodies localize to viral replication centers (Doucas *et al.*, 1996). These proteins include several isoforms of SP100, whereas another (SP100A) becomes localized in the E3 Orf3 tracks (Berscheminski *et al.*, 2014). Overproduction of this isoform stimulated expression of several viral genes in infected cells. Much remains to be learned about the consequences of relocalization of specific PML body components to E4 Orf3 tracks or viral replication centers. However, it is clear that the disruption of PML bodies is important for blocking inhibition of viral reproduction and genome replication by interferon α or γ (Ullman *et al.*, 2007; see also Chapter 9).

6.4 Viral Proteins that Prevent Inhibition of Genome Replication

Viral DNA synthesis is a major target of the antiviral mechanisms induced by interferons, but this innate defense system is counteracted by several gene products (Chapter 9). Several intrinsic cellular defenses can also inhibit viral genome replication directly or indirectly. Not surprisingly, these responses to infection are also blocked or impeded by viral proteins.

6.4.1 *Inhibition of DNA repair responses*

The termini of the linear viral genome, which increase in concentration at a rapid rate once viral DNA synthesis commences, resemble ds breaks in DNA. When the viral measures that block cellular ds break repair pathways cannot operate in infected cells, this response is activated and viral genomes become joined end-to-end (in random orientation) to form large concatemers (J.L. Boyer *et al.*, 1999; Stracker *et al.*, 2002; Weiden & Ginsberg, 1994). Formation of these structures sequesters the terminal origins of viral genome replication, thereby inhibiting viral DNA synthesis

Table 6.1.　HAdV-C5 proteins that prevent activation of double-stranded (ds) DNA break repair.

Viral Proteins	Cellular Target	Mechanism
E1B 55 kDa, E4 Orf6 (+ CUL5, ELO B, and C, and RBX1)	MRN components (MRE11, RAD50, Nibrin)	Ubiquitinylated; targeted for proteasomal degradation
	DNA ligase 4	Ubiquitinylated; targeted for proteasomal degradation
	Bloom syndrome protein	Ubiquitinylated; targeted for proteasomal degradation
	PHO finger protein 13	Ubiquitinylated; targeted for proteasomal degradation
E4 Orf 3	MRE11	Sequestered in Orf3 tracks; relocalized to cytoplasmic aggresomes
	Nibrin	Sequestered in Orf3 tracks
	DNA-PKCs	?
E4 Orf6	DNA-PKCs	?

indirectly. Deletion of terminal sequences during end joining also removes the origins (Karen *et al.*, 2009). The actions of several viral proteins block this deleterious cellular response (Table 6.1).

The virus-specific E3 ubiquitin ligase formed by assembly of the viral E1B 55 kDa and E4 Orf6 proteins with the cellular proteins cullin 5, elongins B and C, and RBX1 (ring box finger 1) (Harada *et al.*, 2002; Querido *et al.*, 2001b) ubiquitinylates components of the cellular MRN complex (Stracker *et al.*, 2002), which associates with, and initiates signaling from, dsDNA breaks (Lavin, 2007; van den Bosch *et al.*, 2003; Williams *et al.*, 2007). Consequently, the modified cellular proteins are targeted for proteasomal degradation and their concentrations decrease rapidly once the viral proteins are made in infected cells (Karen *et al.*, 2009; Stracker *et al.*, 2002). The virus-specific ligase ubiquitinylates and targets for proteasomal degradation additional repair proteins, including DNA ligase IV, which is necessary for genome concatemerization (J.L. Boyer *et al.*, 1999), as well as Bloom helicase (Orazio *et al.*, 2011), and SPOC1 (Schreiner *et al.*, 2013). In addition, MRN proteins are sequestered in the nuclear track-like structure formed by viral E4 Orf3 protein (Evans & Hearing, 2005). Such relocalization depends on sumoylation of MRE11 and NBS1 induced by the E4 Orf3 protein (Sohn & Hearing, 2012). The E4 Orf3 protein has also been reported to induce association of MRE11 with cytoplasmic structures that resemble aggresomes (Araujo *et al.*, 2005; Y. Liu *et al.*, 2005). One consequence of these alterations in MRE11 localization is suppression of signaling via the cellular repair protein kinases ATM (ataxia telangiectasia mutated) and ATR (ataxia telangiectasia and RAD3 related) (Carson *et al.*, 2003; Carson *et al.*, 2009), which are normally activated following recruitment of MRN complexes to ds breaks in DNA (Lavin, 2007; Paull, 2015; Williams *et al.*, 2010). Both the E4 Orf3 and E4 Orf6 proteins have been observed to inhibit DNA-PK (DNA-dependent protein kinase), which also participates in repair signaling and is required for concatemerization of viral DNA (Baker *et al.*, 2007). In addition to these non-structural proteins, the major core protein, protein VII, which remains associated with the viral genome during the early phase of infection (Chatterjee *et al.*, 1986b; Xue *et al.*, 2005), appears to block induction dsDNA break repair during this period (Karen & Hearing, 2011).

The importance of inactivation of ds break repair to successful adenovirus reproduction is emphasized by the observations that the reduced yields and large decreases in viral DNA synthesis characteristic of HAdV-C5 mutants defective for synthesis of E4 Orf3 and either the E1B 55 kDa or E4 Orf6 proteins are largely reversed in cells defective for production of MRE11 or NBS1 (T.G. Boyer *et al.*, 1999; Evans & Hearing, 2005; Huang & Hearing, 1989; Lakdawala *et al.*, 2008; Mathew & Bridge, 2007; Shepard & Ornelles, 2004; Stracker *et al.*, 2002). In normal cells infected by such double mutant viruses, inhibition of viral DNA synthesis is evident before genome concatemers can be detected (Karen *et al.*, 2009). Furthermore, inhibition of DNA damage repair signaling via ATM and ATR prevents formation of genome concatemers, but does not rescue the viral DNA synthesis defects of these mutant viruses. Replication of the mutant genomes is also impaired in cells in which concatemerization is prevented by alteration of DNA ligase IV or the absence of the catalytic subunit of DNA-PK (Lakdawala *et al.*, 2008; Mathew *et al.*, 2007). These observations indicate that, when not mislocalized or degraded, MRN inhibits viral DNA synthesis directly. Such inhibition is not the result of removal of TP from the parental genome (Karen *et al.*, 2009), nor does it depend on the nuclease activity of MRE11 (Karen & Hearing, 2011). It therefore appears that rather than altering the viral origins of replication, the ds damage response proteins sequester or block sequences essential for initiation of viral DNA synthesis. This hypothesis is consistent with the inverse temporal relationship between the loss of core protein VII-containing foci and the increase in appearance of phosphorylated ATM-containing foci (with a constant total number of foci) in cells infected by an E1B 55 kDa and E4 Orf3 protein-null double mutant virus (Karen & Hearing, 2011), as well as with the localization of MRE11, RAD50, and NBS1 to replication centers (Evans & Hearing, 2003; Evans & Hearing, 2005; Stracker *et al.*, 2005). Furthermore, binding of MRN proteins to viral DNA has been detected by ChIP in such mutant virus-infected cells (Mathew *et al.*, 2007; Shah & O'Shea, 2015), although whether the origins of replication are MRN-associated was not examined.

Despite the redundant viral mechanisms to block induction of the ds DNA break repair response, HAdV-C5-infected cells contain increased concentrations of phosphorylated forms of the alternative histone H2AX (pH2AX) (Blackford *et al.*, 2008; Cuconati *et al.*, 2003; Nichols *et al.*, 2009), which is recruited to ds breaks and modified by ATM and DNA-PK very early in the repair response (Miller *et al.*, 2007; Stucki *et al.*, 2006). The concentration of pH2AX detected in infected cells was much higher than seen in mock infected cells exposed to ionizing radiation, and pH2AX was localized to neither replication foci nor discrete foci like those induced by ionizing radiation; instead it spread throughout the nucleus (Nichols *et al.*, 2009). The effects of various kinase inhibitors argue that ATR, which has been reported to be activated early after infection and to phosphorylate known substrates such as replication protein A (RPA) (Blackford *et al.*, 2008; Carson *et al.*, 2009), phosphorylates H2AX (Nichols *et al.*, 2009). Furthermore, phosphorylation of H2AX requires viral DNA synthesis. It therefore appears to be triggered by some feature of replicating genomes, such as the ssDNA or ss ends that are produced during each cycle of viral genome replication (see Fig. 6.2). Consistent with this possibility, the cellular single-stranded DNA-binding protein RPA, as well as the ATR-interacting protein, have been reported to become localized to the centers of viral replication compartments (Blackford *et al.*, 2008), where, as discussed above, ss viral DNA resides. Indeed, it has been reported recently that formation of viral replication centers triggers MRN-independent activation of ATM at these sites and subsequent diffusion of the modified kinase throughout the nucleus (Shah & O'Shea, 2015). This response does not interfere with viral DNA synthesis in cells in culture (Shah & O'Shea, 2015).

6.4.2 *Inhibition of apoptosis*

The viral E1A proteins induce apoptosis via a p53-dependent mechanism associated with stimulation of cell proliferation as well as a second, p53-independent mechanism (Debbas & White, 1993; Lowe & Ruley, 1993; Querido *et al.*, 1997b). Transient activation of p53 may lead to increased production or activation of BCL homology domain 3 (BH3)-only proapoptotic proteins, such as BAX and PUMA (a.k.a. Bcl-2-binding component 3) (Fridman & Lowe, 2003; Kannan *et al.*, 2001; Menendez *et al.*, 2009; Subramanian *et al.*, 2007). The tumor suppressor can also induce apoptosis more directly, following association with mitochondria (Green & Kroemer, 2009; Moll *et al.*, 2005). The p53-independent induction of apoptosis is mediated by activation of expression of the viral E4 gene by the largest E1A protein (Chapter 5) and the E4 Orf1 protein (Lavoie *et al.*, 1998; Marcellus *et al.*, 1996; Marcellus *et al.*, 1998; Teodoro *et al.*, 1995) via its binding the cellular phosphatase PP2A (Marcellus *et al.*, 2000). Apoptosis, an anti-viral defense of last resort, would limit adenovirus reproduction as a result of untimely death of the host cell, and degradation of viral DNA. Consequently, programmed death of infected cells is circumvented by the actions of several viral proteins.

Mutants of HAdV-C2 defective for production of the E1B 19 kDa protein exhibit a characteristic large plaque phenotype, with enhanced cytopathic effect accompanied by reduced accumulation of viral genomes and extensive degradation of both viral and cellular DNA (Subramanian *et al.*, 1984a; Subramanian *et al.*, 1984b). The E1B 19 kDa protein is sufficient to block apoptosis induced by expression of the viral E1A gene (Boulakia *et al.*, 1996; Debbas & White, 1993; Rao *et al.*, 1992) or the exogenous activator of apoptosis TNFα (tumor necrosis factor α) (White *et al.*, 1992). It is a viral equivalent of the cellular anti-apoptotic protein BCL-2 and a functional analogue of cellular MCL-1 (induced myeloid leukemia cell differentiation protein-1) (Cuconati *et al.*, 2002). When overproduced in HeLa cells, BCL-2 complements the phenotypes of E1B 19 kDa-null mutant viruses described above, and the BCL-2 and the E1B 19 kDa protein exhibit limited sequence homology in their central domains (Chiou *et al.*, 1994; Subramanian *et al.*, 1995; Tarodi *et al.*, 1993). Like BCL2, the E1B 19 kDa protein interacts with the pro-apoptotic cellular proteins BAX and BAD (BCL-3-associated agonist of cell death) (Farrow *et al.*, 1995; Han *et al.*, 1996). The BH3 domain of BAX is both necessary and sufficient for association with the viral protein (Han *et al.*, 1996; Han *et al.*, 1998), but only after exposure of cells to an apoptosis-activating signal that leads to conformational change to expose the BH3 domain (Perez & White, 2000; Sundararajan *et al.*, 2001). In uninfected cells exposed to such signals, conformational changes in BAX, as well as BAK, triggers their assembly into high-molecular weight complexes that form pores in the mitochondrial membrane to allow release to cytochrome c and other proteins that activate the cytoplasmic cascade (Cai *et al.*, 1998; Halestrap *et al.*, 2000). By sequestering these proteins, the viral E1B 19 kDa protein circumvents activation of this cascade. The observations that defects in the reproduction of E1B 19 kDa-null mutant viruses are eliminated in human BAX$^{-/-}$ cells (Lomonosova *et al.*, 2002) and BAK-deficient semi-permissive murine cells (Cuconati *et al.*, 2002) establish that prevention of the BAX–BAK interaction is the major mechanism by which the E1B 19 kDa protein blocks apoptosis. However, a fraction of this viral protein has also been reported to localize to the mitochondria, where it interacts with p53 and may block the pro-apoptotic activity of the tumor suppressor at this organelle (Lomonosova *et al.*, 2005).

In contrast to the E1B 19 kDa protein, which acts at terminal steps in activation of apoptosis, other viral proteins interfere with transmission of internal or external pro-apoptotic signals. The activation of transcription by the p53 protein and accumulation of the tumor suppressor are blocked by the viral E1B 55 kDa, E4 Orf6, and E4 Orf3 proteins (see Chapter 5) and likely others (DeHart *et al.*, 2014). The E1B 55 kDa or the E4 Orf3 proteins is also necessary to prevent translocation of apoptosis-inducing factor into the nucleus and nuclear fragmentation (Turner *et al.*, 2014). Several E3 proteins prevent receipt or transmission of external signals that turn on apoptosis (Wold & Ison, 2013).

The heterotrimer comprising two molecules of the E3 14.5 kDa protein (RIDα) and one of the E4 10.4 kDa protein (RID-β), the so-called receptor internalization and degradation (RID) complex, localizes to the plasma membrane and causes endocytosis and subsequent lysosomal degradation of the TNFα, TRAIL (TNF-related apoptosis-inducing ligand) and FAS receptors (Benedict *et al.*, 2001; Elsing *et al.*, 1998; Shisler *et al.*, 1997; Tollefson *et al.*, 1998; Tollefson *et al.*, 2001). In conjunction with the E3 6.7 kDa protein, RID also removes TRAIL receptor 2 from the cell surface (Benedict *et al.*, 2001; Lichtenstein *et al.*, 2004a). The molecular mechanism, which is thought to resemble the normal internalization and turnover of receptors, is not well understood, but requires endosomal sorting signals in the cytoplasmic domains of the E3 RIDα and RIDβ proteins (Chin & Horwitz, 2005; Hilgendorf *et al.*, 2003; Lichtenstein *et al.*, 2002). A fourth E3 gene product, the 14.7 kDa protein also inhibits apoptosis induced by TNFα, TRAIL, and FAS ligand (Gooding *et al.* 1988; Horton *et al.*, 1991). This viral protein has been reported to block internalization of the TNF receptor in response to ligand binding, and hence formation of the internal death-inducing signaling complex that comprises the receptor, TNF receptor-associated death domain (TRADD), FAS-associated death domain (FAD), and caspase 8 (Schneider-Brachert *et al.*, 2006). In these experiments, no effect on signaling via NF-κB (nuclear factor κB), which can also induce apoptosis in response to TNFα (Baud *et al.*, 2001; Wallach *et al.*, 1999), was observed, but in others the E3 14.7 kDa protein was reported to inhibit activation of NF-κB (Carmody *et al.*, 2006; Friedman & Horwitz, 2002). Such inhibition has been attributed to interaction of the viral protein with the cellular FIP3/NEMO protein, a subunit of the kinase that regulates IKβ and related proteins (Li *et al.*, 1997; Li *et al.*, 1998; Li *et al.*, 1999), and to inhibition of DNA binding by NF-κB upon interaction with the viral protein (Carmody *et al.*, 2006).

6.5 What Next?

Replication of the HAdV genome by protein priming was one of the first molecular processes in the viral infectious cycle to be elucidated. Although unprecedented at the time, this parsimonious mechanism proved to be a more general strategy for replication of the genomes of several other classes of DNA and certain RNA viruses. Synthesis of species C HAdV DNA in extracts of infected cells established the first mammalian system for *in vitro* studies of DNA replication, and paved the way for identification of viral replication proteins and analysis of the mechanisms of origin recognition and initiation of DNA synthesis. We now possess a moderately detailed picture of the organization and function of the viral replication proteins and how they cooperate with cellular proteins during initiation. Nevertheless, the dearth of structural information is a major impediment to obtaining further insights into the elaborate and dynamic protein–DNA and protein–protein interactions that mediate viral DNA synthesis. Such information is available only for the C-terminal replication domain of the DBP, and the much greater size of the HAdV-C5 Pol limits the degree to which detailed information can be extrapolated from X-ray crystal structures of the analogous phage ø29 enzyme. It is to be hoped that the increased interest in development of antiviral drugs against HAdVs will spur structural studies of the complex of Pol and preTP. The important issue of how viral replication centers form and function in infected cell nuclei also remains poorly understood. These issues are receiving increasing attention, and the application of newer technologies like sensitive mass spectrometry, high-resolution cryoelectron microscopy (Banerjee *et al.*, 2016) and advanced live-cell imaging should be informative.

A number of viral early protein-induced alterations in cellular gene expression and other processes that either redirect the physiological state of the host cell to provide the cellular proteins and substrates needed for viral DNA synthesis, or impair response to infection that would limit genome replication have been described in some detail. The latter processes include apoptosis and the dsDNA break repair response, which is triggered by the linear viral genome. It is now clear that each of these responses is blocked by multiple viral early gene products, a property that provides redundancy, and that several early proteins, including the E1B

55 kDa, E4 Orf6, and E4 Orf3 proteins, inhibit activation of multiple pathways deleterious for viral genome replication. Interactions of the viral with cellular proteins necessary for these functions have been defined, and the first structural information of the E4 Orf3 protein reported recently. Nevertheless, we have little appreciation of the importance of targeting of the various host responses to successful virus reproduction, particularly in normal human cells resembling those naturally infected, as detailed mutational studies of these viral proteins have not been reported.

Capsid Assembly, Maturation, and Egress   7

7.1 Nuclear Import of Newly Synthesized Capsid Proteins for Assembly

An essential step in adenovirus assembly of progeny virions is the nuclear import of newly synthesized capsid proteins from the cytosol of infected cells. Nuclear import of capsid protein is via the cell trafficking pathways used for signal-mediated import of cellular proteins through the nucleus pore complex (NPC) (Whittaker *et al.* 2000). Binding and transport through the NPC is mediated by nuclear localization signals (NLSs), which generally consist of short stretches of basic amino acid residues recognized by the karyopherin/importin β superfamily of proteins (Bednenko *et al.*, 2003; Kuersten *et al.*, 2001; Macara, 2001). NLSs are present in a number of the adenoviral capsid proteins, but surprisingly, the major capsid protein, hexon, does not contain a recognizable NLS and is not imported by itself (Russell & Kemp, 1995). Instead, protein VI, following its cleavage by the viral cysteine protease (AVP) at its C-terminus, acts as a chaperone for hexon nuclear import via an importin α/β (Wodrich *et al.*, 2003). Strong evidence for hexon-VI import was obtained by studying a temperature sensitive mutant virus designated human adenovirus (HAdV)-C5 *ts*147 that exhibits a profound loss of hexon import at the non-permissive temperature and is also linked to instability in protein VI

(Kauffman & Ginsberg, 1976). One of the five mutations discovered in the *ts*147 hexon, G776D (Wodrich *et al.*, 2003), is located near the base of the hexon trimer where protein VI interacts (Reddy *et al.*, 2014b).

An important question that has not yet been fully addressed is how the viral genome enters the nucleus to initiate the entire process of viral replication. A prevailing hypothesis is that the hundreds of copies of one of the three core protein known as protein VII that are tightly associated with the viral genome (Chatterjee *et al.*, 1986a) promote DNA nuclear import. In support of this hypothesis, protein VII was found to contain multiple NLS that bind to importins α/β (Wodrich *et al.*, 2006). However, the precise mechanism of DNA–protein VII translocation through the nuclear pore and its eventual arrival at replication centers in the nucleus remains to be elucidated.

Once inside the nucleus, the exact mechanism of virus assembly from newly imported capsid proteins is still an enigma, as we lack precise information on the order in which the structural proteins are incorporated into new particles or even how DNA is encapsidated. Some of these partially resolved mechanisms and remaining questions are discussed in this chapter.

7.2 Formation of a Metastable Adenovirus Capsid

The mechanisms comprising the late steps in the adenovirus replication cycle such as capsid assembly and maturation have yet to be fully elucidated. In other viruses such as certain bacteriophages and herpsesviruses, the process of capsid maturation begins with the dual actions of a virus protease and DNA encapsidation (Veesler & Johnson, 2012). However, unlike other viruses that undergo dramatic rearrangements of their outer capsid proteins during maturation (Johnson, 2010), this mechanism does not appear to be a hallmark of adenovirus capsid maturation. One hypothesis that has emerged from ongoing studies of adenovirus is that production of new capsids begins with assembly of so-called immature particles that lack most of the viral DNA and have unprocessed preproteins. These immature particles are unable to mediate host cell infection. Immature particles then undergo protease cleavages of multiple capsid proteins concomitant with DNA encapsidation, resulting in the generation of mature particles capable of infecting host cells. Immature particles can be separated from mature particles that have DNA by cesium chloride density gradient ultracentrifugation: 1.29–1.30 g/cm^3 and 1.34 g/cm^3, respectively (Ishibashi *et al.*, 1974; Sundquist *et al.*, 1973). It still remains to be determined whether immature adenovirus particles are true assembly intermediates or whether they represent dead end byproducts.

The mature virus particle is often referred to as being in a metastable state, i.e. stable enough to withstand the harsh extracellular environment (very low pH, etc.) yet conformationally dynamic and able to respond to host cell stimuli that lead to disassembly in the right place and appropriate time during cell infection. The final product of the maturation process is an adenovirus particle with its double-stranded (ds) DNA cargo fully protected from potentially damaging host and environmental factors until it can be delivered to the host cell nucleus to begin a new infectious cycle. As detailed below, several virally encoded proteins play crucial roles in the processes of capsid assembly/maturation as well as packaging of the viral genome and its incorporation into the mature virus particle.

7.2.1 *The viral protease, AVP, plays a crucial role in maturation of virus particles*

A hallmark of maturation of adenovirus particles is the development of the metastable state described above. Critical to this event is the enzymatic processing of six structural proteins (IIIa, VI, VII, VIII, TP, and μ) from their preprotein forms to the fully cleaved mature polypeptides by the viral protease (AVP) (Fig. 7.1). Failure to achieve maturation cleavage of these capsid proteins in a temperature-sensitive HAdV-C2 known as *ts1* (Imelli *et al.*, 2009; Weber, 1976) produces a hyperstable particle that cannot undergo partial disassembly in the endosome and thus is non-infectious (see below). The viral protease is present in approximately seven copies per virion (Benevento *et al.*, 2014), and cleaves at one of two highly conserved amino acid sequence motifs [(M/I/L)XGG>X or (M/I/L)XGX>G] in the adenoviral capsid preproteins (Diouri *et al.*, 1996). AVP, a 23-KDa cysteine protease encoded by the HAdV-C2 L3 gene, has only low basal enzymatic activity. However, its association with several cofactors increases its rate of substrate cleavage dramatically (Mangel & San Martin, 2014). One important cofactor is an 11-amino-acid peptide, designated pVIc

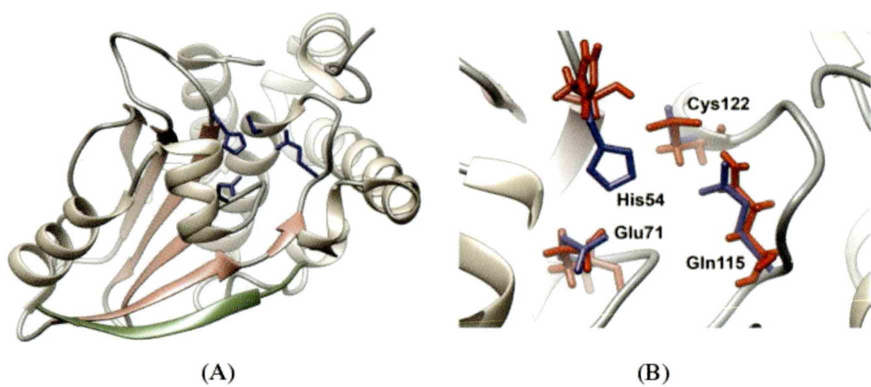

Figure 7.1. Crystal structure of the AVP–pVIc complex and locations of the four amino acid residues involved in catalysis in AVP and in AVP–pVIc. (**A**) Secondary structure of the AVP–pVIc complex with the four amino acid residues involved in catalysis in blue and the pVIc peptide in green. (**B**) The four amino acids involved in catalysis in AVP–pVIc (blue) and in AVP (red) are juxtaposed. Only His54 is in a different position in the two structures. Figure created with UCSF Chimera.

(GVQSLKRRRCF) that is derived from cleavage of the C-terminus of preprotein VI. The basal level of AVP activity without chaperone/cofactors presumably liberates the pVIc peptide. The pVIc peptide then acts as a molecular strap that becomes covalently bound to AVP and holds the two domains of the enzyme in a conformation that is optimal for substrate binding and cleavage (McGrath *et al.*, 2003). Another important AVP cofactor is the viral nucleic acid (Mangel *et al.*, 1993). The interaction of AVP with dsDNA is not nucleotide sequence specific, but association with a high density of negative charges seems to be important for accelerating enzymatic activity. Based on *in vitro* cleavage studies, pVIc and dsDNA together have been estimated to increase the activity of AVP by 34,000 fold (Mangel *et al.*, 1996).

Another molecule that appears to be a substrate of AVP is cellular actin. A C-terminal segment of actin contains a similar sequence (SGPSIVHRKCF) to that of the pVIc C-terminal peptide. In fact, actin binds to AVP with low nanomolar Kd affinity (Brown *et al.*, 2002) and accelerates substrate cleavage. The physiological relevance of actin binding to AVP is still unclear. However, cleavage of actin might weaken the cell cytoskeleton, thereby favoring the egress of progeny virions late in the infection cycle.

The precise molecular dynamics of how AVP cleaves multiple cement proteins on the inside of the virus particle is still poorly understood. Recent studies using total internal reflection fluorescence microscopy (Graziano *et al.*, 2013; Mangel *et al.*, 2016) have led to a proposed model in which the complex of AVP–pVIc slides along the viral dsDNA on the inside of the capsid during particle maturation. As it encounters each precursor protein, AVP cleaves them in rapid succession as it travels along the viral genome. The mean one-dimensional diffusion constant for DNA sliding by the AVP–pVIc complex has been estimated to be ~21×10^6 base pair^2 / second and exhibits Brownian motion (Blainey *et al.*, 2013). The molecular details underlying these processes are likely to be quite complicated, and require further investigation. As AVP-induced virus maturation represents such an important step of the virus infectious cycle, this enzyme might serve as targets for antiviral therapy, as suggested by the discovery of an *in vitro* inhibitor of AVP, 2,4,5,7-tetranitro-9-fluorenone (Pang *et al.*, 2001).

7.2.2 *AdV2-ts1 function provides clues to the process of capsid maturation*

The temperature-sensitive mutant adenovirus known as *ts1* has provided useful information about the process of virus capsid maturation. When grown at non-permissive temperatures of 39–40°C, a specific amino acid substitution (P137L) in the *ts1* protease prevents it from being incorporated into the capsid (Rancourt *et al.*, 1995; Yeh-Kai *et al.*, 1983). As a result, the preproteins of the mutant virus fail to undergo enzymatic cleavage as revealed by SDS-PAGE analyses. More importantly, *ts1* particles are non-infectious; they are able to recognize and utilize host cell receptors and enter cells normally, but they fail to escape the endosome and are either recycled out of the cell or are degraded in lysosomes (Greber *et al.*, 1996; Moyer *et al.*, 2011). The underlying defect of the *ts1* capsid is that it is hyperstable and fails to undergo disassembly during cell infection (Nguyen *et al.*, 2010). As a consequence, protein VI cannot be released from the interior of the capsid to facilitate endosome disruption. The hyperstability of *ts1* capsids has been demonstrated using biochemical and biophysical approaches such as heat-disassembly assays (Perez-Berna *et al.*, 2012; Wiethoff *et al.*, 2005) and atomic force microscopy (AFM) (Perez-Berna *et al.*, 2012). Cryo-electron microscopy (cryoEM) structural analyses have also suggested that the DNA core architecture of *ts1* capsids is distinct from that of native virus (Perez-Berna *et al.*, 2009; Silvestry *et al.*, 2009). In general, protein VI cryoEM density is greater in *ts1* particles than in wild-type virus and the core of the *ts1* capsid also appears to be more connected to the outer capsid.

It has also been suggested that the DNA core of *ts1* is more condensed (Perez-Berna *et al.*, 2009), perhaps due to the actions of precursor forms of the DNA-binding proteins VII and μ. Using biophysical methods such as nanoindentation (AFM), Perez-Berna *et al.* (2012) showed that immature adenovirus particles (*ts1*) are more stable than mature virions and also appear to be more flexible. They found that the core of adenovirus particles undergoes compaction as maturation proceeds, perhaps creating increased internal pressure in the capsid that helps trigger disassembly. Overall, the data suggest that the process of maturation primes the virus capsid to undergo disassembly during cell infection. Additional molecular details that define the precise structural differences between *ts1* and native virus are still lacking.

7.2.3 *Role of the L1 52/55K packaging scaffold protein in adenovirus maturation*

Immature adenovirus particles exhibit differences in protein composition in addition to the absence of mature proteins. One difference is exemplified by a phosphoprotein encoded by the L1 region of the viral genome known as the 52/55K protein that is present exclusively in immature capsids. This protein was originally thought to be a scaffold molecule as it was present in immature particles that lack DNA but absent in mature virions (Hasson *et al.*, 1992). However, the 52/55K protein is not required for capsid protein assembly *per se* but instead promotes encapsidation of the viral nucleic acid (Gustin & Imperiale, 1998). The absence of the 52/55K protein leads to the formation of empty virus particles that lack DNA (Perez-Romero *et al.*, 2006). The 52/55K protein forms a multimeric assembly and associates with the viral DNA packaging signal either directly or indirectly via interactions with other DNA-binding proteins (Perez-Romero *et al.*, 2005; Zhang & Imperiale, 2000). One of the remaining questions related to the 52/55K protein is: why is this molecule present only in immature virus particles? Recently, it was discovered that the 52/55K protein is a substrate for AVP and is cleaved at numerous locations (Perez-Berna *et al.*, 2014). The authors of this study suggested that AVP mediates extensive proteolysis of the 52/55K protein thus facilitating the release of the cleaved fragments from immature particles during maturation.

7.3 Genome Encapsidation and Recognition of the Adenoviral DNA Packaging Signal

Incomplete adenovirus capsids can be isolated from infected cells in a range of cesium chloride densities between that of immature particles and fully mature virions. This range in densities corresponds to the sizes of the DNA present in the particles, which have increasing nucleic acid lengths starting from the left end of the genome (Daniell & Mullebach, 1978; Tibbetts, 1977). This observation is in keeping with the left end of the viral genome as the leading segment of nucleic acid encapsidation, but the exact mechanism of DNA import into the capsid still remains a mystery. Whether it occurs through a unique vertex (i.e. portal), as occurs in an energy-dependent process during assembly of some bacteriophage and herpesviruses, or whether the capsid is assembled around the DNA has yet to be determined.

The left end of the HAdV-C5 genome between base pairs 200–400 contains a series of AT-rich repeat regions (A repeats) that comprise the DNA packaging signal (Grable & Hearing, 1990; Hearing *et al.*, 1987; Parks *et al.*, 1996). Along with the 52/55K protein, two other virally encoded proteins, IVa2 and the late region (L4) 22K protein, recognize this packaging signal and promote genome encapsidation (Ewing *et al.*, 2007; Ostapchuk *et al.*, 2005; Ostapchuk *et al.*, 2006; Zhang *et al.*, 2000). Mutant viruses lacking IVa2 are unable to incorporate DNA into mature virions (Ostapchuk *et al.*, 2011; Zhang & Imperiale, 2003). The role of IVa2 may be more elaborate as there has been a suggestion that it not only mediates recognition of the adenoviral DNA packaging signal but also it participates actively in the process of DNA encapsidation. This protein contains the Walker A/B box motif that is generally associated with ATP binding and hydrolysis (Koonin *et al.*, 1993). However, the predicted secondary structure of IVa2 is distinct from that of the ATPases comprising bacteriophage motor proteins that mediate nucleic acid encapsidation (Rao & Feiss, 2008). Mutations introduced into the Walker A/B box of IVa2 inhibit ATP binding (Ostapchuk & Hearing, 2008), but there is currently no evidence that IVa2 can catalyze ATP hydrolysis on its own, although recent studies reported that hydrolysis of ATP is accomplished by IVa2 in the presence of the L4 33K protein (see below) (Ahi *et al.*, 2015). IVa2 has been reported to be localized at one vertex of the adenovirus particle (Christensen *et al.*, 2008), although it has not been demonstrated that this is a unique vertex representing the long-sought after portal.

The IVa2 protein also associates with the L4 33K protein that is thought to be important for DNA packaging because it forms a ring-like macromolecular structure which appears to stimulate IVa2-dependent ATPase activity (Ahi *et al.*, 2015). The 33K protein is present in immature (empty) but not mature virus particles suggesting a possible role as a scaffold protein (Ahi *et al.*, 2013). Either C-terminal truncation of the 33K protein (Finnen *et al.*, 2001) or the absence of the entire protein blocks capsid assembly, and the latter mutations reduce the yield of complete capsids (Fessler & Young, 1999). If the role of the 33K protein in stimulating IVa2 ATPase activity can be substantiated and explored further, it might reveal additional insights into how the complex of these two proteins (and perhaps others) facilitates viral genome encapsidation.

In addition to DNA-binding proteins that participate in genome encapsidation, adenovirus also encodes the L4 100K protein that plays an essential role in virus assembly (Cepko & Sharp, 1982; Cepko & Sharp, 1983; Gambke & Deppert, 1983). The main function of this protein is as a chaperone required for the proper folding and trimerization of the hexon, the major outer capsid protein. The 100K viral chaperone protein accumulates to high concentration in host cells late in infection, but it is not incorporated into the mature capsids of progeny virus particles. It has also been suggested that the 100K is also a scaffold protein for virus assembly (Morin & Boulanger, 1986; Oosterom-Dragon & Ginsberg, 1981).

7.4 Virus Egress

Newly assembled virus particles face one final challenge following their assembly. They must find their way out of the nucleus in order to gain access to new host cells. For most HAdV types, egress is not a very efficient process. In fact, virus isolation from tissue culture cells usually requires multiple freeze-thaw cycles to liberate progeny particles that remain trapped inside the nucleus. Thus, cell death (apoptosis or necrosis) caused by virus replication is probably the major contributor of adenovirus egress. Previous studies have indicated that several gene products encoded by the viral early region 3 (E3) promote cell death as well as modulate host immune functions (Horwitz, 2004; Lichtenstein *et al.*, 2004b). One of these E3-encoded molecules, the 11.6K protein, also known as the adenovirus death protein (ADP), is produced late in infection via the major late promoter and facilitates apoptosis (Tollefson *et al.*, 1996a). The concentration of ADP attained in certain host cells can be a determining factor for whether the virus establishes persistent infection or enters the pathway for productive infection (Murali *et al.* 2014; see also Chapter 8). Viral mutants that overexpress ADP show early cytopathic effect and cell lysis, produce larger plaques than wild-type virus, and have increased cell to cell transmission, thereby supporting the notion that ADP facilitates virus egress (Doronin *et al.*, 2003). Given that the adenovirus ADP protein is unrelated to other pro-apoptotic proteins, its precise mode of action is yet to be determined.

Although not fully investigated, it is also possible that adenovirus can be transmitted between host cells by mechanisms other than apoptosis. For example, new virus particles might be released from infected cells after envelopment inside cell exosomes, small membranous vesicles of 50–150 nm (Meckes, 2015) that promote cell to cell communication. Another non-enveloped virus was previously reported to spread from cell to cell via exosomes (Feng *et al.*, 2013). More recent studies showed that neural stem cell-derived exosomes were capable of mediating HAdV-C5 dissemination to receptor (CAR)-deficient cells (Sims *et al.*, 2014). An advantage of exosome-mediated virus transmission is that the exosome membrane would protect the virus from recognition by neutralizing antibodies. Further studies are needed to determine whether the exosome route of transmission is an authentic trafficking pathway in adenovirus infection.

7.4.1 *Autophagy may facilitate release from infected cells*

Autophagy (literally 'self-eating') allows recycling of proteins and other materials from damaged organelles and proteins, and facilitates survival in response to nutritional starvation (see Levine & Kroemer, 2009; Yang & Klionsky, 2010). Autophagy can also be induced by virus infection, either to facilitate specific reactions in the infectious cycle or to provide antiviral defense, by triggering of a form of programmed cell death (Baehrecke, 2005; Orvedahl & Levine, 2009). Autophagy has been studied quite extensively in the context of oncolytic adenoviruses, and various approaches taken to promote tumor cell killing by this mechanism. There have also been some reports that this process is induced in cells infected by wild-type HAdV-C5.

Infection of human lung cancer cells (A549) and normal lung fibroblasts by HAdV-C5 was observed to induce the formation of autophagosomes that carry the hallmarks of these specialized vesicles, such as the membrane protein LC3-II, during the late phase of infection (Rodriguez-Rocha *et al.*, 2011). Activation of autophagy via cleavage of the LC3-II precursor and association of the proteins ATG5 (autophagy 5) and ATG12 with one another was reduced when the E1B gene was deleted from the viral genome, and this process has been reported to depend on E1B 19-kDa protein-mediated release of the pro-autophagic protein Beclin1 from its association with BCL-2 (Piya *et al.*, 2011; Rodriguez-Rocha *et al.*, 2011). Inhibition of autophagy by exposure of HAdV-C5 infected cells to drugs that target this process specifically reduced release of infectious virus particles and accumulation of viral late proteins (Jiang *et al.*, 2011; Rodriguez-Rocha *et al.*, 2011). Because of the latter effect, it is not clear whether autophagy facilitates HAdV release from infected cells directly. Furthermore, it has also been reported that induction of autophagy in A549 and normal human bronchial epithelial cells prior to HAdV-C2 infections modestly stimulates expression of the viral E1A gene and viral DNA synthesis, by facilitating entry of virus into the cytoplasm (Zeng & Carlin, 2013).

7.5 What Next?

An accurate picture of how the adenovirus capsid is assembled has been stymied by the large and intricate architecture of the virus. Detailed structural information of mature virus particles has been obtained using high-resolution X-ray diffraction and cryoEM techniques, but has provided only a few clues as to how the complete outer capsid is assembled. In particular, we still lack detailed structural information on the virion core and its associated nucleic acid genome. One of the earliest approaches used to analyze virus assembly employed a chemical solvent, pyridine, to disassemble mature virus particles to reveal their sub-structures (Crowther & Franklin, 1972; Prage *et al.*, 1970). A prominent substructure identified comprises a group of nine hexons (GON) that have a central threefold symmetry axis and were predicted to arise from one facet of the virus capsid. The GON hexon complex was primarily examined by negative staining electron microscopy and therefore its potential association with one or more cement proteins could not be easily determined. It is still unknown whether the GON is a true intermediate of capsid assembly or whether it is a fortuitous byproduct of capsid disassembly.

Apart from their functions in stabilizing the outer and inner surface of the adenovirus capsid, there is relatively little information about the potential roles of cement proteins in the assembly process itself. However, a temperature-sensitive mutant adenovirus with multiple nucleotide substitutions in the region encoding protein IIIa (mutant *ts*112) results in a virus with defective capsid maturation (Chroboczek *et al.*, 1986). The *ts*112 mutant accumulates light precursor capsids that do not mature into complete particles with a full genome (D'Halluin *et al.*, 1978). As protein IIIa plays a prominent role in bridging adjacent hexon trimers in the outer capsid, it is possible that alterations of this protein could impede hexon assembly, but this model has yet to be verified. A substitution in protein VIII, a cement protein on the inner capsid surface that adjoins neighboring hexon), was reported to have decreased virus maturation/ assembly (Liu *et al.*, 1985). Thus, it seems possible that cement proteins not only help stabilize the capsid, but may also contribute to the assembly/maturation of virus particles.

Finally, a major unresolved question is how the viral nucleic acid together with several DNA-binding proteins V, VII, and μ are encapsidated in maturing adenovirus capsids. One observation derived from structural studies is that the DNA and its associated proteins are not icosahedrally ordered in adenovirus as in other viruses that import their genome through a unique vertex (portal). To reveal the architecture of the HAdV DNA core, earlier studies used low energy etching of adenovirus capsids with

Ar$^+$ ions (Newcomb *et al.*, 1984) or treated virus particles with detergent and proteases (Brown *et al.*, 1975). Electron microscopy revealed that that upon removal of the outer capsid, it was possible to discern 12 distinct spheres each with a diameter of ~20 nm. These spheres, termed adenosomes, appeared to have two-, three-, and fivefold rotational symmetry axes, and thus it was proposed that each adenosome might be positioned close to the fivefold vertex of the outer capsid. Consistent with this idea, the close packing of the entire ensemble of adenosomes could be accommodated in the interior of the capsid shell when the individual spheres were positioned near each vertex (Newcomb *et al.*, 1984). It remains to be determined if these early findings demonstrate the authentic arrangement of the DNA core of adenovirus or whether represent artifacts stemming from the approaches used to "peel" away the outer capsid created artifacts and thus do not reflect the arrangement of fully encapsidated nucleic acid. One implication of these earlier studies is that DNA encapsidation might not occur through a unique portal at one vertex. Instead, the outer capsid might assemble around the nucleic acid (adenosomes) using these core structures as scaffolds for assembly. However, a formal possibility that still must be considered is that adenovirus does possess a portal complex for DNA encapsidation. A unique portal, located at a single vertex among the 11 other penton complexes, would likely be obscured in high-resolution cryoEM and X-ray diffraction techniques due to data averaging. Future progress into solving the problem of genome encapsidation may stem from recent developments in electron cryo-tomography (Dai *et al.*, 2013), and computational methods (Q. Wang *et al.*, 2013), as well as the combined use of correlative fluorescence and electron microscopy (Abdelhakim *et al.* 2014; Lanman *et al.*, 2008). These new techniques may help reveal the answers to questions on the mechanisms of assembly, maturation and egress that persist even more than 60 years after the discovery of adenoviruses.

Pathogenesis 8

8.1 Introduction

Most human adenoviruses (HAdVs) are widespread in the human population. The pathology of infection by these viruses varies with adenovirus type, but is even more strongly influenced by the immune status of the infected individual. Most acute HAdV infections are relatively mild and self-limiting in immunocompetent infants, children, and adults, but associated with severe disease, and quite often fatal, in immunosuppressed patients. The mechanisms of pathogenesis have been investigated in a variety of animal models, although most of these models provide only limited facsimiles of the adenovirus–human interaction because they are non- or semi- permissive for virus reproduction. Adenoviruses can also establish long-term, persistent infections in humans, which are thought to be an origin of disseminated acute infection in susceptible patients. Despite their likely clinical importance, the mechanisms underlying the establishment and maintenance of persistent adenovirus infections are not well understood.

8.2 Epidemiology and Pathology

8.2.1 *Diseases caused by HAdVs in immunocompetent hosts*

8.2.1.1 *Respiratory infections in the young*

HAdVs cause a variety of acute diseases in immunocompetent children and adults, usually as a result of infection of epithelial cells. The majority of HAdV types appear to circulate worldwide, but the most predominant change as a function of time and may vary with geographical location (Ampuero *et al.*, 2012; Kajon *et al.*, 2010; Lin *et al.*, 2004; Lin *et al.*, 2010). The HAdVs most commonly detected in human patients are members of species B, C, E, and F (Lion, 2014). They are responsible for some 5–7% of upper respiratory tract infections characterized by nasal secretion, cough, and sometimes tonsillitis in infants and young children (Luksic *et al.* 2013; Pavia, 2011). Such upper respiratory tract disease is typically caused by species C HAdVs and occurs early in life. For example, in one study of infants and children in a day-care center, 298 HAdV isolates were obtained from respiratory secretions, with HAdV-C1, -C2, and -C5 accounting for 92% of the isolates, and were most prevalent in children from 6–12 months of age (Pacini *et al.*, 1987). The incidence of these HAdVs in the general population is high, but geographically variable; based on the presence of antibodies against it, HAdV-C5 has been detected in 30–70% of the population in the U.S., 50–60% in Europe, 85% in China, and in nearly 100% in Northern India (Wold & Ison, 2013). The available data indicate that neonates carry maternal antiviral antibodies, which decline with age and are replaced as children become infected. As HAdVs are contagious, infections are spread with relative ease, particularly in confined and isolated communities, such as day-care centers and military training camps. These HAdVs can be transmitted by the respiratory route, but the results from a long-term study in the U.S. indicated that fecal–oral transmission is much more common in young children (Fox *et al.*, 1977). As indicated by the large difference between the high proportion of the population seropositive for these HAdVs and the much lower frequency of respiratory disease, many infections by the species C HAdVs are asymptomatic in children. However, in contrast to the typical mild or asymptomatic infections by these viruses in children, recent studies suggest that neonates are at greater risk for serious disease (Civardi *et al.*, 2013; Ronchi *et al.*, 2014). For example, in a retrospective study of neonates hospitalized in Dallas, Texas, between 1995 and 2012, 26 had adenoviral infection, and in five cases the infection was disseminated and resulted in death of four of the infants (Ronchi *et al.*, 2014).

Adenoviruses also cause lower respiratory tract disease, and have been reported to be responsible for about 10% of childhood pneumonias (Mull, 1966; Wold & Ison, 2013). This syndrome is associated with infection by HAdV-C1 and -C2 and also HAdV-B3 and HAdV-B7. In some instances, epidemics of lower respiratory tract infections by these adenoviruses result in significant morbidity in children, in particular when patients are infected by HAdV-B7 or -B3 (Hong *et al.*, 2001; Lee *et al.*, 2010; C. Liu *et al.*, 2015). Severe respiratory illness in adults has also been reported. In an outbreak associated with significant morbidity in Oregon, 60% of the patients studied were infected by HAdV-B14 and of these 76% required hospitalization and 78% died (Lewis *et al.*, 2009). Older age and chronic conditions are likely risk factors for severe illness. The more recently arising HAdV-B55 (a recombinant of HAdV-B14 and HAdV-B11) (Walsh *et al.*, 2010; Zhang *et al.*, 2012) has also been increasingly associated with severe respiratory diseases in various parts of China (Cao *et al.*, 2014; Li & Chen, 2014), and in some cases reported to be fatal even in healthy adults (Zhang *et al.*, 2016). Species B HAdVs, as well as HAdV-E4 also cause epidemics of a similar syndrome, acute respiratory disease (ARD), in military recruits (Gaydos & Gaydos, 1995).

8.2.1.2 *Acute respiratory infections in the military*

During the basic phase of U.S. military training, new recruits experience conditions that increase the rate of adenovirus respiratory infections, including high-density living conditions, lower emphasis on personal hygiene, and physical and environmental stress (Broderick *et al.*, 2008; Russell *et al.*, 2006). Before the advent of routine vaccinations, as many as 80% of recruits may have become infected with HAdVs, with 40% having significant respiratory illness (Hilleman, 1957). Among the many viruses associated with respiratory infections in the military, types HAdV-E4, -B7, -B3, and -B21 are the most prevalent (Gaydos & Gaydos, 1995; Ludwig *et al.*, 1998; Metzgar *et al.*, 2007) and these can sometimes lead to fatal pneumonia (Dudding *et al.*, 1972). These HAdV types were therefore considered to be prime candidates for a vaccine program for the military that was eventually initiated in 1971 under a contract with Wyeth Pharmaceuticals. Administration of a vaccine consisting of the live, oral, enteric-coated HAdV-E4 and HAdV-B7 was found to be safe and effective in reducing the incidence of adenovirus respiratory disease in recruit training centers (Top *et al.*, 1971). Unfortunately, in 1996, the sole manufacturer of the vaccine decided to discontinue its production, in part because of cost considerations. This loss resulted in a dramatic increase in the number of cases of acute adenovirus respiratory disease in military recruits (Hendrix *et al.*, 1999). In addition to increased morbidity and mortality, such respiratory diseases substantially increased the cost of basic training. Consequently, the U.S. Department of Defense instituted a new adenovirus vaccine program via a contract with Barr Pharmaceuticals. A dual HAdV-E4–HAdV-B7 vaccine was administered to new military recruits in 2011, and it also proved to be safe and effective, as demonstrated by a rise in the blood concentrations of antiviral antibodies in (Kuschner *et al.*, 2013).

The Naval Health Research Center (NHRC) in San Diego, California, performed long-term surveillance of the HAdV types appearing at eight different recruit training centers across the U.S. beginning in 1996 and ending in 2013 (Radin *et al.*, 2014). Although a number of infections with types such as HAdV-B14 were detected, respiratory diseases due to HAdV-E4, and to a lesser extent, HAdV-B7 continued to predominate during this period. Notably, the re-introduction of the HAdV-E4/B7 vaccine resulted in a 100-fold decrease in adenovirus respiratory infections between 2011 and 2013 (Radin *et al.*, 2014). These authors estimated that the current vaccine prevents approximately one death, 1,000–3,000 hospitalizations, and 13,000 febrile cases each year among military trainees. These numbers translate into an estimated net cost savings for the Department of Defense of roughly $20 million dollars. Continued surveillance of the military, as well as of the general population, for emerging HAdV types such as HAdV-B14p1 (Kajon *et al.*, 2010) and HAdV-B21 (Hage *et al.*, 2014; Kajon *et al.*, 2015) that are associated with ARD, may be necessary to guide modification of the adenovirus vaccine program over the coming years. These vaccines have never been made available to the general public, perhaps because of their development under the auspices of the Department of Defense, and consequent dedication of the supply to military personnel.

8.2.1.3 *Ocular infections*

8.2.1.3.1 Epidemiology and pathogenesis

Conjunctivitis, also known as viral pink eye, is caused by many different HAdV types (Roba *et al.*, 1995). Such infections, although associated with some morbidity, are usually self-limiting and do not cause permanent damage to the cornea. In contrast, infections with certain species D adenoviruses HAdV-D8, 19 (reclassified as 64; see below), 37, 53, 54, and 56 result in much more severe forms of ocular disease known as epidemic keratoconjunctivitis (EKC) (G. Huang *et al.*, 2014; Robinson *et al.*, 2008; Robinson *et al.*, 2009; Walsh *et al.*, 2009; Zhou *et al.*, 2012). As noted in Chapter 3, many of these EKC-associated species D viruses bind to cell surface molecules displaying sialic acid moieties, including the GD1a glycan that is present on human corneal cells (Nilsson *et al.*, 2011). As sialic-acid-linked glycans are widely distributed on many cell types, it is unlikely that binding to these molecules by species D viruses explains their ocular tropism. Other host or viral components that may contribute to ocular tropism, such as lipid rafts and caveolin-1, promote virus internalization via cell signaling and have been reported to be preferentially recognized by type D viruses on corneal cells (Yousuf *et al.*, 2013).

EKC is characterized by severe and painful irritation of the eye, photophobia, blurred vision, and swelling at the site of inflammation (conjunctiva). Such severe symptoms usually persist for a few weeks and then begin to recede, but reduced visual acuity can extend for additional weeks to months. This ocular disease is known to spread rapidly in patients in hospital clinics throughout the world (Calkavur *et al.*, 2012). For example, between 2008 and 2010, six independent adenovirus EKC outbreaks occurred in four different states in the U.S. with a total of 411 cases (Centers for Disease Control and Prevention, 2013). Nosocomial infections from contaminated eye instruments or eye drops have also been reported (Klapper & Cleator, 1995).

One of the hallmarks of EKC pathology is the subepithelial infiltration of immune cells that migrate into and underneath the corneal epithelium (Butt & Chodosh, 2006). Conjunctival exudates obtained from patients with EKC contain a variety of cell types staining positive for CD4, CD8, CD20, CD68, and CD31 markers, indicating the presence of T and B lymphocytes as well as of macrophages and monocytes (Chintakuntlawar & Chodosh, 2010). EKC ocular tissue explants also contain a number of angiogenic and proliferative factors, such as vascular endothelial growth factor, the angiopoietin receptor, TIE-2, and transforming factor-β, which may promote immune cell infiltration as well as endothelial cell proliferation. Early subepithelial corneal infiltrates (three days post-infection with

HAdV-D64) also contain polymorphonuclear neutrophils characterized by cytoplasmic granules that contain degradative enzymes and oxidative molecules. These cells are capable of migrating along gradients of the chemokine interleukin 8 (IL-8), which is produced by corneal stromal fibroblasts in response to virus infection by (Chodosh *et al.*, 2000). Evidence was also obtained indicating that that IL-8 production results from activation of the p38 mitogen-activated protein kinase (MAPK) and its downstream targets HSP-27 and ATF-2 (Rajaiya *et al.*, 2008).

Until recently, the lack of an appropriate animal model that recapitulates the pathology of human EKC hindered systematic investigations of HAdV-induced ocular disease. In an attempt to remedy this situation, the corneas of C57BL/6J mice were microinjected with HAdV-D37 using a glass micropipette needle and examined by histopathological methods after increasing periods (Chintakuntlawar *et al.*, 2007). Compared to those from control, uninfected mice, corneas from mice injected with the virus showed signs of infection and inflammation within one day and this response continued for up to several months. Viral early mRNA (E1A and E1B 19K) synthesis, but not late gene (protein IIIa) expression, was detected at four days post-injection. Virus-infected corneas also showed signs of inflammation as indicated by increased levels of the cytokines IL-6 and MCP-1, and the presence of neutrophils. Additional studies of HAdV-D37 in mice demonstrated that the chemokine CXCL1 and its receptor (CXCR2) contributed to neutrophil infiltration in the cornea, but this process did not appear to affect the development of keratitis suggesting that other parameters were involved (Chintakuntlawar & Chodosh, 2009). One of these virus-host interactions may be the ligation of cellular α_v integrins by the penton base; empty HAdB-B37 capsids devoid of nucleic acid also induced ocular keratitis, and treatment of mouse corneas with a peptide containing the integrin-binding RGD sequence partially blocked virus induced leukocyte infiltration as well as chemokine expression (Chintakuntlawar *et al.*, 2010). Recent transmission electron microscopy analyses of HAdV-D37-infected mouse corneas have revealed a block in capsid disassembly as well as viral DNA nuclear entry, properties that could explain why HAdV-D37 fails to replicate in mouse tissues (Mukherjee *et al.*, 2015).

8.2.1.3.2 Molecular genetics of species D adenoviruses

As noted in Chapter 1, recombination between homologous sequences (largely in the penton base, hexon and fiber genes and the E3 region) is common among species D HAdVs, and critical in their evolution (Adhikary & Banik, 2014; Kaneko *et al.*, 2009; Robinson *et al.*, 2008; Robinson *et al.*, 2009; Robinson *et al.*, 2011a; Robinson *et al.*, 2013b; Singh *et al.*, 2015; Walsh *et al.*, 2009). The evolution of type D19 viruses in particular has shed some light on recombination events and the consequences for ocular disease. HAdV-D19p (prototype strain) was originally isolated in 1955 (Bell *et al.*, 1959), but is not associated with ocular infection and does not cause infection of corneal epithelial cells (Zhou *et al.*, 2012). This virus is similar, but not identical to HAdV-D19c, isolated several decades later and which is associated with EKC (Desmyter *et al.*, 1974; Hierholzer *et al.*, 1974; Wadell *et al.*, 1980). Further genomic sequence analyses established that all EKC-associated HAdV-D19 strains have likely undergone DNA recombination among types D37, D22, and D19p, the latter with the original 19p hexon region that preserves the serological identity among these strains (Zhou *et al.*, 2012). Based on the new typing criteria, established in consultation with GenBank (Seto *et al.*, 2011) and with the Human Adenovirus Working Group, these authors have now reclassified all EKC-D19 associated viruses as D64. An intriguing question that remains to be answered is how recombination of the E3 region and penton base, fiber, and hexon genes contributes to the etiology of HAdV-induced EKC. It may be the case that certain combinations of these gene products, or even other uncharacterized viral molecules, are required for the development of EKC. The continuing emergence of novel recombinant species D adenoviruses (J. Huang *et al.*, 2014; Singh *et al.*, 2015) with sequence variations in their penton base genes (Robinson *et al.*, 2013b) or other E3 and late genes might provide further clues to the origin of EKC-associated adenoviruses.

8.2.1.3.3 Antiviral treatments for HAdV ocular infections

Until recently, the number of antivirals specifically targeting adenovirus ocular infections has been quite limited (Romanowski & Gordon, 2008).

However, compounds that dampen the immune response generated by virus infection have been useful. For example, cyclosporine, delivered as a 0.05% topical solution, was reported to reduce the severity of ocular HAdV disease (Okumus *et al.*, 2012). Similarly, oral administration of steroids has shown some efficacy in pediatric cases of HAdV EKC (Kim *et al.*, 2015). While helpful in treating the underlying immunopathology associated with this disease, these anti-inflammatory drugs do not prevent infection by the virus.

On the other hand, topical administration of cidofovir, a viral DNA polymerase inhibitor (see Chapter 10) reduced the severity of HAdV-EKC (Hillenkamp *et al.*, 2002). Other antivirals have also been explored in clinical trials to treat HAdV-EKC. One of these antivirals is N-chlorotaurine (NCT) (Teuchner *et al.* 2005), a derivative of taurine, produced by polymorphonuclear neutrophils and monocytes during infection. NCT and another taurine derivative known as NVC-422 (Yoon *et al.*, 2011) cause oxidation of thiol and amino groups of proteins, thereby neutralizing pathogens non-specifically (Romanowski *et al.*, 2006). Another compound that has been analyzed in preclinical animal models is doxovir (CTC-96, Redox Pharmaceuticals). This cobalt-chelating agent causes protein denaturation and has been shown to be effective against HAdV-C5, although its effectiveness against EKC-inducing virus types has yet to be established (Epstein *et al.*, 2006). More recently, ganciclovir, an inhibitor of DNA polymerases of adeno-viruses and multiple human herpesviruses, showed significant antiviral activity against EKC adenovirus types 8, 64, and 37 at non-toxic doses in cell culture systems (G. Huang *et al.*, 2014). Whether this compound can be used as an effective antiviral agent for adenovirus-induced ocular disease in humans remains to be established. An interesting approach for more selective antiviral therapy for EKC is the use of multivalent sialic acid conjugates, referred to as molecular wipes (Aplander *et al.*, 2011). These molecules have the capacity to aggregate adenovirus particles, thereby preventing their binding to cell surface sialic acid residues. While these sialic-acid-based compounds are likely to be safer than other more commonly used remedies, their antiviral effectiveness *in vivo* is unknown.

8.2.1.4 *Gastrointestinal disease in children*

HAdVs are also an important cause of gastroenteritis and diarrhea in children under four years of age, but not adults. Adenovirus-associated gastroenteritis occurs in all parts of the world, although the reported incidences varies widely, from 1% or less of gastroenteritis cases to 40% (e.g. Aminu *et al.*, 2007; Cunliffe *et al.*, 2002; Fajfr *et al.*, 2014; Iturriza Gomara *et al.*, 2008; Jarecki-Khan *et al.*, 1993; Lekana-Douki *et al.*, 2015; L. Li *et al.*, 2005; Lin *et al.*, 2000; X. Liu *et al.*, 2015; Magwalivha *et al.*, 2010; Mayindou *et al.*, 2015; Motamedifar *et al.*, 2013; Moyo *et al.*, 2014; Oh *et al.*, 2003; Saderi *et al.*, 2002; Silva *et al.*, 2008; Subekti *et al.*, 2002; Thongprachum *et al.*, 2015; Uhnoo *et al.*, 1986; Van *et al.*, 1992; Zhang *et al.*, 2011). Although many HAdVs can be shed in the stool, gastroenteric disease is associated with the so-called enteric adenoviruses, HAdV-F40 and -F41 first identified in 1983 (de Jong *et al.*, 1983; Uhnoo *et al.*, 1983), and also with HAdV-G52 and some members of species D (Echavarria, 2008; Jones *et al.*, 2007; Lynch *et al.*, 2011; Matsushima *et al.*, 2012; Matsushima *et al.*, 2013). In some populations, such as children in urban areas of Kenya, species D HAdVs were the most common (Magwalivha *et al.*, 2010). In many of these studies, gastroenteritis caused by infection by multiple viruses was not uncommon. Although hospitalization can be necessary, for example because of dehydration, adenovirus gastroenteritis is typically associated with self-limiting watery diarrhea. Infrequent complications of HAdV infections of the gastrointestinal tract include colitis, pancreatitis and hepatitis (Lynch *et al.*, 2011; Wold & Ison, 2013).

8.2.2 *Disease caused by HAdVs in immunocompromised hosts*

With the emergence of the AIDS epidemic, the increasing number of cancer patients receiving immunosuppressive chemotherapies, and a large rise in organ and cell transplants, the number of immunocompromised persons in the human population has increased substantially in the past 4 to 5 decades. Adenovirus infection is a cause of serious illness and death in such populations.

In the early days of the AIDS epidemic before effective therapies were available, severe, and sometimes fatal, cases of HAdV-associated pneumonia, nephritis, hepatitis, meningeoencephalitis, gastroenteritis, and disseminated disease were reported (Khoo *et al.*, 1995; Schnurr *et al.*, 1995). The risk of HAdV infection in such patients was high, 17–28% depending on the clinical stage of AIDS (Khoo *et al.*, 1995). However, with the availability of effective anti-retroviral therapies, such severe outcomes have become relatively rare, and HAdV infection in patients receiving these therapies is generally asymptomatic or associated with diarrhea (Adeyemi *et al.*, 2008; Thomas *et al.*, 1999; Nebbia *et al.*, 2005). Some fatalities in AIDS patients were associated with HAdV-B3 and HAdV-C1 and -C2, but most types detected in these patients are members of species D (Echavarria, 2008; Khoo *et al.*, 1995). As discussed in Chapter 1, many of the most recently identified HAdVs are members of species D, and were first detected in HIV-infected patients. These immunosuppressed individuals might carry multiple HAdV types for extended periods and hence provide increased opportunities for recombination between viral DNA genomes.

Infants with inherited immunodeficiencies, particularly those suffering from severe combined immunodeficiency, who lack both cell-mediated and antibody-mediated immunity, are highly vulnerable to HAdV infection. In these children, infection by a variety of HAdV types results in pneumonia, bronchiolitis, hepatitis, or gastroenteritis, as well as disseminated disease that can be fatal (Echavarria, 2008; Lion, 2014). Although only limited incidence data are available, the case fatality rate may be >50% in these children.

Pediatric patients who receive solid organ or hematopoietic stem cell transplants are also at greater risk for severe HAdV-associated disease than adult transplant recipients. It has been estimated that HAdV DNA is detectable in <10% of such adult patients, and from 4–57% of pediatric transplant patients, depending in part on the organ transplanted (de Mezerville *et al.*, 2006; Engelmann *et al.*, 2009; Florescu *et al.*, 2010; Florescu *et al.*, 2013; Hoffman, 2009; Humar *et al.*, 2005; M. Liu *et al.*, 2009; Liu *et al.*, 2010; Pinchoff *et al.*, 2003; Shirali *et al.*, 2001). The clinical manifestations of HAdV infection in solid organ transplant patients are often evident in the transplanted organ, for example, jaundice and hepatitis observed in liver transplant recipients. In such adult patients,

HAdV infection can correlate with increased rates of organ rejection and morbidity (Bridges *et al.*, 1998; Clark *et al.*, 2013; Vu *et al.*, 2011), but severe complications are more common in pediatric patients, with case fatality rates as high as 50% (Kojaoghlanian *et al.*, 2003).

In children who receive an allogeneic hematopoietic stem cell transplant, mortality from HAdV infection can be as much as 6% (Lion *et al.*, 2010; Lion *et al.*, 2003) and is associated with disseminated disease and multi-organ failure (Forstmeyer *et al.*, 2008; Kalpoe *et al.*, 2007; Lion *et al.*, 2010). HAdVs may infect transplant patients *de novo*, be introduced in the transplanted organ, or result from reactivation from a latent state in either the recipient or the transplanted organ when an immunosuppressive regimen is imposed (Lion, 2014). It can be difficult to distinguish among these alternatives. However, the observation that 30% of stem cell transplant recipients carried multiple HAdV types (Kroes *et al.*, 2007) suggested that reactivation (or this phenomenon in combination with primary infection) makes an important contribution. This hypothesis is strongly supported by the observation that the HAdV type detected in allogeneic stem cell transplant patients prior to treatment is generally that identified after the transplant (Veltrop-Duits *et al.*, 2011).

The high morbidity and mortality of HAdV infections in pediatric transplant patients has stimulated development of therapies and anti-adenoviral drugs (Chapters 9 and 10), and also clinically predictive screening methods. Postmortem analysis of HSCT patients indicated that organ failure was associated with extensive HAdV reproduction leading to cell lysis and release of virus particles into the blood (viremia) (Forstmeyer *et al.*, 2008). The magnitude of such viremia correlates with the degree of organ damage (Heim, 2011). Furthermore, it has been reported that surveillance of stool for HAdVs and evidence of their replication can be of value in predicting disseminated HAdV infection and disease (Jeulin *et al.*, 2011; Legoff *et al.*, 2013; Lion *et al.*, 2010), with HAdV loads of less than 10^6 copies/gram associated with low risk of HAdV viremia. Although HAdV DNA can be detected by PCR in multiple organs of allogeneic HSCT patients (Baldwin *et al.*, 2000; Legrand *et al.*, 2001; Lion *et al.*, 2003), it is its presence (DNAemia) and concentration in peripheral blood that are predictive of disseminated disease (Echavarria *et al.*, 2001; Erard *et al.*, 2007; Ganzenmueller *et al.*, 2011;

Lankester *et al.*, 2002; Lion *et al.*, 2003). There is no consensus threshold value for HAdV DNAemia for increased risk, and hence initiation of drug treatment, in part because different techniques are employed and have not been standardized among laboratories (Lion, 2014). However, it has been reported that the viral DNA load in peripheral blood increases rapidly prior to the onset of symptoms of HAdV disease, suggesting that the kinetics of DNAemia may be the most relevant predictive parameter (Lion *et al.*, 2003; Seidemann *et al.*, 2004). Some guidelines recommend screening peripheral blood for HAdV DNA in allogeneic HSCT patients with specific risk factors (e.g. grade III or IV graft versus host disease), but it remains unclear whether treatment once viremia is detected may already be too late, and whether earlier preemptive therapy is of greater benefit (Lion, 2014).

8.2.3 *Diseases that may be associated with HAdV infection*

8.2.3.1 *Acute diseases*

Infection by HAdVs has also been linked to a variety of other syndromes, notably myocarditis, interssusception, and celiac disease, largely by retrospective surveys of patients for the presence of virus particles, anti-adenovirus antibodies, or viral DNA. In no case, however, is evidence of HAdV infection found in all patients, indicating that other parameters are also important for the development of these diseases. These associations have been described in detail previously (Horwitz, 2001; Wold & Ison, 2013) and, as no new information has come to light, are described only briefly here.

Adenovirus DNA has been detected by PCR analyses in cardiac biopsies from both children and adults with myocarditis, with from 23–40% of samples positive (Bowles *et al.*, 1996; Bowles *et al.*, 2003; Calabrese *et al.*, 2002; Kuhl *et al.*, 2005). Sequencing of PCR products established that the majority were HAdV-C2 and most of the remainder HAdV-C5 (Bowles *et al.*, 2003). In this study, none of the control patients tested positive for HAdV, whereas 22% of those diagnosed with myocar-ditis and 12% of patients with dilated cardiomyopathy (a late consequence of myocarditis) did. This substantial difference, as well as a report that clearance of the virus from patients with these conditions was associated with improved symptoms, indicates that HAdV infection may contribute to development of such cardiac diseases.

Intussesception is a severe intestinal syndrome in which one segment of the intestine slides into another. The association of HAdV with this syndrome rests on the results of similar epidemiological surveys, with substantial differences between test and control patients (see Hurwitz, 2003). The mechanisms by which HAdV infection might contribute to development of this syndrome or of myocarditis are not known, although inflammatory responses to infection seem likely to contribute. In contrast, it has been proposed that antibodies against the HAdV-A12 E1B 55-kDa protein contribute to the development of celiac disease. This common autoimmune syndrome is triggered by ingestion of gluten proteins present in cereals such as wheat and barley. The major gluten component A-gliadin, which is known to cause celiac disease, and the HAdV-A12 E1B protein share a short region of homology. Furthermore, antibodies against the viral protein cross react with A-gliadin, as well as with a synthetic A-gliadin peptide corresponding to the region of homology (Kagnoff *et al.*, 1984). Consistent with a role for HAdV-A12 infection, neutralizing antibodies against this virus or those that recognize the synthetic peptide were detected in a significantly greater fraction of celiac disease patients than in control groups (Kagnoff *et al,*. 1984; Lahdeaho *et al.*, 1993a; Lahdeaho *et al.*, 1993b).

8.2.3.2 *Chronic obstructive pulmonary disease*

Cigarette smoking is the major risk factor for development of chronic obstructive pulmonary disease (COPD), but epidemiological studies indicate that respiratory illness in children increases the risk of COPD, which arises in only a minority of smokers. As noted, HAdV infections of the respiratory tract are common in children, and the virus can persist in tonsils and T-lymphocytes (see below). These properties led to the hypothesis that latent HAdV infection might increase inflammatory responses in the lungs and hence contribute to the development of COPD (Hayashi, 2002). Indeed, higher concentration of species C HAdV DNA (that of the E1A gene) was detected in lung tissue from COPD patients than in matched control subjects (Matsuse *et al.*, 1992). In a guinea pig model, E1A DNA was observed to persist in a few alveolar and airway epithelial cells after the initial acute infection and was associated with an increased inflammatory response to a single exposure to cigarette smoke (Vitalis *et al.*, 1996). In addition, the production of pro-inflammatory molecules like IL-8 and ICAM-1 in response to lipopolysaccharide stimulation, and of their mRNAs, was reported to be increased in A549 and primary human bronchial epithelial cells in which the viral E1A protein is present (Higashimoto *et al.*, 1999; Keicho *et al.*, 1997; Keicho *et al.*, 1999). In the primary cells that synthesize viral E1A gene products, expression of genes that encode growth factors such as TGF-β1 is also increased, as did synthesis of mesenchymal markers (Frickmann *et al.*, 2012; Ogawa *et al.*, 2004). While there have been reports that respiratory HAdV infections in childhood increase the risk of COPD in smokers (Matsuse *et al.*, 1992), the data from COPD patients are variable, with only infrequent detection of HAdV-C5 E1A DNA in more recent studies (McManus *et al.*, 2007).

8.2.3.3 *Obesity*

Initial indications of a link between adenovirus infection and obesity came from studies in animals, and were spurred by the observation that a fowl adenovirus increased obesity in experimentally infected chickens and uninfected chickens housed with them (Dhurandhar *et al.*, 1992). Several HAdVs, including HAdV-C5, -D36, and -D37, similarly increased obesity in infected mice, rats, and marmosets, whereas HAdV-C2 and HAdV-A31 did not (Dhurandhar *et al.*, 2000; Dhurandhar *et al.*, 2002; Pasarica *et al.*, 2006; So *et al.*, 2005; Whigham *et al.*, 2006). For example, in one longitudinal study of marmosets that become naturally infected with HAdV-D36, 15–30% increases in body weight were recorded with concomitant reductions in serum cholesterol concentrations (Dhurandhar *et al.*, 2002). Subsequent studies focused on HAdV-D36, in part because it is readily distinguished from other HAdVs. Furthermore, comparison of the sequence of the HAdV-D36 genome with those of other HAdVs revealed non-conserved sequences in several genes, including the fiber gene, and in combination with molecular modeling of the HAdV-D36 fiber suggested that this species D member may have a unique, or unusual, tissue tropism (Arnold *et al.*, 2010).

In HAdV-D36-infected rodents, such symptoms of obesity as weight gain, increased insulin sensitivity, and glucose uptake are accompanied by enhanced differentiation of the pre-adipocytes into adipocytes and by expression of a master transcriptional regulator of this program, C/EBP (Pasarica *et al.*, 2008; Rathod *et al.*, 2009; Vangipuram *et al.*, 2007). These observations suggest that HAdV-D36 infection increases obesity primarily by stimulation of adipogenesis, and accumulation of fat stores within adipocytes. Expression of the viral E4 Orf1 gene has been reported to be both necessary and sufficient to induce differentiation of pre-adipocytes in culture (Dhurandhar *et al.*, 2011). This viral gene product also leads to increased glucose uptake in primary human skeletal muscle cells and adipose tissue, as a result of activation of signaling via RAS and PI3K and consequent increased expression of the genes encoding the glucose transporters GLUT1 and GLUT4 (Rogers *et al.*, 2008; Wang *et al.* 2008). Increased glycolysis would expand supplies of acetyl CoA, the substrate for synthesis of fatty acids. The differentiation of HAdV-D36 infected cells is also associated with reduced production and secretion of leptin (Wang *et al.*, 2008), an important hormone that regulates eating behavior and signals satiation. This change could therefore result in increased appetite and feeding in infected animals, and hence contribute to weight gain.

While there is growing appreciation of the mechanistic foundation of the adipogenicity of HAdV-D36 in animals, an association between the virus infection and development of obesity in humans has not been clearly established. The prevalence of HAdV-D36 has been examined in obese adults and children, usually but not invariably with non-obese control groups, and in populations varying in such parameters as body mass index in North and South America, Europe, and Asia. Infection has been assessed by serum neutralization tests or sensitive PCR assays for the presence of viral DNA, for example in fat tissue (Dhurandhar *et al.*, 2011; Ponterio & Gnessi, 2015). In some studies, HAdV-D36 prevalence was notably higher in obese individuals, for example 30% compared to 11% in non-obese controls in a study of adults in the U.S. (Atkinson *et al.*, 2005), and 65% and 35% presence in obese and non-obese cohorts, respectively, in Italy (Trovato *et al.*, 2009). However, in other studies, the differences were minimal or somewhat higher prevalence was recorded in the non-obese populations (Broderick *et al.*, 2010; Na *et al.*, 2012;

Ponterio *et al.*, 2015). Obesity is a complicated, multifactorial syndrome that is governed by genetic and socioeconomic parameters, as well as diet, fitness (exercise), and age. Variation in these parameters among the different populations studied could well confound their comparison and interpretation. Two meta-analyses have indicated that HAdV-D36 infection increases the risk for development of obesity (but is not associated with markers of this condition such as waist size, or blood glucose and cholesterol levels) (Shang *et al.*, 2014; Yamada *et al.*, 2012). Nevertheless, additional studies are clearly needed to establish the contribution of HAdV-D36 infection, in particular relative to those of other risk factors, to the development of obesity in human adults and children. Prospective, longitudinal studies in which large cohorts of subjects are tracked for multiple risk factors would seem likely to be particularly informative.

8.2.3.4 *Cancer*

The differential ability of HAdVs to induce tumors in rodents was an important early criterion for their classification, with A, B, and C types defined as highly, weakly, and non-oncogenic, respectively (Table 1.1) (Huebner, 1968). The identification of HAdV-A12 as the first human virus with tumorigenic activity inspired the investigations of transformation described in Chapter 1, and also led to the identification of mechanisms by which species A HAdV E1A gene render transformed cells refractory to detection and destructive by cells of the innate and adaptive immune systems (Chapter 9). They also raised the question of whether HAdVs contributed to the development of any human cancer.

Extensive early studies to assess the association of HAdVs with cancer in humans, for example, sponsored by the U.S. National Cancer Institute, found no evidence for the increased presence of adenoviral genetic information in human tumors compared to normal tissues (Green & Wold, 1976; Green *et al.*, 1979; Mackey *et al.*, 1976; Wold *et al.*, 1976). The one early report of such an association in neuronal tumors (Ibelgaufts *et al.*, 1982) has not been confirmed. It might be argued that this question should be reexamined using the far more sensitive methods for detection of nucleic acids based on PCR developed subsequently. Those few studies that have employed such methods to survey the association of HAdV nucleic acids with various types of cancer have proved inconclusive (leukemias and lymphomas; Fernandez-Soria *et al.*, 2002), revealed no differences between malignant and non-malignant tissues (pediatric brain tumors; Kosulin *et al.*, 2007), or detected no viral DNA in tumor samples (pediatric leukemias, lymphomas, and solid tumors; Kosulin *et al.*, 2007). In a study of lung cancer, viral E1A sequences were detected in 31% of 35 small cell lung cancer samples, but in none of the 40 non-small cells lung cancer samples (Kuwano *et al.*, 1997). In the positive samples, viral DNA was observed by *in situ* hybridization in tumor cells in 8 of 11 cases, and in normal epithelial cells in the remaining three. It was therefore proposed that HAdV infection might result in induction of the mutations in the p53 and RB tumor suppressor genes that are known to be associated with small cell lung cancer (Kuwano *et al.*, 1997). However, this hypothesis has not been confirmed, and there is, to our knowledge, no compelling evidence that HAdVs transform human cells in natural infections and hence contribute to the development of cancers.

More recently, a causative role for prenatal HAdV infection in the development of childhood acute lymphoblastic anemia (ALL) has been considered. This disease is associated with specific chromosomal translocations of unknown origin (Wiemels, 2012), but thought, on the basis of epidemiological evidence, to be induced by infection with a common pathogen *in utero* (Eden, 2010; Greaves & White, 2006). Adenoviruses might be candidates for such infectious agents: viral gene products inactivate DNA repair pathways (Chapter 6), and expression of the viral oncogenes leads to chromosomal damage like that seen in the absence of timely repair of double-stranded DNA breaks (Hart *et al.*, 2005; Nevels *et al.*, 2001). It has therefore been proposed that HAdV infection could be oncogenic in a "hit-and-run" fashion; expression of viral genes results in permanent changes to the host genome that promote oncogenesis (Nevels *et al.*, 2001).

An initial study to assess the association of prenatal HAdV infection with ALL reported a statistically significant greater prevalence of viral DNA in Guthrie cards (blood samples from neonates dried onto filter paper for long-term storage) from children who developed ALL than in subjects who did not (26.5% versus 6.4%) (Gustafsson *et al.*, 2007). In a follow-up study with larger numbers of samples from both ALL and control patients, viral DNA could be detected at only very low frequencies in both populations (2 of 727 total samples) (Honkaniemi *et al.*, 2010), and similar values were reported by other investigators (Vasconcelos *et al.*, 2008). These differences could be the result of loss of DNA during Guthrie card storage under non-optimal conditions (false negatives), or contamination of stored samples with HAdV during handling (false positives). However, screening of cord blood lymphocytes collected from infants within a day of birth (Ornelles *et al.*, 2015) or of amniotic fluid (Baschat *et al.*, 2003; Reddy *et al.*, 2005; Van den Veyver *et al.*, 1998; Wenstrom *et al.*, 1998) has consistently indicated prenatal HAdV infection in 3.7–5.4% of the population. The most common genetic abnormality in ALL, an ETV6-RUNX translocation, was also detected in a greater percentage of HAdV-positive blood lymphocyte samples, although the number of such samples was too small to assess statistical significance (Ornelles *et al.*, 2015). Examination of larger numbers of infants by these methods, and prospective studies of their health, should allow the contribution of prenatal HAdV infection to ALL to be assessed.

8.3 Pathogenesis

8.3.1 *Animal models and their limitations*

The clinical manifestations of HAdV infection provide important clues about such questions as whether and how tissue tropism and host responses determine the diseases associated with these viruses. However, as with any human virus, detailed mechanistic understanding, for example, of the contributions of specific viral gene products, can come only from studies in animals. Experimental animals can be infected with the human viruses of interest, or related viruses of other animals can be studied in their natural hosts. In the case of HAdVs (and many other human viruses), each approach has limitations.

8.3.1.1 *HAdV infection of experimental animals*

Research in the experimental animal most closely related to humans, the chimpanzee (*Pan troglodytes*), has been declining for a considerable period, and essentially ceased with the June 2015 classification of captive chimpanzees as an endangered species by the U.S. Fish and Wildlife Service. Other primates, such as rhesus monkeys and macaques, have been used widely to assess the efficacy of adenovirus-based vaccines and investigate immune responses to adenovirus vectors (Chapter 12), but in few studies of pathogenesis (Calcedo *et al.*, 2009; Dhurandhar *et al.*, 2002). It is not clear that the expense and more importantly, the invasive procedures and sacrifice necessary for studies of pathogenesis in these animals would be justified. Cells from the African green monkey (an Old World species) are not fully permissive for reproduction of the HAdVs that have been examined (members of species C): virus yield is approximately three orders of magnitude lower than obtained from human cells, as a result of defects in synthesis and processing of viral mRNAs, and hence in production of viral proteins (especially fiber) during the late phase of infection (Anderson & Klessig, 1984; Klessig & Anderson, 1975; Johnston *et al.*, 1985; Ross & Ziff, 1992; Ross & Ziff, 1994).

Mice (*Mus musculus*), the "work horse" of experimental studies in animals, are non-permissive for HAdVs: few or no infectious progeny virus particles can be recovered, even when mice are intravenously injected with high doses (>10^9 pfu) (Duncan *et al.*, 1978; Ginsberg *et al.*, 1991; Oualikene *et al.*, 1994; Ying *et al.*, 2009). Nevertheless, immuno-competent mice are routinely employed in pre-clinical testing of HAdV vectors (Chapters 11 and 12) and the impact of HAdV infection on host responses has been examined in such mice as well as various knock-out strains (Chapter 9). Differences in human and murine immune systems and especially the lack of viral replication and its associated cytotoxicity in mice severely limit the value of these laboratory animals for studies of HAdV pathogenesis.

One species of rat, the cotton rat (*Sigmodon trispidus*), was shown early on to develop pneumonia with pathology similar to that seen in humans when HAdV-C5 was administered intranasally (Pacini *et al.*, 1984). In this model, virus replication was detected in epithelial cells of the upper and lower respiratory tracts, the degree of pathology was proportional to virus dose, and was primarily the result of host responses, such as infiltration of infected tissue by monocytes, macrophages and lymphocytes (Ginsberg *et al.*, 1989; Prince *et al.*, 1993). This system has been exploited to demonstrate that the coding sequence for the E1B

55 kDa protein is necessary for the development of HAdV-C5 induced pneumonia and inflammatory responses, independently of any effects on the efficiency of virus reproduction (Ginsberg *et al.*, 1999), and that deletion of the E3 gp19K coding sequence increases the inflammatory response (Ginsberg *et al.*, 1989). These animals have also been used to compare the efficiencies with which replicating and non-replicating HAdV-C5 vectors induce immune responses (Eloit & Adam, 1995; Kim *et al.*, 2014) and to evaluate the efficacy of oncolytic HAdVs (e.g. Steel *et al.*, 2007; Toth *et al.*, 2005). However, as cotton rats are very difficult to handle and breed in large numbers (Wold & Toth, 2012), their use as a routine model is severely limited.

Syrian hamsters (*Mesocricetas auratas*) offer a number of practical advantages as experimental animals including extreme docility (Wold & Toth, 2012), and are quite permissive for HAdVs. Several lines of tumor cells from these animals support efficient production of species C HAdV virus particles, with yields only 10- to 100-fold lower than obtained from very permissive human A549 cells (Bortolanza *et al.*, 2009; Spencer *et al.*, 2009; Thomas *et al.*, 2006; Ying *et al.*, 2009). When introduced intranasally, HAdV-C5 replicates in the lung of young and adult animals, which develop pneumonia and anti-adenovirus antibodies (Hjorth *et al.*, 1988; Thomas *et al.*, 2006). Enteric infection of infant hamsters leads to shedding of infectious virus in the feces (Hjorth *et al.*, 1988), as also occurs in humans. When inoculated intravenously, a route that results in disseminated spread, HAdV-C5 replicates in the liver, but also various other organs, including lung, heart, and kidney, with greatest damage to the liver (Ying *et al.*, 2009). These properties indicate that Syrian hamsters represent useful models for investigation of the mechanisms of HAdV pathogenesis. These animals have, in fact, been used to evaluate compounds that inhibit adenovirus replication (Diaconu *et al.*, 2010; Toth *et al.*, 2008; Zarubalev *et al.*, 2007; see also Chapter 10), as well as in preclinical studies of oncolytic derivatives of HAdV-C5 (Wold & Toth, 2012; see also Chapter 12). They have also been exploited in studies to investigate the basis for the increased pathogenesis of a more virulent, and recently appearing strain derived from HAdV-B14, termed 14p1. This virus, in contrast to HAdV-B14, failed to repress synthesis of pro-inflammatory cytokines in infected human alveolar macrophages in culture, and induced increased immunopathology and bronchopneumonia in the lungs of Syrian hamsters (Radke *et al.*, 2015).

8.3.1.2 *Adenoviruses of other animals infecting their native hosts*

Adenoviruses have been isolated from a wide range of mammals, including dogs, horses, cows, and sheep, in addition to non-human primates. Several are under investigation as potential vectors for gene delivery (e.g. canine viruses) or vaccines (e.g. chimpanzee viruses) (Chapters 11 and 12). Not surprisingly, studies of the pathogenesis of these mastadenoviruses of non-human animals are rare, for both ethical reasons and the high costs of working with large animals. However, two murine adenoviruses have been identified, and the pathogenesis of MAdV1 in mice has been investigated quite extensively.

Inbred strains of laboratory mice vary considerably in their susceptibility to MAdV1 (Charles *et al.*, 1998; Guida *et al.*, 1995; Kring *et al.*, 1995; Spindler *et al.*, 2001), and the outcome of infection is also age-dependent (Guida *et al.*, 1995; Moore *et al.*, 2004; Procario *et al.*, 2012). Intraperitoneal inoculation of susceptible animals results in a severe central nervous system disease, fatal hemorrhagic encephalitis with disruption of the blood brain barrier (Gralinski *et al.*, 2009; Guida *et al.*, 1995; Kring *et al.*, 1995). This severe pathology has been attributed to MAdV1 replication in vascular endothelial cells (Charles *et al.*, 1998). In immuno-compromised mice, MAdV1 infection leads to disseminated disease with pneumonia, hepatitis, encephalitis, and gastroenteritis (Charles *et al.*, 1998; Moore *et al.*, 2003; Moore *et al.*, 2004; Pirofski *et al.*, 1991). When inoculated intravenously, MAdV1 gene products and DNA can be detected in the respiratory epithelium, concomitant with the increased synthesis of chemokine mRNAs and recruitment of infiltrating cells (Procario *et al.*, 2012; Weinberg *et al.*, 2005). The viral E1A protein is not required for inflammatory response in the lungs, or dissemination of MAdV1 to the brain, even though viral gene expression was diminished in its absence (Weinberg *et al.*, 2007).

The various pathologies that MAdV1 infection can cause in mice bear some resemblance to those observed in HAdV-infected humans described previously. However, it is important to keep in mind that in naturally infected mice, MAdV1 replicates in endothelial cells throughout the animal (Charles *et al.*, 1998; Guida *et al.*, 1995; Kajon *et al.*, 1998; Kring *et al.*, 1995; Moore *et al.*, 2003), in contrast to the epithelial cell tropism of HAdVs. It has also been reported that in immunodeficient mice, MAdV1 does not exhibit the strong tropism for the liver displayed by HAdV-C5 (Lenaerts *et al.*, 2009). Furthermore, the coding sequences and protein functions of MAdV1 and HAdVs differ considerably. For example, the murine viral genome encodes a single E1A protein, analogous to the HAdV-C5 289R E1A protein (Ball *et al.*, 1989), but this protein is dispensable for replication of MAdV1 in murine fibroblasts (Ying *et al.*, 1998). Nor does the MAdV1 genome include a VA-RNA gene (Meissner *et al.*, 1997) and only slight similarities to the HAdV-C5 E3 gp19K protein could be identified (Beard *et al.*, 1990). The important contributions to these HAdV-C5 gene products to shaping and modulating the host responses of host cells to infection, including the interferon antiviral defense (Chapter 9), underscore the limitation of the MAdV1-infected mice model for studies of the molecular basis of HAdV pathogenesis.

8.3.1.3 *Cross-species transmission of primate adenoviruses*

In general, infections by adenovirus are usually considered to be species specific based on their replication patterns in tissue culture (Rowe & Hartley, 1962 and in animal model systems (Ginsberg *et al.*, 1991). Nonetheless, serological analyses have detected antibodies to Old and New World monkey adenoviruses in normal human sera in geographic areas in which the monkeys are endemic (Ersching *et al.*, 2010; Xiang *et al.*, 2006). Further, the substantial sequence relatedness between some adenoviruses from the great apes and human species B viruses suggest that monkey adenoviruses could cross over into humans.

Evidence for a potential zoonotic infection was obtained at a regional primate facility in which a high percentage of New World monkeys (titi monkeys; *Callicebus cuperus)* were found to be infected with a monkey adenovirus, designated TMAdV. These animals developed respiratory infections that ultimately progressed into a more serious disease (pneumonia and hepatitis) that either killed the animals or which required euthanasia

(Chen *et al.*, 2011). During the course of TMAdV infections, a researcher with close contact with the titi monkey colony as well as a family member, developed mild respiratory infections and their sera tested positive for antibodies to TMAdV. A small but significant number of randomly collected human sera from the western U.S. (~2.5%) were also found to contain antibodies to TMAdV suggesting that cross-species transmission of TMAdV is possible. Further evidence that cross transmission of monkey adenoviruses is possible was obtained in experimental studies in which nasal inoculation of TMAdV into cotton top tamarins was shown to produce a mild respiratory disease in these animals that was similar to that observed in the putative human-TMAdV infections (Yu *et al.*, 2013).

More recently, human sera that were found to contain neutralizing antibodies to adenoviruses isolated from infected baboons, provisionally designated simian adenovirus C (SAdV-C), suggesting prior exposure to SAdV-C as a result of cross-species transmission (Chiu *et al.*, 2013). These studies emphasize the potential of zoonotic infections between emerging primate adenoviruses and humans. This situation could lead to potentially serious respiratory diseases and thus more vigilant procedures may be necessary to prevent such occurrences, especially in regional primate facilities where there is a high density of animals that are in potential contact with humans.

8.3.2 *Virulence*

8.3.2.1 *Tissue tropism*

As noted previously, HAdVs of different species are associated with diseases of different organs. This specificity for infection of different types of epithelial cells cannot be ascribed to recognition of distinct, cell-type-specific receptors by different HAdV species (Chapter 3). In addition, the species C HAdVs that are ubiquitous in young children and associated with upper respiratory tract disease can be shed for months in the feces (Fox *et al.*, 1977), indicating that they can reproduce in gastrointestinal epithelial cells, at least sporadically, but cause no symptoms. It therefore appears that intracellular host components govern the ability of HAdV species to reproduce in particular cell types, or that host responses to infection that contribute to pathology vary from site to site in the host.

8.3.2.2 *Spread within the host*

Although adenovirus particles are spread in the blood in immunosuppressed human patients, viremia is quite uncommon in normal, immunocompetent adults and children, indicating that in most infections, the virus can spread only locally. This process appears to be facilitated by viral proteins that bind to cell surface HAdV receptors.

Adenoviral fibers that bind to desmoglein 2 (DSG2) (Chapter 3) have the ability to induce transient opening of the intercellular DSG2 junctions between epithelial cells (Wang *et al.*, 2011a). This alteration in cell–cell contact is the result of activation via MAPK signaling of the metalloprotease disintegrin and metalloproteinase domain-containing protein 17 (ADAM17), which liberates the extracellular domain of DSG2 (Wang *et al.*, 2015). Reduction of cell–cell contact in this way was proposed to facilitate the spread of the species B HAdVs that bind to DSG2 (3, 7, 11, and 14) locally and perhaps into deeper layers of the lower respiratory epithelium (Wang *et al.*, 2011b). Consistent with this hypothesis, pretreatment of epithelial cells with free pentons or recombinant HAdV-B3 fiber knob increased access to receptors masked in intercellular junctions and stimulated spread of HAdV-B3 particles in a polarized epithelial cell line (Lu *et al.*, 2013; Wang *et al.*, 2011a; Wang *et al.*, 2011b). The more pathogenic strain of HAdV-B14 (p1) was observed to spread more efficiently than its parent in a human cancer xenograft model (Wang *et al.*, 2015). This property was not the result of increased cleavage and shedding of DSG2 (Wang *et al.*, 2015), indicating that other viral components modulate spread in epithelial cell layers.

The ability of excess fiber produced in and released from cells infected by CAR-binding HAdVs (e.g. 1, 2, 5, and 6) to disrupt tight junctions and the presence of a specific CAR isoform on the apical surfaces of polarized epithelial cells described in Chapter 3 may also facilitate local spread of these HAdV types in the host. On the other hand, it has been reported that excess production of fiber in a small number of cells in transformed human cells in culture infected by species C and B HAdVs leads to receptor masking in neighboring cells and hence inefficient spread of particles (Rebetz *et al.*, 2009). This phenomenon could limit spread *in vivo*, but whether normal human cells also produce excess fiber is not clear.

8.3.2.3 *Cytotoxicity*

Destructive effects of viral replication and viral gene products

Anyone who has monitored human cells permissive for HAdV infection will be familiar with the direct cytotoxicity of these viruses. With increasing time after infection, nuclei appear darker and become enlarged, a phenomenon particularly evident in flat and tightly adherent normal human fibroblasts, the cells begin to pull up from the substratum, become rounded and retractile and eventually detach from the surface as the die. Such death does not exhibit the characteristic features of apoptosis, which is blocked by several viral gene products (Chapter 6), but rather is considered necrotic. It results, in part, from the inhibition of normal cell processes: when proliferating cells in culture are infected by HAdV-C2, -C5, or -A12 cellular DNA synthesis is inhibited (Ledinko & Fong, 1969; Pina & Green, 1969); in infected normal cells alterations in cellular gene expression, with many inhibited, increase in number and magnitude as the infectious cycle progresses (Ferrari *et al.*, 2012; Miller *et al.*, 2007; Zhao *et al.*, 2012) and, during the late phase of infection, cellular protein synthesis declines steadily until it becomes largely undetectable (Anderson *et al.*, 1973; Beltz & Flint, 1979). HAdV-C5 infection of human cell lines in which cellular protein synthesis is resistant to inhibition results in considerably reduced cytopathic effect, as does treatment of infected 293 with 2-aminopurine, a drug that prevents shut-off of cellular protein synthesis (Huang & Schneider, 1990; Zhang *et al.*, 1994). These observations provide experimental support for the view that the inability of HAdV-C5-infected cells to make cellular gene products contributes to cytotoxicity. In addition, certain viral proteins perturb host cell integrity more directly.

The first such viral protein to be identified was the L3 protease (AVP). The intermediate network of keratin filaments of HeLa cells becomes disassembled and disorganized from 24 hours. After infection with HAdV-C5, keratin 18 was found to be cleaved between its globular head and α-helical rod domains (Chen *et al.*, 1993). This cleavage occurred at a consensus site for AVP, and neither keratin 18 scission nor disruption of cytoplasmic keratin filaments took place when cells infected with a temperature-sensitive

AVP mutant of HAdV-C2 produced at a non-permissive temperature (Chen *et al.*, 1993). Destruction of the intermediate filament network is much less pronounced when inhibition of protein synthesis is blocked by 2-aminopurine, presumably because newly synthesized keratins are available to repair filaments, leading to poor cell lysis and 2 orders of magnitude reduction on release of virus particles (Zhang *et al.*, 1994). Whether this mechanism of disruption of host cell integrity promotes cell lysis and contributes to tissue damage and pathogenesis *in vivo*, as seems likely, has not been examined.

As described previously (Chapter 7), the viral E3 membrane protein adenovirus death protein, ADP, is required late in infection for efficient cell lysis and releases of virus particles (Tollefson *et al.*, 1996a; Tollefson *et al.*, 1996b). It also enhances up to a 100-fold the release of HAdV-F41, which is difficult to propagate in established lines of human cells in culture (Lu *et al.*, 2013). Furthermore, when the HAdV-C5 genome is engineered for overproduction of E3 ADP from the major late (ML) promoter in the absence of other E3 proteins, cytopathic effects and cell lysis appear considerably earlier than in cells infected by wild-type HAdV-C5 (Doronin *et al.*, 2003). This mutant virus and HAdV-C5 exhibited very similar toxicities and biodistribution in permissive Syrian hamsters (Lichtenstein *et al.*, 2009; Ying *et al.*, 2009), but the viruses were introduced by intravenous inoculation, a route that, in effect, bypasses the local infection and spread that normally initiate infection. Viruses that direct overproduction of ADP and are targeted via various additional modifications for tumor-cell-specific replication destroyed tumor cells efficiently in xenograft models and prevented tumor growth (Doronin *et al.*, 2000; Toth *et al.*, 2004; Kuppuswamy *et al.*, 2005; Shashkova *et al.*, 2007). Most significantly, such ADP-overexpressing viruses suppressed the growth of subcutaneous A549 human lung cell tumors in mice and prolonged survival considerably more effectively than did HAdV-C5 or another oncolytic virus that did not direct ADP overproduction (Toth *et al.*, 2010). These properties are consistent with an important role for the ADP in promoting cell lysis at initial sites of natural infection, although this hypothesis has not been tested directly.

Immunotoxicity

Characterization of HAdV pathology in humans and the studies in animal models, especially those permissive for reproduction of the human viruses, both summarized in previous sections, underscore the importance of inflammatory responses and consequent infiltration of cells of the innate and adaptive immune systems into infected organs and tissues. A dramatic, tragic and unintended demonstration of the role of immunotoxicity was provided by the fatal systemic inflammatory response to a HAdV-C5 gene therapy vector trial (Raper *et al.*, 2003), described in detail in Chapter 11.

8.3.3 *Persistence and latency*

Within a few years of their discovery, it was reported that species C HAdVs competent to replicate could persist, for example, in tonsils and adenoids (Evans, 1958; Israel, 1962). Furthermore, genomes of HAdVs that initially infect such respiratory tissues could be found in the feces long after they could no longer be detected in nasopharyngeal washing, with intermittent shedding of virus particles in stool (Fox *et al.*, 1969; Fox *et al.*, 1977). The mechanisms by which species C HAdVs establish such long-lasting infections are of considerable interest and clinical relevance in view of the high morbidity and mortality of these viruses in immunosuppressed patients, particularly children, described previously.

The frequent detection of viral DNA in the absence of infectious particles, first reported nearly 30 years ago (Neumann *et al.*, 1987), initially indicated that a latent infection in which the genome is maintained but no progeny particles are produced can be established in humans. Subsequent studies have shown that T-lymphocytes from a large fraction (some 80%) of adenoids and tonsils from young children harbor species C HAdV DNA in the absence of viral gene expression or genome replication (Garnett *et al.*, 2002; Garnett *et al.*, 2009). In samples examined in detail, the numbers of viral genomes per cell and the numbers of positive cells varied considerably, for example, the former parameter from 3 to 3×10^3 genomes/10^7 lymphocytes. Although infectious virus particles could be recovered from only ~12% of viral DNA-positive tissue samples, viral gene expression and genome replication were induced when tissue-derived lymphocytes maintained in culture were activated, in most cases with production of virus particles able to reproduce in permissive human epithelial cells (A549 cells) (Garnett *et al.*, 2009).

The observations are consistent with the conclusion that species C HAdV genomes are maintained in a latent, quiescent state in T-lymphocytes from naturally infected children, but in the absence of studies of single cells, the possibility that at any one time a very small number of cells in the populations support a productive infectious cycle cannot be excluded. However, certain established lines of human T-lymphocytes reproduce such latency and viral genome maintenance when infected by HAdV-C5 or HAdV-C2 (Zhang *et al.*, 2010) or by HAdV-A31 or HAdV-B3 or -B11 (Markel *et al.*, 2014). Remarkably, the persistently infected cells proliferate at the same rate as uninfected cells, and viral genomes could be detected at ~100 copies/cell for months to up to a year after the initial infection in the absence of expression or assembly of infectious virus particles (Markel *et al.*, 2014; Zhang *et al.*, 2010). HAdV-C5 viral genomes were shown to persist as non-integrated viral linear monomers (Zhang *et al.*, 2010). The mechanism by which they are replicated in the absence of significant expression of the genes that encode the viral replication remains a mystery.

The long-term persistence of HAdV-C5 genomes in established lines of T-lymphocytes correlated with rapid loss (within 24 hours) of cell surface Coxsackie and adenovirus receptor (CAR) (Markel *et al.*, 2014; Zhang *et al.*, 2010). This process initially required viral late gene expression and a fiber protein that can bind CAR, and was followed by loss of CAR mRNA from some 30 days after infection (Zhang *et al.*, 2010). Cells that reached this state could not be re-infected, even when the CAR gene was expressed via a retroviral vector, suggesting that additional changes restrict viral reproduction (Zhang *et al.*, 2010). The loss of CAR would be expected to limit spread of virus particles from the lymphocytes initially infected. Perhaps of greater relevance to the long-term viability of the T-lymphocytes in which HAdV-C5 genomes persist is the reduced production of the E3 ADP (and it is mRNA) (Murali *et al.*, 2014), which facilitates the release of virus particles from epithelial cells (Chapter 7). A mutant virus that does not direct synthesis of ADP established a persistent infection in a T-lymphocyte line that supports productive infection of wild-type parent (HAdV-C5), illustrating the importance of the decreased production of this protein for development of latent infection.

8.4 What Next?

Infection by adenoviruses is common in humans, particularly infants and children, in all parts of the world. In immunocompetent individuals, these infections are generally mild and self-limiting, or asymptomatic. Nevertheless, specific HAdV types can cause more serious disease, notably acute respiratory disease and EKC. The significant morbidity, and consequently cost, associated with ARD in military recruits resulted in the development of effective vaccines against the relevant HAdV types, underwritten by the U.S Department of Defense. Infection by HAdVs has also been associated with other more serious syndromes, including myocarditis and obesity, but the available evidence indicates that the infection may contribute to, but is not necessary for, the development of these diseases. HAdVs are a much more serious problem in immunosuppressed patients, especially pediatric transplant patients, when they cause disseminated diseases that affect multiple organs and are often fatal. The increasing number of such patients has spurred the development of anti-HAdV drugs (Chapter 10) and other therapeutic approaches (Chapter 9), although further development and testing will be needed.

Clinical experience with HAdVs and studies in animal models underscore the important roles of host immune responses not only in controlling and clearing infection but also in pathogenesis. Some viral gene products that contribute to pathogenesis in animal models, for example, by promoting release and spread of virus particles, have also been identified. However, our understanding of HAdV pathogenesis remains quite limited. The major impediment to progress in this area has been the lack, until quite recently, of experimentally accessible animal models that reproduce the characteristics of HAdV infection and disease in humans. It has now been established that the Syrian hamster is quite permissive for HAdV replication and exhibits pathology similar to that seen in humans when infected by different routes. It is to be hoped that further studies in this animal model will help address such outstanding issues as the relative importance (and roles) of specific viral gene products and host immune system components in governing pathogenesis, and the molecular basis of the difference on tropism among HAdVs. As Syrian hamster are rodents that differ considerably from humans in multiple respects, including their immune system, it seems likely that further development of this animal model will be needed, for example, humanization of immune systems components.

Host Defenses   9

9.1 Introduction

Adenovirus infections are contained, controlled and eventually cleared by elaborate antiviral host responses. Front-line defense is provided by the intrinsic (cell-autonomous) mechanisms and innate immune responses that are triggered rapidly upon infection of a host cell, and in some cases, even before an adenovirus particle gains entry. Engagement of these initial countermeasures to infection leads to activation of adaptive immune responses, and the production and recruitment to sites of infection of T cells and antibodies that specifically recognize adenovirus particles or infected cell that are synthesizing viral proteins. During development of these more slowly acting adaptive defenses, immune memory is established to confer protection against subsequent infections.

Such immune defenses are very effective, as illustrated, for example, by the high prevalence of antibodies that recognize species C human adenoviruses (HAdVs) in children and the severe disease and high mortality associated with HAdV infections of immunosuppressed patients (Chapter 8). However, the very properties of the immune responses that confer effective protection against these viruses can hamper therapeutic applications (Chapters 11 and 12). Indeed, the need to foil, or at least limit, the immune responses to HAdV vectors inspired many of the efforts to elucidate the innate and adaptive responses to adenovirus described in this chapter.

9.2 Cell Autonomous Defenses

Mammalian cells possess a surprising variety of intrinsic mechanisms for interfering with viral reproduction independently of the extracellular signaling molecules that activate and coordinate innate and adaptive immune responses (Flint *et al.*, 2015). Such immediate defenses not only confer antiviral protection, but can also promote subsequent immune responses. Several are induced following HAdV infection, but can also be minimized by the actions of viral gene products.

9.2.1 *Apoptosis*

Induction of apoptosis in virus-infected cells represents a defense of last resort, in which the suicide of the infected cell limits virus reproduction and spread to nearby cells. Activation of the apoptotic caspase cascade leads to characteristic nuclear fragmentation, membrane blebbing and the release of subcellular apoptotic bodies as the host cell undergoes programmed cell death. So-called sentinel cells, such as plasmacytoid dendritic cells and macrophages, which patrol common sites of infection, can engulf the debris of a virus-infected cell produced by apoptosis. Activation of these sentinel cells leads to release of signals that amplify apoptosis (TNF-α) or the stimulation of the innate and adaptive immune systems (e.g. type 1 interferons, IFNs).

The HAdV-C5 E1A and E4 proteins can activate apoptosis by p53-dependent and independent, mechanisms, respectively. However, the intrinsic and extrinsic pathways that induce this process, as well as activation of the caspase cascade, are blocked by the multiple mechanisms described in Chapter 6. At least seven viral proteins confer resistance to induction of apoptosis, in some cases with apparent redundancy (Table 9.1). The targeting of this response by such a plethora of viral proteins suggests that apoptosis has the potential to limit severely the reproduction and propagation of HAdVs. However, a recent report indicates that inhibition of apoptotic death of HAdV-C5-infected cells can also prevent inflammatory responses; when A549 cells infected by a mutant virus deleted for the E1B 19 kDa gene died by apoptosis, inflammatory responses were induced in macrophages that came into contact with the infected cell corpses (Radke *et al.*, 2014). However, such responses were repressed when

Table 9.1. HAdV-C5 proteins that avert apoptosis.

Viral Protein	Cellular Target(s)	Mechanism
E1B 19 kDa	Proapoptotic proteins, e.g. BAX, BAD, BAK	Sequestered
55 kDa[°]	AIF	Translocation to nucleus blocked
55 kDa[*]	p53	Ubiquitinylated; targeted for proteasomal degradation
E3 RIDα ⎱ RID RIDβ ⎰	TNFαR, TRAIL1R, FAS	Removed from cell surface; degraded
6.7k + RID	TRAIL2R	Removed from cell surface; degraded
14.7k	TNFαR	Internalization ligand-bound receptor blocked
	NF-κB	Activation and DNA binding prevented
E4 Orf6[*]	p53	Ubiquitinylated; targeted for proteasomal degradation
Orf3	p53	Inhibition p53-dependent transcription

[°]When E4 Orf3 coding sequence also deleted.
[*]Mediated by the virus-specific E3 ubiquitin ligase formed with the cellular proteins CUL5, ELO B and C, and RBX1.

wild-type virus-infected cells underwent non-apoptotic death (Radke *et al.*, 2014). These observations indicate that, by blocking apoptosis, the viral E1B 19 kDa protein may indirectly limit inflammatory responses at sites of infection.

9.2.2 *Tripartite motif family proteins*

The large family of (tripartite motif) TRIM proteins includes several initially implicated in antiviral defense as exemplified by the restricted reproduction of retroviruses (Nisole *et al.*, 2005; Rajsbaum *et al.*, 2014; Yap *et al.*, 2012). It also includes proteins that confer resistance to HAdV infection unless they are inactivated or destroyed by the actions of viral gene products, namely PML (TRIM19) and members of the transcriptional intermediary factor (TIF) group. Disruption of PML bodies and reorganization of PML protein into the track-like structures formed by the viral E4 Orf3 protein is necessary for timely and efficient viral DNA synthesis (Doucas *et al.*, 1996; see Chapter 6), and also facilitates viral gene expression (Chapter 4). In addition, PML and the associated protein DAXX (death domain-associated protein) have the potential to inhibit viral genome replication in response to type I IFN (see below).

More recently, TRIM24 (TIF1α) and TRIM33 (TIF1γ) were shown to bind to the HAdV-C5 E4 Orf3 protein and to be recruited to the nuclear tracks formed by the viral protein (Vink *et al.*, 2012; Yondola & Hearing, 2007; see also Chapter 6). The ability to reorganize TRIM24 in this way is conserved among members of HAdV species A, C, D. and E, suggesting that it is important for successful HAdV reproduction. The E4 Orf3 protein is sufficient for reorganization of both TRIM24 and TRIM33 (Forrester *et al.*, 2012; Vink *et al.*, 2012; Yondola & Hearing, 2007). However, only TRIM33 is directed for proteasomal degradation by the E4 Orf3 protein (of species A, B, and C HAdVs) by a cullin RING-finger ligase-independent mechanism (Forrester *et al.*, 2012). Knockdown of this cellular protein by RNAi resulted in earlier and somewhat increased synthesis of viral early and late proteins, whereas its overproduction delayed viral early gene expression (Forrester *et al.*, 2012). These observations indicate that TRIM33 is a restriction factor for HAdV reproduction, although its mechanism of action is not yet known. In addition, both TRIM24 and TRIM28 (TIF1β) interact with the E1B 55 kDa protein in infected cells (Forrester *et al.* 2012), but whether these multifunctional regulators of transcription (Cammas *et al.*, 2012; Hatakeyama, 2011) also limit viral gene expression if not inactivated remains to be determined.

The ability of the E4 Orf3 protein to assemble into intranuclear tracks seems likely to influence the course of the viral infectious cycle by additional mechanisms. In particular, this process leads to altered expression of over 400 cellular genes, including genes associated with inflammatory and immune responses, in a human osteosarcoma cell line by TIF protein-independent mechanisms (Vink *et al.*, 2015).

9.3 Innate Immune Responses to HAdVs

The innate immune system embraces a large repertoire of secreted cytokines, including antiviral interferons and pro-inflammatory cytokines, chemokines that are important for recruiting immune system cells to sites of infection, defensins, and the serum complement proteins. The majority of mammalian cells can produce type I IFNs and other cytokines upon virus infection, but in animals these soluble mediators of innate immunity are produced predominantly by patrolling sentinel cells. Such macrophages and dendritic cells, which can detect infection indirectly by phagocytosis of infected cell corpses or themselves become infected, and cells that destroy infected cells directly, such as natural killer (NK) cells, are critical cellular components of the innate immune response. In addition to furnishing non-specific and broadly acting anti-viral defenses that can be deployed rapidly (within minutes to hours of infection) components of the innate immune system trigger and shape the subsequent adaptive immune responses that are tailored to each pathogen.

The innate immune response to HAdV infection and the mechanisms by which they are subverted by viral gene products have received increasingly intense scrutiny over the past 20–25 years. Such investigations have been spurred in large part by the recognition that even HAdV vectors from which all coding sequences have been deleted elicit strong innate immune responses that can result in inflammatory responses or, conversely, increase the potency of oncolytic HAdVs (Aldhamen *et al.*, 2011; Hendrickx *et al.*, 2014; see also Chapters 11 and 12).

9.3.1 *Sensors of HAdV-C5 infection*

Regardless of whether they act locally or systemically, antiviral defenses are powerful, and frequently result in death and destruction of infected cells. Consequently, deployment of innate immune defenses is tightly coupled to recognition of viral infection. In the case of HAdVs, multiple molecular signs of infection, from the initial interactions of virus particles with cell-surface receptors to expression of viral genes can trigger signal transduction and transcription of genes that encode interferons and other cytokines, and initiate pro-inflammatory responses. Although the primary sites of HAdV infection in humans are various epithelial cells, many studies of sensors of infection have employed murine cells, particularly sentinel cells (macrophages and dendritic cells), which are critical in the detection and amplification of initial signals of infection. Murine cells typically support viral early gene expression, but not genome replication and subsequent steps in the infectious cycle, and may also differ from human cells in their susceptibility to HAdVs.

One important mechanism for initiation of innate responses to infection is recognition of distinctive molecular features of bacteria, viruses and other pathogens. These pathogen-associated molecular patterns (PAMPs) are detected by specialized proteins termed pattern-recognition receptors (PRRs). The first PRRs to be recognized were the Toll-like receptors (TLRs), which reside in the plasma or endosomal membranes, but it is now clear that a variety of detectors of microbial RNA or DNA are present in the cytosol and nucleus, and that the initial interactions between a virus particle and the host cell can also serve as early warning detectors of infection.

9.3.1.1 *Signaling from the cell surface: immune responses elicited by adenovirus receptor association*

The initial association of HAdV particles with cell-surface receptors induces immune responses that are governed by the identity of both the receptor and the infected cells. Interaction of the HAdV-C5 fiber with the CAR receptor on A549 epithelial cells in culture was reported to induce a modest and transient production of certain pro-inflammatory cytokines such as IL-8 via triggering of ERK1/2 and JNK mitogen activated kinases (Tamanini *et al.*, 2006). It is striking that the expression of CAR on certain cell types is influenced by immune activation itself. An alternative isoform of CAR, designated CAR-Ex8, is located on the apical side of polarized epithelial cells, a property that allows infection from this membrane location (Kotha *et al.*, 2015). IL-8, a pro-inflammatory cytokine, was reported to stimulate the accumulation of CAR-Ex8 as a result of the activation of the signaling molecules AKT and S6K.

Interactions of adenovirus particles with other attachment receptors may also modulate immune responses, although relatively little is known with regard to desmoglein 2 (DSG2) and GD1a. However, purified GD1a can impair the maturation of human dendritic cells in response to lipopolysaccharide (LPS) (Shen & Ladisch, 2002), and in this way attenuate differentiation of T helper cells (Shen *et al.*, 2005). Thus, it is possible that adenovirus interactions with GD1a might induce similar immune impairments, but this possibility needs further investigation.

CD46, a receptor for cell attachment of certain species B adenoviruses (e.g. HAdV-B35, -B16), activates specific immune responses following its ligation by different microbial pathogens (Cattaneo, 2004; Hay *et al.*, 2014) or by association with its normal cellular ligand, the Notch family protein jagged1 (Le Friec *et al.*, 2012). CD46 is a strong co-stimulator of IFNγ production in effector T helper (TH)-1 cells and mediates the conversion to interleukin 10 (IL-10) secreting regulatory T cells (Cardone *et al.*, 2010; Kemper *et al.*, 2003). Interactions of CD46 with diverse microbial pathogens such as measles virus and HHV-6 are known to limit the capacity of immune cells to produce IL-12, a cytokine that modulates both innate and adaptive immune responses (Karp *et al.*, 1996; Smith *et al.*, 2003). Adenoviruses that bind to CD46 (HAdV-B35), but not those that attach to CAR (HAdV-C5), also reduce IL-12 production in human peripheral blood mononuclear cells as a result of interference with IFNγ signaling mediated by the transcription factor C/EBPβ (Iacobelli-Martinez *et al.*, 2005).

Recent studies in cells in culture showed that the binding of recombinant HAdV-B35 fiber knob proteins to CD46 either enhances or diminishes human T cell co-stimulation, depending on the overall receptor-binding affinity (Hay *et al.*, 2014). Recombinant forms of the HAdV-B35 fiber knob selected for high receptor binding affinity caused the loss of CD46 from the surface of T cells, resulting in diminished co-stimulatory activity. In contrast, an arginine substitution mutant of this fiber knob that decreased CD46 binding (Wang *et al.*, 2010) provided co-stimulatory activity. The mechanism by which high affinity fiber knob binding to CD46 restricts co-stimulatory activity is not known. Whether intact CD46-binding adenovirus particles can influence normal T cell functions *in vivo* also remains to be determined. CD46-binding HAdVs have been used as vaccine vectors in mice (DiPaolo *et al.*, 2006) or rhesus monkeys (H. Li *et al.*, 2012) and did not exhibit significantly altered T cell responses to test antigens or to the viral vectors themselves. The ability of HAdV-B35 fiber knob to reduce CD46 expression on tumor cells has also been explored as a potential therapy to render these cells susceptible to complement-mediated destruction (Beyer *et al.*, 2013; Wang *et al.*, 2010).

As discussed in Chapter 4 and summarized in Fig. 9.1, subsequent interactions of the adenovirus penton base with α_v integrins not only promote virus internalization but also induce signal transduction events that modulate host immune responses.

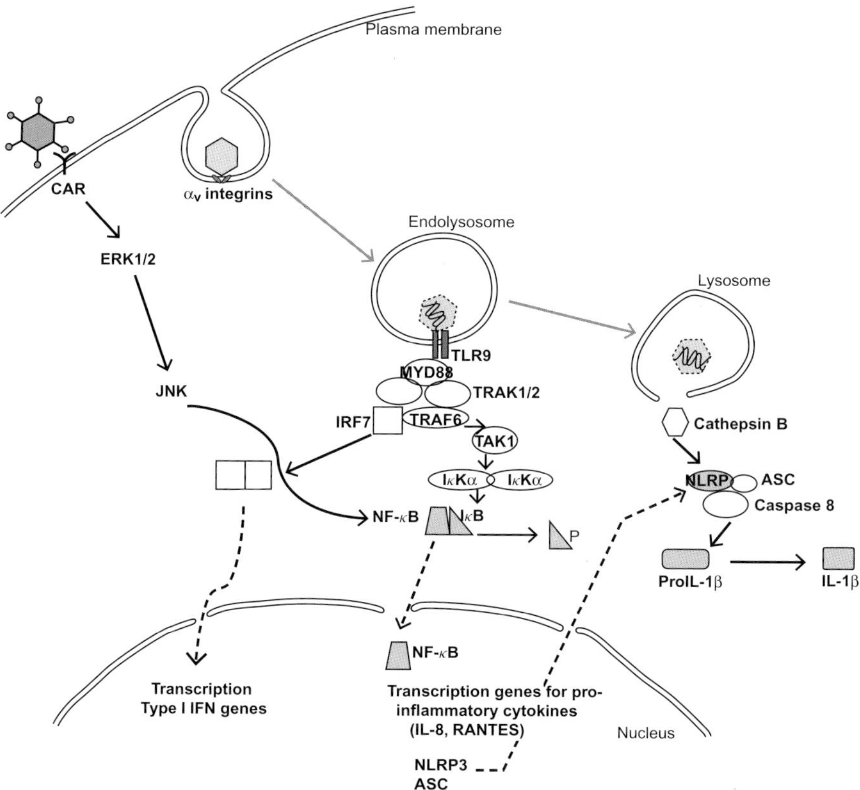

Figure 9.1. Induction of innate immune responses during HAdV attachment and entry. The pathways shown for activation of transcription of genes that encode pro-inflammatory cytokines, type I interferons (IFNs) and components of the inflammasome (NLRP3, ASC) have been identified in different cell types and appear to be determined by the identity of the initial attachment receptor. Coxsackie virus and adenovirus receptor (CAR)-binding HAdVs induce stimulation of interleukin (IL)-8 production in transformed human epithelial cells, recognition of CD46-binding HAdVs in endolysosmes by TLR-9 initiates signaling

Figure 9.1. (Continued) that culminates in activation of NF-κB and IRF-7 and transcription of genes controlled by these activators in murine plasmacytoid dendritic cells, and, production of the pro-inflammatory cytokine IL-8 occurs in human but not murine macrophages infected by species A, B, and D HAdVs that traffic to lysosomes leading to membrane disruption, release of cathepsin B, and activation of the NLRP3 inflammasome. Additional mechanisms are described in the text. Activation of proteins and movement of proteins into or out of the nucleus are indicated solids and dashed, respectively, black arrows, while grey arrows denote the entry pathway of adenovirus particles.

ASC, apoptosis-associated speck-like protein containing a CARD; ERK1/2, extracellular signal-regulated kinase 1 and 2; IRF7, interferon regulatory factor 7; IκB, inhibitor of NF-κB; IκK, inhibitor of NF-κB kinase; IRAK1/2, IL-1 receptor-associated kinase 1 and 2; JNK, Janus kinase; MYD88, myeloid differentiation primary response protein; NFκB, nuclear factor κB; NLRP3, NSACHT, LRR, and PYD domains-containing protein 3; TAK1, transforming growth factor-beta-activated kinase 1; TLR, Toll-like receptor; TRAK1/2, trafficking kinesin-binding protein 1 and 2.

9.3.1.2 *Cytosolic detectors*

In addition to TLR-dependent mechanisms, initiation of innate immune responses to infection by HAdV-C5 (and typically by "first-generation" vectors lacking at least the E1A, E1B, and E3 genes) was observed to be TLR-independent in various murine cells of the myeloid lineage, such as macrophages and conventional dendritic cells (Appledorn *et al.*, 2008b; Cerullo *et al.*, 2007; Fejer *et al.*, 2008; Nociari *et al.*, 2007; Yamaguchi *et al.*, 2007; Zhu *et al.*, 2007a): as indicated by such read-outs as synthesis of IFNα or other cytokines like IL-6, these responses were impaired only partially or not at all in cells from TLR9$^{-/-}$ mice. Subsequent experiments using murine macrophages, a macrophage-like cell line or bone-marrow-derived dendritic cells established that the viral DNA genome is recognized in the cytosol to trigger phosphorylation of IFN regulatory factor (IRF3), a protein that, as its name implies, is critical for initiation of the IFN response (Nociari *et al.*, 2007; Nociari *et al.*, 2009; Stein & Falck-Pedersen, 2012). Of the various cytosolic PRRs, the DNA sensor cyclic GMP–AMP synthase (cGAS) is critical for induction of phosphorylation (activation) of IRF3 in murine macrophage or endothelial cell lines infected by a first-generation HAdV-C5 vector: RNAi-mediated knockdown of cGAS or of proteins via which it signals, stimulator of IFN genes (STING) and Tank-binding kinase 1 (TBK1) (Fig. 9.2), resulted in greatly reduced phosphorylation of IRF3 and reversal of the infection-induced expression of the IFNβ gene and various IFN-sensitive genes (ISGs) (Lam & Falck-Pedersen, 2014).

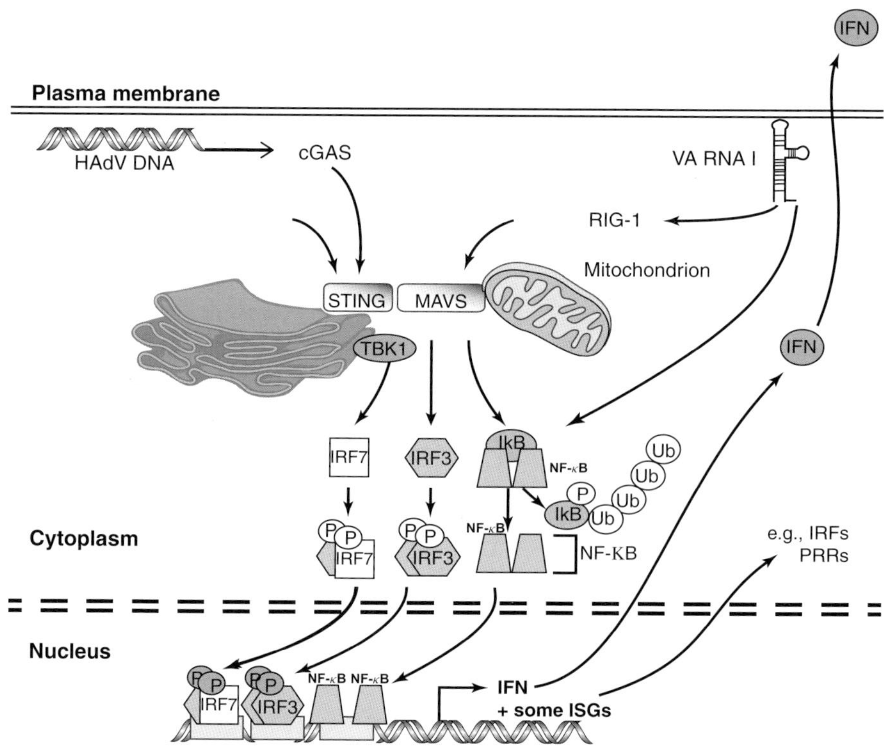

Figure 9.2. Cytosolic detectors of HAdV infection. Both viral DNA that enters the cytoplasm at the beginning of infection and VA RNA I made later in the infectious cycle are bound by cellular PRRs, cGAS, and RIG-1, respectively, and the signal transduced to transcriptional activators such as IRF7 and NF-KB to induce synthesis of IFNα and -β and transcription of some interferon-sensitive genes (ISGs). cGAS, cyclic GMP–AMP synthase; IRF3/7, interferon regulatory factor 3/7; MAVS, mitochondrial antiviral signaling protein; NF-κB, nuclear factor κB; RIG-1, retinoic acid-inducible gene protein 1; STING, stimulator of interferon genes protein.

Sentinel cells such as the macrophages and dendritic cells examined in the studies summarized in the previous paragraph are critical for initiation of both innate and adaptive responses to adenoviral vectors in mice (Elkon *et al.*, 1997; Muruve *et al.*, 1999; Trevejo *et al.*, 2001; Yang *et al.*, 1995). However, such cells are not typically those initially infected in humans, but rather may become infected once virus particles are released and spread from epithelial cells at sites of primary infection. Infection of established human cell lines that can mount an IFN response by an HAdV-C5 vector has also been observed to activate the cGAS-STING pathway early after infection and IRF3-dependent expression of genes that encode IFNβ, and ISGs (Lam *et al.*, 2014). Activation of the cGAS–STING–IRF3 signaling pathway was also induced in HeLa cells infected by wild-type HAdV-C2, but greater phosphorylation of IRF3 was observed in cells infected by species B HAdVs that enter cells via binding to CD46 (HAdV-B35) or DSG2 (HAdV-B7A) (Lam *et al.*, 2014). As cGAS recognizes DNA without sequence specificity (Cai *et al.*, 2014), this finding may suggest that the efficiency of delivery of HAdV to the cytoplasm, or access to other PRRs, is governed by the route of entry. However, it has been reported recently that HAdV-C5 E1A gene products alone reduce (by some fourfold) type I IFNŠ production in response to signaling via GAS–STING, as a result of interaction of the viral proteins with STING (Lau *et al.*, 2015).

Viral DNA entering the cytoplasm of HAdV-infected cells of various species and types clearly serves as an important PAMP, but whether it sensed by DNA-recognizing PRRs other than cGAS, such as IFI16 (Dempsey & Bowie, 2015) is not yet known. On the other hand, additional PRRs that recognize RNA have been implicated in induction of the IFN response. Introduction of a plasmid carrying the HAdV-C2 VA genes was observed to activate the IFNβ promoter and greatly increase the accumulation of IFNα and IFNβ mRNAs (Weber *et al.*, 2006). Activation of IRF3 in HAdV-C5-infected cells was also reported to require transcription by RNA polymerase III (Chattopadhyay *et al.*, 2011). Similarly, it was observed that introduction of virus-associated (VA) RNAs I and II into a human carcinoma-derived line stimulated accumulation of these IFN mRNAs, as well as those of various ISGs (Minamitani *et al.*, 2011). In this system, RNAi-mediated knockdown of the RNA-detector retinoic acid-inducible gene 1 (RIG1) (Fig. 9.1) greatly reduced activation of the IFN response, as did that of the downstream signaling protein IRF3 (Minamitani *et al.*, 2011). Although somewhat artificial, these systems appear to reflect a function of VA RNAs during the infectious cycle. In a human cell line infected with an E1-deleted HAdV-C5 vector, knockdown of RIGI or IRF3 impaired the second of two phases of increased accumulation of $IFN\beta$ and ISG mRNAs, and secretion of $IFN\beta$, which occurred from 48 hours after infection, but not the early induction (from 12 hours p.i.) (Minamitani *et al.*, 2011). The late-phase accumulation of $IFN\beta$ mRNA and protein was not induced by infection with UV-inactivated virus, that is, required viral gene expression or genome replication, and coincided with the accumulation of VA RNAs.

These observations are consistent with a mechanism in which the VA RNAs amplify the initial interferon response to species C HAdV infection. Whether these viral RNAs govern the later (or secondary) induction of IFN remains to be demonstrated directly. Furthermore, there are some indications that the mechanism of VA RNA I detection may be cell-type specific. In murine bone-marrow-derived dendritic cells and mouse embryo fibroblasts (MEFs), inhibition of RNA polymerase III transcription or deletion of the VA RNA genes impaired induction of $IFN\beta$ release, but in MEFs this response required mitochondrial antiviral signaling (MAVS) protein, but not RIG1 (Yamaguchi *et al.*, 2010).

Our understanding of the initiation of antiviral responses to HAdV infection has improved enormously in the past 10–15 years, with the identification and characterization of many components of the various pathways by which infection is registered by host cell components to activate signal transduction. However, numerous observations argue that PRRs for HAdVs operate in cell-type-specific fashion, and that the route of entry governs both the detection pathway and the magnitude of the response. At this juncture, direct comparisons among published studies are hampered by use of different vectors, of different cell types or of cells of different species. Clearly, systematic evaluation of the contributions of the various PRRs in individual cell types infected with wild-type HAdV, and comparison to infection with a typical E1- and E3-deleted vector would be of great benefit.

9.3.2 *Anti-adenoviral actions of the innate immune system*

9.3.2.1 *Defensins*

As members of a family of small antimicrobial proteins, defensins are made in epithelial cells in the skin and at mucosal surfaces such as those of the gastrointestinal tract and the eye (Selsted & Ouellette, 2005). Defensins are perhaps most recognized for their anti-bacterial activities (Lehrer *et al.*, 1989). However, these innate immune molecules are also capable of neutralizing fungi and viruses (Buck *et al.*, 2006; Wilson *et al.*, 2013). Defensins are small (18–45 amino acids), cationic, amphiphilic peptides with mostly β-sheet secondary structure that is stabilized by three pairs of disulfide bonds. Human α-defensins are one of the major classes of the antimicrobial peptides that are produced in large quantities (up to low mM local concentrations) in the small intestine (Ayabe *et al.*, 2000) or at lower levels in neutrophils and other immune cells.

In one of the earliest investigations of the effects of defensins on viruses, human α-defensin, HNP1, was demonstrated to inhibit infection by HAdV-C5, although the mechanism of inhibition was not explored (Bastian & Schafer, 2001). Subsequently, human α-defensins HD5 and HNP1 were shown to block HAdV-C5 infection of cultured cells by preventing capsid disassembly during cell entry (Smith & Nemerow, 2008). Defensin neutralization occurred with HAdV-C5 vectors displaying the type 5, 16, or 35 fiber and importantly, antimicrobial peptides selectively targeted virus escape from the early endosome. α-defensins bind directly to the capsid and prevent release of the vertex region and the associated protein VI molecules that mediate membrane disruption (Nguyen *et al.*, 2010). Further evidence of virus capsid stabilization by α-defensin HD5 was obtained using atomic force microscopy (Snijder *et al.*, 2013). These biophysical analyses demonstrated that association of the defensin with HAdV-C5 particles strengthens the fivefold axis, consistent with the earlier biochemical and functional studies. To gain further insights into the mechanism of defensin-mediated HAdV neutralization, Smith and colleagues (2010) created several adenovirus type 5/19 chimeric particles and examined their ability to be neutralized by α-HD5. These virus chimeras consisted of a HAdV-C5 (defensin-sensitive) capsid in which the type 5 penton base or fiber protein or both proteins was replaced by the corresponding protein(s) from the defensin-resistant HAdV-D19 virus. The replacement with the type 19 fiber and penton base together conferred nearly complete protection against defensin neutralization. Moreover, a small amino sequence, GYAR, in the fiber protein of type D viruses, also conferred defensin resistance. These findings suggested that the penton base, together with the N-terminal fiber tail, is the site of defensin binding to and neutralization of species C HAdV-C5. Subsequent cryo-electron microscopy (cryoEM) studies, combined with molecular dynamic modeling analyses, revealed prominent defensin HD5 density associated with a region spanning the fiber and penton base RGD loop of HAdV-C5 but this defensin density was absent on HAdV-D19, indicating that HD5 inactivates certain adenoviruses by stabilizing a specific region in the virus vertex region (Flatt *et al.*, 2013). By analyzing a series of alanine-scanning mutants of HD5, several hydrophobic amino acids (L26, L16) and a charged residue (R28) on one external face of the defensin molecule were found to be crucial for HAdV-C5 neutralization (Tenge *et al.*, 2014). Neutralization of HAdV-C5 by HD5 also depends on a relatively high stoichiometry (B_{max} = 3,050) rather than overall affinity (K_D = 1.8 µM).

Although these studies have provided additional insights into how innate immune molecules restrict adenovirus infection *in vitro*, further investigations are needed to determine whether defensins also play a prominent antiviral role *in vivo*. In this regard, a new small intestinal organoid model has been developed to facilitate further investigations (Wilson *et al.*, 2015).

9.3.2.2 *Complement*

The complement system, comprising 30 different proteins, is an important part of the host innate immune response to invading microbial pathogens (Walport, 2001). There are three distinct pathways of complement activation: classical (antibody-mediated), alternative (antigen specified or spontaneous), and lectin (mannan-binding lectin or ficolin-directed) pathways. The classical and alternative complement pathways are well known for their ability to inactivate enveloped or nonenveloped viruses either by aggregation, membrane lysis, or coating of particles with complement proteins that prevent cell attachment or virion disassembly (Cooper & Nemerow, 1983). Opsonization (phagocytosis) of viruses and bacteria via covalent attachment of fragments of the third component of complement (C3) can also lead to pathogen uptake and destruction by cells of the reticuloendotheial system, which possess the complement receptors CR1 and CR3 (Merle *et al.*, 2015).

Bacterial membranes and viral envelopes are major targets for the complement membrane attack complex C5b-9, and thus the lack of a lipid envelope raised the question of whether complement played any role in neutralizing adenovirus. One of the earliest studies focusing on this question showed that incubation of different HAdVs (Ad1, Ad3, Ad4, Ad5, and Ad9) with human plasma, containing either neutralizing or non-neutralizing antibodies, activated complement based on the generation of the C3a fragment, and that activation appeared to be mostly antibody dependent (Cichon *et al.*, 2001). However, a subsequent investigation reported that incubation of HAdV-E4 or HAdV-C5 in agammaglobulinemic human serum (lacking antibodies) was also capable of activating C3 via the alternative complement pathway, and that C3 was deposited on these virus particles, although the consequences were not investigated further (H. Jiang *et al.*, 2004). Other investigators have reported that adenovirus-mediated complement activation *in vitro* (cell-free systems) assays requires the presence of anti-viral antibodies while activation by the virus *in vivo* occurs via the classical (antibody) and non-classical (alternative) pathways and by the reticuloendothelial system in a mouse model (Tian *et al.*, 2009). Coating virus particles with polyethylene glycol (PEG) reduces complement activation both *in vitro* and *in vivo* (Tian *et al.*, 2009), suggesting that this approach may be helpful for some gene transfer experiments.

More thorough investigation of HAdV–complement interactions was aided by the development of various complement knock out mouse models. In one study, systemic administration of a HAdV-C5 vector into C3-knockout mice produced significantly lower levels of pro-inflammatory cytokines as well as more normal levels of platelets compared to C3-sufficient mice, suggesting that complement activation contributes to blunting the innate immune response to the virus (Kiang *et al.*, 2006).

Evidence has also been obtained exploiting C3 knock out mice that complement activation plays a role in the development of neutralizing antibodies to the virus capsid (Appledorn *et al.*, 2008a).

Deposition of C3 on the surface of adenovirus particles, as well as other non-enveloped viruses, has several important consequences for intracellular signaling events (Tam *et al.*, 2014). When HAdV-C5 is incubated with human immune serum, IgG, C3, and C4 are deposited onto capsid surfaces and the resulting virus–immune complexes can be taken into human 293 cells. HAdV-C/IgG complexes activated nuclear factor κB, AP-1, and the IRF3 transcription pathways, leading to a strong pro-inflammatory cytokine response, and the production of IFNβ. In a parallel pathway to that mediating pro-inflammatory cytokine production, Ig-complement-coated HAdV particles are degraded by the AAA-adenosine triphosphate (ATPase) valosin-containing protein (VCP) and the proteasome, as indicated by the ability of specific inhibitors of VCP or the proteasome to override this process (Tam *et al.*, 2014). Thus, this broadly acting and complement-dependent response induces an antiviral state, and also provides a direct pathway for destroying invading pathogens such as HAdV.

Systemic administration of HAdV-C5 can also result in the delivery of this virus to the liver in a coagulation factor X (FX)-dependent bridging mechanism (Kalyuzhniy *et al.*, 2008; Shayakhmetov *et al.*, 2005; Waddington *et al.*, 2008). However, FX was found to be dispensable for HAdV-C5 liver transduction in mice deficient in antibodies or the complement components C1q or C4 (classical pathway mediators) (Xu *et al.*, 2013), suggesting that another role for FX association with capsids is to protect them from assault by antibodies and complement. Thus, the associations of HAdV particles with distinct plasma proteins can have varying effects on adenovirus tissue tropism as well as virus neutralization.

9.3.2.3 *TRIM21*

Recent studies have revealed a fascinating new mode of antibody neutralization of viruses including adenovirus, inactivation inside host cells via the action of a cytosolic antibody receptor known as TRIM21 (Mallery *et al.* 2010; Yoshimi *et al.*, 2009). TRIM21 is made in most cell types and binds with micromolar to subnanomolar affinity to the Fc portion of IgG, IgM, or IgA molecules via its PRYSPRY domain (Bidgood *et al.*, 2014; James *et al.*, 2007; Mallery *et al.*, 2010). Association of TRIM21 with virion-bound antibody marks the complex for proteasomal degradation via a process known as antibody-dependent intracellular neutralization (McEwan *et al.*, 2012). In addition to this degradative pathway of antibody-mediated virus inactivation, TRIM21 can also activate IRF3, IRF5, IRF7, and AP-1, thereby triggering an antiviral state (McEwan *et al.*, 2013). More recently, TRIM21 has been shown to help present double-stranded (ds) DNA viruses such as adenovirus to the cytosolic DNA sensor cGAS thereby generating a more rapid immune response upon infection (Watkinson *et al.*, 2015).

9.3.2.4 *Interferons*

9.3.2.4.1 The antiviral state induced by IFNα and -β

Humans possess 13 IFNα genes and a single IFNβ gene that can be activated in response to viral infection. Although it is not clear whether the multiple IFNα proteins fulfill identical functions or operate in specific cell types or circumstances, α and β IFNs bind to a common, heterodimeric receptor (IFNAR) to initiate signaling via the relatively simple JAK–STAT (Janus kinase–signal transducer and activation of transcription) pathway (Fig. 9.3) (Ivashkiv & Donlin, 2014; Schneider *et al.*, 2014). Such signaling may be autocrine, leading to a second wave of IFN synthesis in an infected cell, or paracrine. In either case, the net result is assembly of the transcriptional regulator ISGF3 (IFN-stimulated gene factor 3), which enters the nucleus and activates a response that is far from simple, altered transcription (increased in the great majority of cases) of over 3,000 human genes (Rusinova *et al.*, 2013). The large set of ISGs include the IFNα and IFNβ genes themselves, genes for proteins that participate in or regulate inflammatory, innate and adaptive immune responses, such as PRRs and components of antigen presentation pathways, in addition to genes associated with the inhibition of virus reproduction (Gonzalez-Navajas *et al.*, 2012; Wang & Fish, 2012). The products of the latter class of ISGs can act at essentially every step in viral infectious cycles from entry to release of infectious virus particles

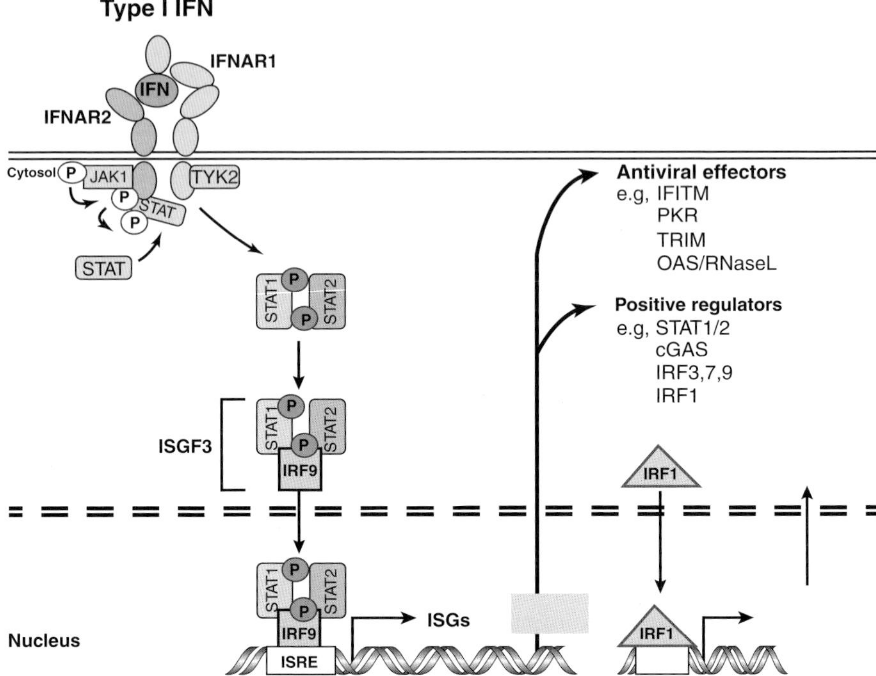

Figure 9.3. Activation of transcription of ISGs. Binding IFNα and -β to the common receptor IFNAR activates the associated Janus kinase (JAK) and tyrosine kinase 2 (TYK2), which phosphorylate the receptor, allowing recruitment and phosphorylation of signal transducer and activator of transcription STAT1 and STAT2. The modified proteins dimerize, associate with IRF9, and translocate to the nucleus to stimulate transcription of ISGs. As indicated, ISGs encode not only proteins that impair virus reproduction, but also proteins that amplify the innate response, including IRFs that initiate a second burst of IFN synthesis.

(Schneider *et al.*, 2014; Schoggins, 2014; Schoggins & Rice, 2011). However, the molecular functions and mechanisms of action of only a relatively small number of antiviral ISGs have been elucidated. In the case of HAdVs, only a few that impair progression through the infectious cycle have been identified, as a result of the identification of viral gene products that confer resistance to IFNs.

9.3.2.4.2 Viral gene products that impair induction or execution of the IFN response

There is considerable evidence indicating that type 1 IFNs influence the responses to HAdV infection *in vivo* (see next section). Nevertheless, the reproduction of all HAdVs in permissive cell lines in culture that have been examined (representatives of species A–F) is more resistant to inhibition by mammalian IFNα or IFNβ than that of many other viruses (Gallagher & Khoobyarian, 1969; Gallagher & Khoobyarian, 1971; Uchio *et al.*, 2011). Such relative insensitivity may be in part a consequence of the genetic changes that accompanied tumor development or transformation of cells in culture, such as the presence and expression of the human papillomavirus type 16 E7 and the HAdV-C5 E1A proteins in HeLa and 293 cells, respectively; these viral proteins impair cytosolic sensing of DNA via the cGAS–STING pathway and activation of IFNβ synthesis (Lau *et al.* 2015). Indeed it was reported recently that pretreatment with IFNβ reduced HAdV-C5 reproduction by some three orders of magnitude in normal human bronchial epithelial cells, but was without effect in an adenocarcinoma cell line (Zheng *et al.*, 2016). In the normal cell, type I IFN was less inhibitory, reducing virus yields from 10- (Chahal *et al.*, 2012) to 100-fold (Zheng *et al.*, 2016), depending on the IFN concentration and perhaps differences in the normal cell populations. Under conditions of substantial inhibition, pre-treatment with IFNα or IFNγ was observed to inhibit expression of the E1A gene by inducing both impaired association of the transcriptional activator GABP with E1A enhancer 1 (see Fig. 5.1) and increased binding of RB or the related protein p107 to a conserved E2F-binding site in this enhancer (Zheng *et al.*, 2016).

There appear to be modest differences in IFN-sensitivity among HAdV species, a property exemplified by the greater sensitivity of HAdV-F40 and HAdV-F41 observed in a direct comparison with HAdV-C2 (Tiemessen & Kidd, 1993). The rate of infection by HAdV-F40 has also been reported to be reduced by IFNα pretreatment in a cell culture model of polarized gastro-epithelial cells that resemble natural host cells (Sherwood *et al.*, 2012). Pre-infection of human cells with HAdV-C2 overcame the inhibition of species F HAdV replication, suggesting that genes for antiviral proteins of species C HAdVs are not well conserved in the gastroenteric HAdV genomes (Tiemessen & Kidd, 1993). Several such HAdV-C5 gene products that interfere at various steps in the type I IFN-mediated antiviral defense have been identified.

The HAdV-C5 E1A proteins not only stimulate transcription of viral and numerous cellular genes to promote virus replication (Chapter 5), but also are important repressors of host cell antiviral responses. This protective function was first suggested by the observation that IFNβ-mediated inhibition of vesicular stomatitis virus reproduction was blocked by infection with HAdV-C5, but not by infection with an E1A-deleted mutant (Anderson and Fenni, 1987). Expression of the E1A gene alone is sufficient to repress expression of IFN-sensitive reporter genes in cells treated with IFNα/β or IFNγ (Ackrill *et al.*, 1991; Gutch & Reich, 1991; Kalvakolanu *et al.*, 1991; Leonard & Sen, 1996; Reich *et al.*, 1988), establishing a direct role for these viral gene products. In cells treated with IFN α/β or IFNγ, the E1A proteins blocked assembly of the crucial activators of transcription of ISGs, ISGF3, and a STAT1 homodimer termed γ-activating factor, respectively, and their translocation to the nucleus (Ackrill *et al.*, 1991; Gutch & Reich, 1991; Kalvakolanu *et al.*, 1991; Leonard & Sen, 1996). These defects have been ascribed to failure to phosphorylate STAT1 because of its reduced concentration in E1A producing cells (Ackrill *et al.*, 1991; Leonard & Sen, 1996). More recently, the E1A gene products were reported to be necessary to restrict phosphorylation of STAT1 (as well as STAT2) in HAdV-C5-infected normal human bronchial epithelial cells (Shi *et al.*, 2007), a response that correlated with decreased synthesis of the mRNA for the relevant kinase, JAK1 (Fig. 9.2).

In reporter assays, disruption of IFN signaling by E1A proteins required an N-terminal sequence common to the two major proteins (Gutch & Reich, 1991; Kalvakolanu *et al.*, 1991). This E1A sequence includes amino acid residues required for binding of E1A proteins to p300/CBP (Fig. 5.6), and E1A proteins were observed to impair IFN-dependent gene expression by competing with STAT1 and STAT2 for binding to their common p300/CBP co-transcriptional activator (Bhattacharya *et al.*, 1996; Gerritsen *et al.*, 1997; Zhang *et al.*, 1996). Interaction of E1A with both p300/CBP and RB has also been implicated in repression of ISG transcription in infected, quiescent human fibroblasts (Ferrari *et al.*, 2014). As altered E1A proteins that cannot bind p300/CBP, interact directly with STAT1 and impair IFNγ induced gene expression (Look *et al.*, 1998), it appears that these viral proteins exert both direct and indirect effects on STAT1 function.

An additional mechanism for inhibition of STAT1 has been described more recently, sequestration of phosphorylated STAT1 at viral replication centers dependent on viral genome replication (Sohn & Hearing, 2011). Furthermore, the ability of E1A gene products to block expression of ISGs upon infection of established lines of human cells has been correlated with binding to an N-terminal E1A sequence (amino acids 4–25 in HAdV-C5 E1A proteins) of hBRE1 (Fonseca *et al.*, 2012). This cellular protein is a component of a heterodimeric enzyme that catalyzes addition of mono-ubiquitin to a single lysine residue in histone H2B, a post-translational modification that is characteristic of very actively transcribed chromatin (Lee *et al.*, 2007; Xiao *et al.*, 2005). Interaction of hBRE1 with E1A proteins reduced this modification at the several ISGs examined, probably by competing for binding to the ubiquitin ligase UBE2B, an impaired induction of expression of these genes by IFNβ (Fonseca *et al.*, 2012).

The antiviral function of the HAdV-C5 E1B 55 kDa protein came to light in a comparison of alterations in cellular gene expression in normal human fibroblasts infected by wild type virus and an E1B 55 kDa-null mutant (Miller *et al.*, 2009). Expression of a set of over 300 genes is increased substantially in the absence of the E1B protein, and bioinformatics analyses revealed remarkable enrichment for genes associated with antiviral defense and ISGs. As predicted by these findings, virus yield is reduced by several orders of magnitude in IFNβ-treated normal human fibroblasts and bronchial epithelial cells infected by E1B mutant viruses, but reproduction of HAdV-C5 is little affected (Chahal *et al.*, 2012). The E1B 55 kDa protein has been demonstrated to block ISG expression at the level of transcription, a function that requires the previously described (Yew *et al.*, 1994) repression domain, but not formation of the virus-specific, E1B 55 kDa protein-containing E3 ubiquitin ligase (Chahal *et al.*, 2012; Chahal *et al.*, 2013). This protein has been reported to interact with various cellular proteins implicated in repression of gene expression, including DAXX and SIN3A (Punga & Akusjarvi, 2000; Zhao *et al.*, 2003), but the mechanism of repression of transcription of ISGs, and the basics of specificity for this set of molecules are not yet known. The primary defects were substantially reduced viral genome replication and failure to form viral replication centers (Chahal *et al.*, 2012). The latter process is also inhibited in IFN-treated cells infected by a mutant virus defective for synthesis of the viral E4 Orf3 protein (see below). However, relocalization of the PML protein with the E4 Orf3 protein was not perturbed in IFN-treated cells infected with an E1B 55 kDa protein-null mutant virus (Chahal *et al.*, 2012), indicating that the products of one or more ISGs other than PML can interfere with formation of viral replication centers and genome replication. The identity of such ISGs, among the ~130 E1B 55 kDa protein-repressed candidates, is not known.

As described previously (Chapter 6), the HAdV-C5 E4 Orf3 protein assembles into track-like structures in infected cell nuclei, to which it recruits the PML protein (among others). Expression of the genes encoding PML and other PML-NB-associated proteins, such as SP100 and DAXX, increases in cells exposed to IFNα, β, or γ, and PML-NBs increase in size and number (Regad *et al.*, 2001). Treatment of permissive cells infected with E4 Orf3-null mutant virus with IFNα or IFNγ was reported to impair viral genome replication and formation of viral replication centers (Ullman *et al.*, 2007). Such inhibition was rescued by RNAi-mediated knockdown of PML or DAXX, indicating that these proteins block viral DNA synthesis, presumably indirectly, in response to exogenous IFN (Ullman & Hearing, 2008). Expression of viral genes, including those for the viral E2 replication proteins, is little changed in IFN-treated, mutant virus-infected cells (Ullman *et al.*, 2007), and the molecular basis of inhibition of viral DNA synthesis by PML and DAXX has not been established. This function of the E4 Orf3 protein is conserved in species A, B, and C HAdVs (Ullman *et al.*, 2007).

The first antagonist of the IFN antiviral defense against adenovirus to be identified was VA RNA I, when reproduction of a VA RNA I-deleted mutant virus was shown to be reduced some 10-fold by pretreatment of cells with IFNα (Kitajewski *et al.*, 1986). This small RNA is required for maximally efficient synthesis of viral late proteins in untreated host cells, independent of defects in production of viral late mRNAs (Thimmappaya *et al.*, 1982), and this requirement is more pronounced following exposure to IFN. Detailed mutational molecular and biochemical studies established that, in the absence of VA RNA I, phosphorylation of elF2α is increased in infected cells. This modification results in high affinity association of elF2α with its guanine nucleotide exchange factor (elF2B) and inhibition of translation when sufficient elF2α is phosphorylated to trap all the elF2B (Davies *et al.*, 1989; O'Malley *et al.*, 1986; Reichel *et al.*, 1985; Schneider *et al.*, 1985; Siekierka *et al.*, 1985). Subsequently, it was established that VA RNA I targets the dsRNA-activated protein kinase PKR, which was known to phosphorylate the critical serine residue in elF2α (Katze *et al.*, 1986; Kitajewski *et al.*, 1986; O'Malley *et al.*, 1986). Activation of PKR depends on binding of at least two molecules to the same dsRNA and autophosphorylation of specific threonine residues near the active site of the enzyme. Mutational and biochemical experiments indicated that VA RNA I prevents association, and hence activation, of PKR and that this function requires the apical stem of the viral RNA (Fig. 5.4), which is bound by the tandem dsRNA-binding domains of PKR, and the central stem (Clarke & Mathews, 1995; Ghadge *et al.*, 1994; Kitajewski *et al.*, 1986; McKenna *et al.*, 2007; Spanggord & Beal, 2001; Spanggord *et al.*, 2002). The results of analysis of the interactions of PKR with VA RNA and of structural studies of the RNA alone and associated with the RNA-binding domain of the protein by small angle X-ray scattering are consistent with a model in which VA RNA I is perfectly tailored to bind to a single molecule of PKR (Dzananovic *et al.*, 2014; Launer-Felty *et al.*, 2010; Launer-Felty *et al.*, 2015; Wahid *et al.*, 2008; Wilson *et al.*, 2014). Although the interaction of PKR with the apical stem of VA RNA-I is of a relatively low affinity (Launer-Felty *et al.*, 2015), the very high concentration of the RNA that accumulate during the late phase of infection (see Chapter 5) would allow its effective competition with dsRNA for binding to PKR.

As noted in Chapter 4, all mastadenovirus genomes carry at least one VA RNA gene, of the VA-RNA-I type, but there have been few studies of the abilities of these small RNAs of distinct HAdVs either to induce the IFN response or to prevent activation of PKR. However, it has been suggested recently that the slow reproduction of HAdV-A12 in cells in culture and low cytotoxicity may be related to reduced production (or perhaps altered function) of VA RNA I; cells infected by HAdV-A12 were observed to exhibit increased production of IFNβ compared to HAdV-C2-infected cells, accompanied by activation of PKR and elF2α

phosphorylation throughout the infectious cycle, and knockdown of PKR or synthesis of HAdV-C5 VA RNA stimulated HAdV-A12 reproduction and production of the late protein hexon (Wu *et al.*, 2015).

A second ISG gene product that appears to be modulated by VA RNA I is 2'-5'-oligoadenylate synthetase (OAS), which like PKR, is activated by binding to dsRNA (see X.L. Li *et al.*, 2011); the viral RNA binds to OAS, but rather than inhibiting, activates the enzyme (Desai *et al.*, 1995). The results of mutational analyses established that OAS and PKR bind to VA RNA I via different structural elements (Desai *et al.*, 1995). Such activation of an enzyme long thought to be anti-viral as a result of synthesis of 2'-5'-oligoA and consequent activation of RNAse L and other mechanisms (Chakrabarti *et al.*, 2011; Kristiansen *et al.*, 2011) may seem paradoxical. However, the effect, if any, of OAS activation on HAdV reproduction has not been investigated, and the results of recent experiments suggest that the "byproduct" of Dicer processing of VA RNA I which comprises the apical stem and central domain (see Fig. 5.4), can inhibit rather than activate OAS, as it does PKR (Vachon *et al.*, 2015).

9.3.2.4.3 The interferon and cytokine response *in vivo*

Numerous studies of innate immune responses to HAdV vectors administered to mice by different routes have reported increased serum concentrations of IFNs, other cytokines (e.g. IL-6 and TNFα) and chemokines and increased concentrations of the corresponding mRNAs in various cell types, with the details of the response varying with the genetic background of the mice, the inoculation protocol, and the cell types studied (Aldhamen *et al.*, 2011; Hartman *et al.*, 2008). Intraperitoneal injection of mice with wild-type species B and C HAdVs also induces the rapid appearance of IFN α/β in plasma, with concentrations peaking 8 hrs after virus inoculation (Fejer *et al.*, 2008). In this system, dendritic cells in the spleen were primarily responsible for production of the IFNs following HAdV-B3 infection, a response that depended on endosomal escape of virus particles but not TLR9 (Fejer *et al.*, 2008). As murine cells do not support HAdV replication, it is not possible to evaluate whether and how synthesis of IFN α/β in infected mice limits virus reproduction and spread. However, these cytokines have been reported to govern various immune responses to HAdV infection that can be assessed in these animals. For example, infection with a wild type HAdV-C5 with some E3 sequences from the very closely related HAdV-C2 greatly increased sensitivity to induction of TNFα by lipopolysaccharide (Fejer *et al.*, 2005), a pro-inflammatory response that was not observed in mutant mice lacking the receptor for IFN α/β (Fejer *et al.*, 2008). Furthermore, production of IFNs (and likely other cytokines) in mice infected by HAdV-C5 vectors that lack genes for the viral proteins that block the IFN response is critical for protective responses mediated by NK, B, and T cells (e.g. Xu *et al.*, 2004; Zhu *et al.*, 2007a; Zhu *et al.*, 2008).

More recently, the impact of the IFNα/β response on HAdV-C5 reproduction and pathogenicity was examined in Syrian hamsters, a small experimental animal model that is quite permissive for HAdVs (Chapter 8). These studies exploited a STAT2 knock out hamster line, in which activation of transcription of ISGs (Fig. 9.3) in response to infection with HAdV-C5 was not detected (Toth *et al.*, 2015). In these animals, intravenous inoculation of HAdV-C5 results in much more severe pathology, as indicated by more rapid death, increased inflammatory responses in the liver and 2–3 orders of magnitude increases in viral load in that organ (Toth *et al.*, 2015). Neither the production of anti-HAdV-C5 antibodies nor the clearance of the infection in surviving animals was impaired in the mutant hamsters (Toth *et al.*, 2015). These observations emphasize the major contribution of innate immune responses to limiting reproduction and pathogenesis of at least species C HAdVs *in vivo*. Furthermore, in a phase I human clinical trial, systemic administration of an E1-, E3-, and E4-deleted HAdV-C5 vector at 5×10^{12} to 3.8×10^{13} total particles elicited substantial IL-6 and IL-10 production, which was accompanied by a fatal pro-inflammatory response in a subject with an underlying monogenetic deficiency (Raper *et al.*, 2003).

9.3.2.5 *Red blood cells*

Under-appreciated regulators of HAdV-C5 dissemination *in vivo* are human red blood cells (RBCs), which display multiple adenovirus receptors on their surfaces including the Coxsackie and adenovirus receptor (CAR) (Carlisle *et al.*, 2009; Lyons *et al.*, 2006; Seiradake *et al.*, 2009). Consequently, HAdV-C5 is capable of interacting with the CAR, and other HAdVs such as HAdV-D37 with sialic acid, on the plasma membrane of human but not murine RBCs, an association that prolongs the half-life of virus particles when they are administered systemically. Moreover, HAdV–RBC interactions restrict virus transport to the liver or other target organs. The high number of RBCs in the circulation thus present a significant impediment for gene transfer to specific target tissues and may account for the high doses of vector that are needed to effectively reach the desired host cell *in vivo*.

9.3.2.6 *Natural killer cells*

NK cells, which represent a substantial class (>5%) of circulating lymphocytes, can recognize virus-infected cells independently of virus-specific epitopes, and destroy them (Biron *et al.*, 1999; Degli-Esposti & Smyth, 2005; Lanier, 2008). Once activated by interaction with an infected cell, NK cells synthesize and release cytokines, for example TNFα and IFNγ, which promote local inflammatory responses, signal to the components of the adaptive immune system, and stimulate production of antibodies. They also produce cytotoxic molecules like granzymes and perforin, which are introduced into target infected cells (Topham & Hewitt, 2009). A variety of observations indicate that NK cells respond to and help contain HAdV infection *in vivo*.

In adenovirus-infected infants aged 1–70 days, the severity of clinical manifestations of infection was observed to correlate inversely with NK cell number (Mistchenko *et al.*, 1998), implicating these cells in limiting virus reproduction and spread in natural infections. Consistent with this conclusion, infection of primary human fibroblasts or peripheral blood mononuclear cells with wild-type HAdV-C5, -E1, -E3-deleted HAdV-C5 vectors or HAdV-B35 activated lysis by NK cells in culture (Pahl *et al.*, 2012; Tomasec *et al.*, 2007). HAdV-C5 vectors have also been observed to induce sustained activation of NK cells in mice (Ruzek *et al.*, 2002; Zhu *et al.*, 2008). Furthermore, depletion of NK cells led to accumulation of large quantities of viral DNA in the liver of such infected animals, concomitant with increased transgene expression, liver damage and hepatocyte apoptosis (Z.X. Liu *et al.*, 2000; Peng *et al.*, 2001; Zhu *et al.*, 2008).

In mice lacking both copies of the gene for the common receptor for IFNα and IFNβ (IFNAR), NK cells were not activated following HAdV-C5 vector infection, as indicated by the failure to synthesize characteristic proteins like IFNγ (Zhu *et al.*, 2008). As adoptive transfer of wild-type NK cells into the mutant mice restored the ability to reduce viral DNA accumulation and transgene expression in the liver, it was concluded that IFN α/β produced in response to infection acts directly on NK cells to activate them. More recent studies indicate that the efficacy of such IFN α/β activation of NK cells varies with HAdV species, because of differences in the efficiency of infection and hence production of IFN α/β by infected cells (Johnson *et al.*, 2014; Pahl *et al.*, 2012).

Activation of NK cells in response to HAdV infection of cells in culture or in inoculated mice depends on the NK cell receptor NKG2D, as masking the receptor with anti-receptor antibodies blocks the NK cell response (Tomasec *et al.*, 2007; Zhu *et al.*, 2010). Synthesis of ligands recognized by this NK cell receptor has been reported to increase in infected cells *in vivo* (Zhu *et al.*, 2010) and in tumor cells that synthesize HAdV-C5 E1A proteins and that are sensitive to NK cell killing in culture (Routes *et al.*, 2005). Conversely, tumorigenic rodent cells transformed by HAdV-A12 escape NK cell recognition as a result of reduced cell surface accumulation of NKG2D ligands (Heyward *et al.*, 2012). These observations illustrate the importance of modulation of NK receptor ligands for activation of NK cells in response to HAdV infection, but whether the synthesis or cell surface accumulation of these proteins is altered *in vivo*, and whether viral gene products or type I IFNs are responsible, are not known. However, it may be relevant that the expression of genes that encode NKG2D ligands such as RAE1 and MICB (major histocompatibility complex, MHC class I chain-related protein B) is increased 4–8-fold in human cells treated with type 1 IFN (Rusinova *et al.*, 2013).

Adenoviral gene products that might counter the NK-cell-mediated antiviral defense have received relatively little attention. As noted above, E1A proteins of the oncogenic virus HAdV-A12 can lead to decreased cell surface concentrations of NKG2D ligands, because of reduced expression of the genes that encode them (Heyward *et al.*, 2012). Such downregulation was characterized in HAdV-A12 transformed rodent cells, so it is not known whether this phenomenon is also induced in species A HAdV-infected permissive human cells. In fact, the only viral gene products shown to impair recognition of HAdV infected cells by NK cells are E3 proteins. The E3 gp19K protein has been known for some time to interfere with antigen presentation by sequestration of MHC class I proteins in the endoplasmic reticulum (ER) (see next section). More recently, increased synthesis of the NKG2D ligands MICB and MICA was reported to occur in HAdV-C5-infected cells, but the proteins were retained in the ER by E3 gp19K (McSharry *et al.* 2008). This observation suggests that typical HAdV vectors, all of which lack the E3 gene, may be more susceptible to the anti-viral actions of NK cells than the wild-type, but this inference has not been tested directly.

An E3 protein unique to species D HAdVs, E3/49K, (Blusch *et al.*, 2002; Windheim & Burgert, 2002) has been observed to suppress activation and functioning of not only NK cells, but also T cells (Windheim *et al.*, 2013). Like E3 gp19K, the 49-kDa protein enters the ER and is glycosylated, but subsequently is cleaved and the large luminal domain secreted (Windheim & Burgert, 2002; Windheim *et al.*, 2013). This secreted form binds to all isoforms of the cell surface protein tyrosine phosphatase CD45, an antigen common to all lymphocytes (Windheim *et al.*, 2013), but how it blocks activation of NK and T cells has not yet been explored.

9.4 The Antibody Response

9.4.1 *Neutralizing antibodies*

Antibodies are one of the major components of the host adaptive immune response that play a significant role in controlling adenovirus dissemination. Some of the earliest studies of adenovirus neutralization focused on the properties of various animal antisera directed against the fiber, penton base or hexon. Most of these early studies showed that the hexon protein was the major determinant for inactivation by antibody (Kjellen & Pereira, 1968). Further investigations in mouse models reported that the development of anti-hexon antibodies protected mice from subsequent infection with intact virus (Mautner & Willcox, 1974). Moreover, there was some cross protection observed in that vaccination with type 2 hexon afforded some defense against HAdV-C5 infection and vice versa. These studies in mice also showed that prior vaccination with the fiber protein could protect against HAdV-C5 infection, although the fiber did not elicit as potent an antibody response as the hexon. Anti-penton base neutralizing antibodies have also been found in sera from patients previously vaccinated with an oncolytic HAdV-C2 vector (Hong *et al.*, 2003). However, this same study reported that circulating anti-RGD antibodies were not very prevalent and were poorly immunizing. Other reports indicated that the neutralizing capacity of anti-penton base antibodies can be bolstered by the presence of anti-fiber antibodies, suggesting a synergistic effect of these distinct anti-body types (Gahery-Segard *et al.*, 1998). Anti-hexon, and to a much lesser extent, anti-fiber or anti-penton antisera obtained from rabbits mediated type-specific virus neutralization (Wadell, 1972). Fab or F(ab)'2 fragments of anti-hexon or anti-fiber antibodies did not possess neutralizing activity even though they were capable of binding to the virus, suggesting that the Fc portion of the antibody molecule plays a crucial role in neutralization. In subsequent analyses of the kinetics and stoichiometry of anti-hexon neu-tralization, a single-hit mechanism revealed that 1.4 antibody molecules bound to the virus particle was sufficient for neutralization (Wohlfart, 1988). This author showed that anti-hexon antibody prevented conforma-tional changes in the virus hexon as determined by susceptibility to protease digestion at low pH, but this biochemical alteration was not directly linked to an exact neutralization mechanism. One possibility is that antibody attachment to the hexon disrupts subsequent dynein motor interactions with the hexon trimer thereby hindering trafficking to the nucleus (Scherer & Vallee, 2015). However, other neutralization mechanisms, such as intracel-lular destruction via association with TRIM21, discussed previously, could also contribute.

A comparison of the amino acid sequences of the hexon protein from human species A, B, C, D, F as well as from several non-HAdVs, identi-fied seven distinct hypervariable region (HVR) loops displayed on the outer region of this capsid protein (Crawford-Miksza & Schnurr, 1996) and these HVR loops are recognized by type specific, human neutralizing antisera (Bradley *et al.*, 2012; Crawford-Miksza & Schnurr, 1996; Tian *et al.*, 2015). In a detailed study of HAdV-C5 neutralization, an anti-hexon monoclonal antibody, designated 9C12, was shown to form a meshwork-like coating of the outer surface of the virus particle (Varghese *et al.*, 2004). This antibody did not prevent cell attachment or internalization of the virus, but rather restricted a later step in cell entry. In a follow up investigation, it was discovered that the 9C12 antibody blocks microtubule-dependent transport of the virus to the nucleus following endosomal escape (J.G. Smith *et al.*, 2008). Prevention of microtubule dependent transport could be a result of the interference by the antibody with the association of hexon with the dynein motor. In support of this possibility, 240 antibodies per virus particle, or one 9C12 antibody per hexon trimer, are needed to neutralize the virus (Varghese *et al.*, 2004).

9.4.2 *Pre-existing antibodies restrict HAdV vector delivery*

The presence of anti-capsid antibodies in human subjects who have been naturally infected with virus has important consequences for adenovirus-mediated gene transfer. This property is particularly important for recombinant HAdV-C5-based vectors, as the majority of people worldwide have circulating HAdV-C5 antibodies (Sumida *et al.*, 2005). Most human anti-HAdV-C5 neutralizing antibodies are directed against the hexon, while a smaller proportion recognizes the fiber protein (Bradley *et al.*, 2012). In an effort to develop vector vaccines that are not restricted by preexisting anti-HAdV-C5 antibodies, different HAdVs, such as types B11, B35, B50, D26, D48, and D49, have been engineered based on their low seroprevalence throughout the world (Abbink *et al.*, 2007; Barouch *et al.*, 2011). While these new vectors (especially HAdV-D26) appear promising in terms of avoiding preexisting antibody (Chapters 11 and 12), the generation of novel vectors is time consuming and extensive safety testing is required. As an alternative approach, the well-established HAdV-C5 vector has been modified by replacing the immunogenic hexon protein of HAdV-C5 with the hexon from a less immunogenic virus type such as HAdV-C6 (Youil *et al.*, 2002) or HAdV-D48 (Bradley *et al.*, 2012). Replacing only the hexon HVRs has also been explored, but this strategy requires removing each of the seven HVRs of HAdV-C5 and replacing them with the corresponding HVRs of less immunogenic HAdV types, because each HVR is a potential antigenic site for neutralizing antibodies (Bradley *et al.*, 2012).

So-called non-specific antibodies can also limit the efficacy of HAdV vector delivery. In certain strains of mice, natural IgM molecules recognize the virus and contribute to its removal via Kupffer cells of the liver (Qiu *et al.*, 2015). Whether this antiviral defense system is operative in humans remains to be determined but natural antibodies could be a hindrance to efficient systemic delivery of HAdV vectors.

9.4.3 *IgG–Fc receptor interactions*

Not only do anti-capsid antibodies have the capacity to neutralize HAdVs by themselves, but they can also combine with or assist other immune molecules in the destruction of the virus. Systemic delivery of HAdV-C5 vectors in mice revealed that a sizable portion of the input virus (~25%) can be taken up by migratory neutrophils via Fc or complement type 1 receptors, thereby reducing the efficiency of vector transduction of epithelial cells (Cotter *et al.*, 2005). However, the precise contributions of Fc receptors that facilitate the removal of adenovirus *in vivo* remains poorly defined. In contrast, there is ample evidence for IgG-Fc-receptor-mediated enhancement of the uptake of HAdV-antibody complexes in cells in culture (Ebbinghaus *et al.*, 2001; Hong *et al.*, 2003; Korokhov *et al.*, 2003; Leopold *et al.*, 2006). The majority of these studies have been performed using CAR-deficient cells, and show that Fc receptors promote gene delivery (Leopold *et al.*, 2006) via complexes of HAdV vectors containing a recombinant CAR–Ig(Fc) protein (Ebbinghaus *et al.*, 2001) or other Ig(Fc) ligands (Hong *et al.*, 2003; Korokhov *et al.*, 2003). Whether similar Fc-mediated enhancement of HAdV vector transduction also occurs *in vivo* has not been well established. The functions of neutralizing antibodies in cell culture studies do not always correlate with *in vivo* neutralization, suggesting that some antibodies could actually enhance infection rather than reduce host cell invasion (Pichla-Gollon *et al.*, 2009).

9.5 T Cell Responses

Efforts to develop adenoviruses of humans and other mammals for delivery of vaccine components and the rising prevalence of serious diseases associated with HAdV infections have spurred interest in T cell responses to these viruses. Immunosuppressed patients, particularly children who receive hematopoietic stem cell transplants, are at considerable risk for development of disseminated HAdV disease associated with significant mortality (Chapter 8). In such patients, host T cells are ablated prior to treatment and immunosuppressive drugs may be administered subsequently to prevent or treat graft versus host disease. In several studies, the risk of HAdV-associated disease and death has been observed to be an inverse correlate of the recovery of T cell activity (Chakrabarti *et al*,. 2002; Feuchtinger *et al.*, 2005; Heemskerk *et al.*, 2005; Myers *et al.*, 2007; van Tol *et al.*, 2005). Studies in the past 10–15 years have established that HAdV infections lead to both CD4$^+$ helper and CD8$^+$ cytolytic T cell responses, and furthermore, that such cells can be harnessed for treatment of HAdV disease in susceptible patients (see below and Chapter 10).

9.5.1 *The repertoire of T cell responses to HAdV antigens*

Initial studies in healthy human donors established that the majority, from about two-thirds to essentially all, carried HAdV-reactive CD4$^+$ T cells, and, based on the constellation of cytokines secreted, mounted a TH1-type response (Wold & Horwitz, 2006; Wold & Ison, 2013). The ability of individual viral proteins or peptides derived from them to stimulate such T cells in populations of peripheral blood mononuclear cells has been used to identify epitopes recognized by these cells. As is the case for neutralizing antibodies against HAdVs (see previous section), certain hexon epitopes are immunodominant, recognized by the CD4$^+$ T cells in 50–70% of subjects (Heemskerk *et al.*, 2006; Serangeli *et al.*, 2010). These epitopes are MHC class II restricted and the great majority are derived from sequences of the hexon conserved among different HAdVs (Heemskerk *et al.*, 2006; Olive *et al.*, 2002; Onion *et al.*, 2007; Tang *et al.*, 2006; Veltrop-Duits *et al.*, 2006), suggesting that T cells specific for these epitopes could recognize, and protect against, multiple HAdVs. Indeed, CD4$^+$ T cell clones generated against HAdV-C5 in culture recognized cells infected by, and inhibited reproduction of, not only HAdV-C5 and HAdV-C2, but also of species A, B, and D HAdVs (Heemskerk *et al.*, 2006; Veltrop-Duits *et al.*, 2006). Similarly, HAdV-specific CD4$^+$ T cells from all of 17 human subjects tested also recognized two chimpanzee adenoviruses (Hutnick *et al.*, 2010). Conserved hexon sequences also function as epitopes recognized by HAdV-specific and MHC class I-restricted CD8⁺ T cells, but such epitopes have also been identified in other structural proteins (penton and fiber), and the viral DNA polymerase (Joshi *et al.*, 2009; Leen *et al.*, 2004; Leen *et al.*, 2008; Tang *et al.*, 2006). Immunoprecipitation of specific MHC class I proteins from lysates of cells infected with a clinical strain of HAdV-C2 and sequencing of the eluted peptides by mass spectrometry methods established that previously described immunodominant hexon epitopes are naturally presented, and identified novel epitopes from the 13.6-kDa protein encoded in the alternatively spliced i-leader of major-late transcripts and from protein VIII (Gunther *et al.*, 2015). The prevalence of T cells specific for conserved epitopes of HAdV has important implications, both inauspicious and beneficial, for therapeutic applications (see below, and Chapters 11 and 12).

The phenotypic and functional properties of human T cells specific for HAdV epitopes have been investigated to only a limited degree. Such CD4⁺ T cells were reported to be largely mono-functional, producing IFNγ, IL-2, and TNFα (TH1 cells) and cell surface markers characteristic of central memory and effector memory cells in response to stimulation (Hutnick *et al.*, 2010; Makedonas *et al.*, 2010). In contrast, HAdV-specific CD8⁺ T cells have been observed to be more polyfunctional, and mount a strong effector response (production of perforin) when stimulated (Chakupurakal *et al.*, 2010; Hutnick *et al.*, 2010; Makedonas *et al.*, 2010). These T cell phenotypes were maintained in individuals following vaccination with an E1-deleted HAdV-C5 vaccine vector for HIV-1 proteins as the fraction of HAdV-C5-specific CD8⁺ cells present in peripheral blood samples expanded (Hutnick *et al.*, 2010). However, the phenotypes of adenovirus-specific T cells present in different organs and tissues may vary; in naturally-infected humans, CD8⁺ T cells specific for a conserved hexon epitope predominated in peripheral blood and killed HAdV-C5-infected fibroblasts only when treated with IFNγ to stimulate antigen presentation by MHC class I proteins. In contrast, viral DNA-polymerase-specific T cells were more prevalent in tonsils and exhibited a greater proliferative potential and IFNγ-independent killing of HAdV-C5-infected cells (Joshi *et al.*, 2011). At present, little is known about the causes or consequences of such differences in natural populations of HAdV-specific T cells.

9.5.2 *Viral proteins that thwart T cell responses*

MHC class I proteins, in which T cell peptide epitopes (average length: nine amino acids) are presented to T cell receptors, decorate the surfaces of most mammalian cells, and acquire peptides derived from intracellular proteins, including those of viruses, in the ER prior to transport to the cell surface. Viral antigens displayed in this way on the surfaces of infected cells are recognized as non-self by CD8[+] cytotoxic T lymphocytes (CTLs), which, when appropriately engaged via immunological synapses, kill the target cells by induction of apoptosis or the introduction of cytoplasmic granules carrying proteases (granzymes) and the membrane-disrupting protein perforin (Cemerski & Shaw, 2006; Stinchcombe & Griffiths, 2003). The genomes of a variety of viruses encode proteins that blunt the CTL response, a precedent set with the HAdV-E3 19-kDa glycoprotein (E3 gp19K).

The E3 gp19K of HAdV-C2 or -5, a major protein made during the early phase of infection, is a transmembrane glycoprotein with an N-terminal signal sequence that is removed upon entry into the ER and a relatively large luminal domain and a small cytoplasmic tail (Wold *et al.*, 1985). Cells infected by wild-type HAdV-C2 or HAdV-C5 are markedly less sensitive to killing by CTLs in culture than cells infected by mutant viruses that cannot direct synthesis of E3 gp19K (Burgert & Kvist, 1985; Burgert & Kvist, 1987). This protective action of the viral protein was traced to its non-covalent binding to particular domains of MHC class I proteins (Burgert & Kvist, 1987; Cox *et al.*, 1991; Feuerbach *et al.*, 1994; Flomenberg *et al.*, 1987; Kampe *et al.*, 1983; Paabo *et al.*, 1986). The E3 19-kDa glycoprotein resides in the ER, a location determined by an ER retrieval signal in the cytoplasmic domain and a transmembrane ER retention signal (Fu & Bouvier, 2011; Paabo *et al.*, 1986; Sester *et al.*, 2010; Windheim *et al.*, 2004). Consequently, its interaction with MHC class I proteins prevents exit of these cellular proteins (and associated antigens) from the ER and transport to the cell surface, a model confirmed by the observation that the signals needed for ER retention of E3 gp19K are required for intracellular sequestration of MHC class I proteins (Sester *et al.*, 2013). There has been one report that E3 gp19K also binds to a protein that transports antigenic peptide into the ER (TAP) and hence delays maturation and cell surface accumulation of MHC class 1 proteins

(Bennett *et al.*, 1999). Consistent with these properties of the viral protein, inclusion of its coding sequence in an HAdV-C5 vector reduced lysis of infected cells by T cells raised against the E3 region-lacking vector (Lee *et al.*, 1995), and increased vector persistence and transgene expression in murine lung (Bruder *et al.*, 1997). Conversely, mutant viruses that cannot direct synthesis of E3 gp19K replicated as well as wild-type virus in the lungs of cotton rats, but were much more pathogenic.

Mutational analyses indicated that two disulfide bonds and several other residues in the luminal domain of the HAdV-C5 E3 gp19K are required for binding to MHC class I proteins (Fu *et al.*, 2011; Liu & Marmorstein, 2007; Sester & Burgert, 1994; Sester *et al.*, 2010). The X-ray crystal structure of HAdV-C2 E3 gp19K made in insoluble form in *E. coli* and refolded in the presence of a native MHC class I protein bound to a peptide antigen (L. Li *et al.*, 2012) has provided important insights into the specificity of the interaction. The luminal domain of the viral protein forms an elongated V-shaped wedge comprising two anti-parallel β-sheets and a hydrophobic core at the base of the V wedge. E3 gp19K makes contact with two MHC class I chain domains (Fig 9.4A), in fact with those predicted from mutational analyses, far from the peptide-binding groove, and also with the MHC class constant chain, β2-microglobulin (L. Li *et al.*, 2012). This mode of interaction is consistent with the observation that E3 gp19K binds to MHC class I protein regardless of whether the peptide-binding groove is occupied (H. Liu *et al.*, 2005), and renders sequestration of these cellular proteins in the ER independent of the nature of the antigens they carry.

Although their sequences are not highly homologous, binding to MHC class I proteins is a common property of the E3 gp19K proteins of species B and E HAdVs (Burgert & Blusch, 2000; Burgert *et al.*, 2002; Deryckere & Burgert, 1996; Flomenberg *et al.*, 1987; Flomenberg *et al.*, 1992; Hermiston *et al.*, 1993). Mapping of the strictly or high conserved residues of the E3 proteins onto the structure of the HAdV-C2 protein revealed that several were present in the hydrophobic core described above, indicating that they make important contributions to maintaining protein stability (L. Li *et al.*, 2012). Other conserved residues maintained hydrophilic or hydrophobic interactions with the MHC class I protein (Fig 9.4B), accounting for the conservation of this function. The results

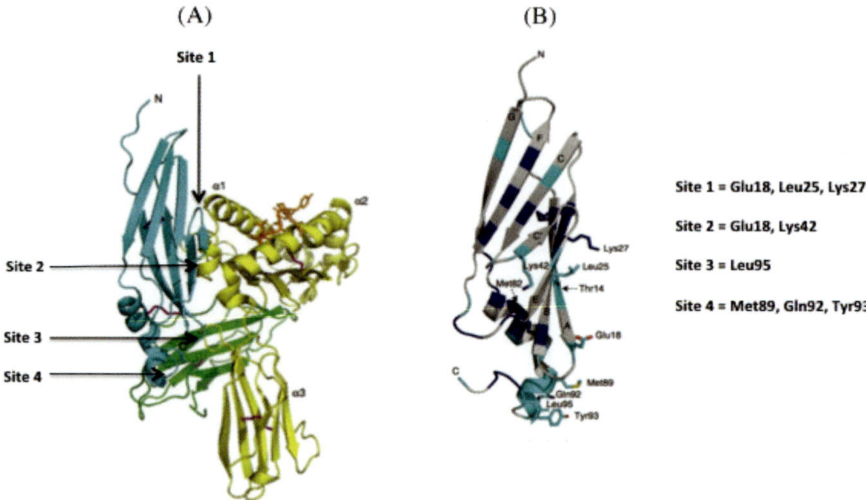

Figure 9.4. Crystal structure of the HAdV-C2 E3 gp19K bound to MHC class I. (**A**) The structure of the viral protein synthesized in E. coli and solubilized and refolded in the presence of the MHCI protein HLA-A2 bound to a peptide from the HTLV1 Tax protein was determined at 1.95-Å resolution by X-ray crystallography. In the ribbon form shown, E3 gp19K is in cyan, the HLA-A2 heavy chain in yellow and the β2-microblobulin invariant chain in green. The Tax protein peptide is shown in orange and disulfide bonds in magenta. (**B**) Amino acids strictly and highly conserved among the E3 proteins of species A–D HAdVs are shown in blue and cyan, respectively, on the HAdV-C2 protein structure. The side chains of these residues in the MHC-interaction sites, which are listed at the right, are shown with N, O, and S atoms indicated in blue, red, and tan, respectively. Adapted from L. Li *et al.* (2012), with permission, images courtesy of M. Bouvier, Univ. of Illinois at Chicago.

of *in vitro* studies indicated that the viral glycoprotein does not bind to the MHC class proteins encoded by all alleles, but exhibits higher affinity for HLA-A than HLA-B proteins and very low or no interaction with HLA-Cw[x] or HLA-E proteins (Fu *et al.*, 2011; L. Li *et al.* 2012). These preferences can be attributed to the contribution of polymorphic amino acids to the interaction with E3 gp19K (L. Li *et al.*, 2012). As discussed previously, this viral protein also sequesters the MHC class I-related proteins and NK receptor ligands MICA and MICB in the ER, but both mutational analyses and structural information indicated that distinct surface of E3 gp19K binds to the two classes of cellular proteins (L. Li *et al.*, 2012; Sester *et al.*, 2010).

An alternative mechanism of downregulation of MHC class I proteins was discovered and elucidated following the observations that rodent cells transformed by oncogenic HAdV-A12 were more tumorigenic *in vivo*, and more resistant by CTL killing in culture, than cells transformed by non-oncogenic species C HAdVs (Gallimore, 1972; Gallimore & Paraskeva, 1980). The observations that HAdV-12 transformed hamster cells induced tumors only before thymic responses develop, or following removal of the thymus, implicated T cells in tumor rejection (Cook *et al.*, 1979a; Cook *et al.*, 1979b). The ability of transformed cells to form tumors in animals correlated with the origin of the viral E1A gene (Bernards *et al.* 1982; Sawada *et al.*, 1985). Expression of the HAdV-A12 E1A gene prevented transformed cell lysis by allogenic CTLs (Bernards *et al.*, 1982; Eager *et al.*, 1985; Yewdell *et al.*, 1988) and led to reduced concentrations of MHC class I proteins on the surface of transformed cells, whereas the HAdV-5 E1A proteins did not exhibit these activities (Ackrill & Blair, 1988a; Eager *et al.*, 1985). The HAdV-12 E1A gene products were shown to repress transcription of MHC class I genes (Ackrill & Blair, 1988a; Ackrill & Blair, 1988b; Vasavada *et al.*, 1986), in part by inducing hypophosphorylation of a subunit of the transcriptional activator NF-κB, thereby preventing its binding to an MHC enhancer (Guan *et al.*, 2008; Jiao *et al.*, 2010; Kushner & Ricciardi, 1999; Kushner *et al.*, 1996; Nielsch *et al.*, 1991). The HAdV-12 E1A protein also promotes binding to a second enhancer of a repressor comprising the nuclear receptor family member COUP-TFII associated with histone deacetylases (Kralli *et al.*, 1992; Liu *et al.*, 1994; Smirnov *et al.*, 2000; Smirnov *et al.*, 2001; Zhao & Ricciardi, 2006; Zhao *et al.*, 2003). Furthermore, production of other components of the antigen presentation machinery, including the peptide transporters TAP1 and TAP2 and specific subunits of the proteasome associated with the generation of peptide epitopes, is inhibited in HAdV-A12-transformed cells (Proffitt & Blair, 1997; Rotem-Yehudar *et al.*, 1994; Rotem-Yehudar *et al.*, 1996; Vertegaal *et al.*, 2003). The ability of the E1A proteins, particularly the largest, to bind to STAT1 and inhibit STAT1-dependent transcription described previously has been implicated in blocking expression of genes encoding these proteasome subunits (Berhane *et al.*, 2011; Chatterjee-Kishore *et al.*, 2000). These mechanisms presumably ensure that the presentation of antigens on the surfaces of

HAdV-A12-transformed cells is diminished, allowing the cells to escape detection and lysis by CTLs, a property not shared with cells transformed by species C HAdVs (see Williams *et al.*, 2004). Whether they also blunt T cell response to HAdV-A12 infection *in vivo* has not been investigated.

9.5.3 *Adoptive T cell therapy*

The need for T cell responses to clear HAdV infection in hematopoietic stem cell transplant patients coupled with the limited armamentarium of anti-adenoviral drugs (Chapter 10) has provided a strong impetus for development of adoptive T cell transfer as a therapeutic option. Consequently, as described above, multiple adenoviral epitopes recognized by CD4$^+$ and CD8$^+$ T cells present in humans have been characterized as have, in many cases, the particular MHC alleles in which they are presented (Gunther *et al.*, 2015; Joshi *et al.*, 2011; Olive *et al.*, 2002; Serangeli *et al.*, 2010). In the context of adoptive T cells therapy, the broad specificity of T cells that recognize hexon epitopes offers considerable advantage, in principle at least, permitting a single preparation of HAdV-specific T cells to offer protection against and clear HAdVs of multiple species.

Protocols have been developed for isolation of HAdV-C5-specific, IFNγ-producing T cells (mixtures of CD4$^+$ and CD8$^+$) under GMP conditions (Aissi-Rothe *et al.*, 2010; Feuchtinger *et al.*, 2006; Feuchtinger *et al.*, 2008). In culture, such cells exhibit high toxicity against autologous cells loaded with adenoviral antigens, but lower activity against allogeneic target cells (Aissi-Rothe *et al.*, 2010; Feuchtinger *et al.*, 2006), and strong cross-reactivity with antigens of other HAdVs (Feuchtinger *et al.*, 2008). Initial studies indicated that adoptive transfer of such T cell preparations into HSCT patients is generally safe and effective in clearing HAdV infection (Feuchtinger *et al.*, 2006; Leen *et al.*, 2009). A more recent larger clinical trial in which hexon-specific CD4$^+$ TH1 cells were administered to 30 patients refractory to treatment with anti-antiviral drugs, reported no increase in graft versus host disease, T cell expansion in 61% of evaluable patients (14/23), and complete clearance of viremia in 67% of the patients responding to the adoptive T cell transfer (Feucht *et al.*, 2015). Furthermore, while all patients who failed to respond to this treatment died (mean 24 days after T cell transfer), 71% of the responders survived. In view of the difficulties in treating HAdV infections in these patients (Chapter 10), these are encouraging results, particularly because it has been calculated that a single isolation and infusion of the virus-specific T cells costs no more than 4.9 doses of the commonly administered drug cidofovir (Feucht *et al.*, 2015).

9.6 What Next?

Antibodies that recognize and neutralize HAdVs were found in human sera very soon after the discovery of these viruses, and exploited in early studies, for example, to aid classification. However, the breadth and depth of the versatile immune armamentarium against HAdVs had been established only more recently, in parallel with the development of HAdV vectors. Investigation of these responses has yielded much fundamental information about the host intrinsic, innate, and adaptive defenses against HAdV infection, including unanticipated mechanisms that restrict infection *in vivo*, and viral genes products that block or minimize these responses. Such knowledge has also guided strategies to develop both HAdV vectors for such therapeutic applications as gene therapy and vaccine delivery, and adoptive T cell therapies for transplant recipients. Nevertheless, important questions remain.

A number of viral proteins are now known to counter the action of antiviral proteins encoded by ISGs, but the identities of these cellular proteins, and consequently how they can disrupt the viral infectious cycle are largely unknown. Perhaps of greater biological and practical significance, the relative importance in permissive hosts of the many mechanisms by which immune components have been shown to help control and clear HAdV infection has not been established. Most investigations of immune responses have been performed in cells in culture or non-permissive mice, and focused on vectors defective for expression of viral immunomodulatory genes. In view of the inherent limitations of experimental studies of human viruses, it is not clear that this issue can be fully resolved. Nevertheless, it seems likely that further development and exploitation of permissive small animal models like the Syrian hamster would be beneficial in this regard.

Antiviral Approaches   **10**

10.1 Development of Antiviral Therapies

While the majority of acute human adenovirus (HAdV) infections in normal individuals are generally self-limiting, adenovirus remains a significant cause of morbidity and mortality in immunocompromised patients (Lion, 2014). As described in Chapter 8, patients with HIV and recipients of hematopoietic stem cell transplants (HSCTs) or solid organ transplants are particularly at risk for either graft rejection and/or developing potentially fatal disseminated HAdV infections (Lion, 2014). Disseminated infection is usually defined as the involvement of two or more organs, excluding viremia (Florescu & Hoffman, 2013). The most common occurrence of HAdV infections arise in allogeneic HSCT recipients with the highest rate being close to 50% of bone marrow transplant patients. The use of potent immunosuppressive agents such as anti-thymocyte globulin, alemtuzumab, or prednisone to protect the recipients of HSCT leads to higher rates of virus-induced morbidity and mortality, particularly in pediatric patients (Myers *et al.*, 2005). Adenoviruses belonging to species B1 and B2 including types 11, 31, 33, 34, 35, and 40 are often associated with disseminated adenoviral disease as well as hemorrhagic cystitis (Echavarria, 2008; Sandkovsky *et al.*, 2014). A diagnosis of disseminated disease is usually made in the clinic by histopathology, as well as by performing qualitative and quantitative polymerase chain reaction (PCR) that is capable of detecting as few as 100-virus genome copies/ml. PCR appears to be the approach most relied on to monitor disease progression as well as treatment efficacy.

Many adenovirus types, and especially those belonging to species D (Robinson *et al.* 2013b), are also a frequent cause of ocular infections. The most severe form is known as epidemic keratoconjunctivitis (EKC). This disease is associated with infections with types 37, 19, and 8 and can lead to cornea scarring with permanent loss of visual acuity. In Japan, adenovirus induced EKC is a particularly important eye disease with very many individuals affected each year (Kaneko *et al.*, 2011b). Adenoviruses have been a constant source of acute respiratory infections in U.S. military recruits (Kajon *et al.*, 2007), and the suspension of a vaccine in the U.S. military in 1999 resulted in a significant rebound in the number of adenovirus infections, some of which resulted in fatalities. Emergence of new adenoviruses by genetic recombination, such as the case with HAdV-B 14p (Seto *et al.*, 2009), is also of concern given the lack of immunity to emerging genetic variants in the general population.

In view of the fact that there are ~70 different adenovirus types, many of which are associated with human diseases, it is unlikely that a broadly neutralizing viral vaccine will be developed in the near future. The current standard of care treatment for HAdVs includes the use of a limited armamentarium of antiviral compounds, detailed below. In general, these agents have proven to be only moderately effective, primarily due to low bioavailability and significant toxicity. Thus, significant effort has been expended toward the development of new and more efficacious antiviral therapies, some of which are being currently investigated in the clinic, while others are on the horizon. This chapter will not detail every antiviral study but instead provides an overview of the most promising current experimental and clinical uses of antiviral therapies for HAdVs.

10.2 Clinical Uses of Current and Emerging Antiviral Compounds

To date, there are no antiviral drugs specifically directed against human adenoviruses approved by the U.S. Food and Drug Administration (FDA). However, there are a number of non-selective antiviral compounds developed for other double-stranded (ds) DNA or even RNA viruses that have been employed in the most severe clinical settings of adenovirus infections of immunocompromised patients. Ribavirin, a guanosine analog that has been used to treat hepatitis C virus and respiratory syncytial virus, is also effective against the majority of HAdV types (Morfin *et al.*, 2009). While this compound has been employed for treating patients with HAdV infections, the lack of randomized controlled trials has made it difficult to determine whether it is truly effective (Shah *et al.*, 2012). Gangciclovir, a drug that restricts the activity mainly of herpes simplex virus DNA polymerases and is approved for treatment of herpesvirus infections, has also been shown to be somewhat effective against adenoviruses in animal models (Ying *et al.*, 2014).

More recently, the nucleoside phosphonate derivative (De Clercq, 2011) known as cidofovir (Fig. 10.1), was clinically approved for the treatment of human cytomegalovirus induced retinitis. Cidofovir was shown to have potent antiviral activity in adenovirus ocular infections in

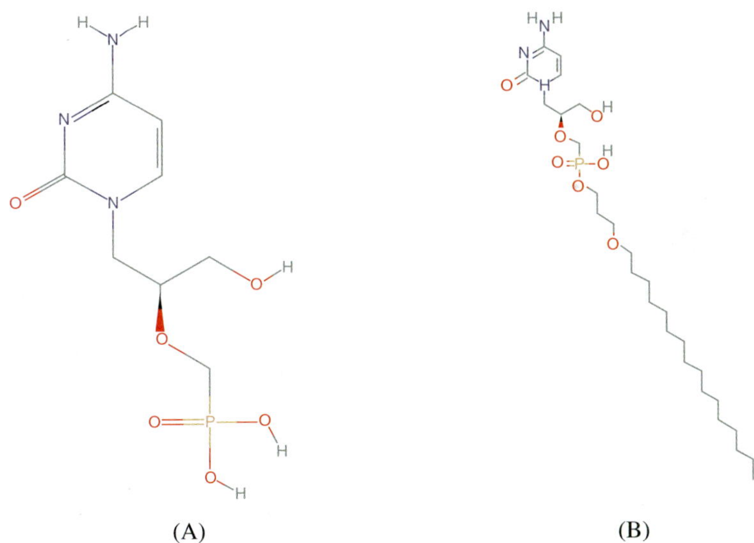

(A)   (B)

Figure 10.1. Chemical two-dimensional structures of cidofovir (**A**) and its derivative brincidofovir (**B**). The hexadecyclopropyl-cidofovir (B) shows enhanced oral bioavailability, lower toxicity, and improved efficacy. Adapted from PubChem (CID 60613, 483477).

rabbits (Romanowski *et al.*, 2001) and cotton rats (Kaneko *et al.*, 2004), as well as in a Syrian hamster model of immunosuppression (Tollefson *et al.*, 2014). Cidofovir has also recently been employed as a treatment for adenovirus infections in immunocompromised patients with some success (Matthes-Martin *et al.*, 2012), although controlled clinical trials have yet to be completed.

A major problem with the systemic use of cidofovir is that this drug exhibits significant nephrotoxicity and limited bioavailability (Izzedine *et al.*, 2005). A lipid conjugate of cidofovir, hexadecyloxypropyl-cidofovir (CMX001, brincidofovir, Chimerex, Durham, NC; Fig. 10.1) was developed to remedy these problems (Ciesla *et al.*, 2003). CMX001 was shown to have increased bioavailability, diminished accumulation in the kidney, and was well tolerated at doses up to 2 mg/kg in normal adult volunteers (Painter *et al.*, 2012). In further animal model studies using the Syrian hamster, CMX001 was shown to protect against HAdV-C5 infection and to reduce viral replication in important target organs (Toth *et al.*, 2008). This compound was also effective at reducing HAdV-C5 infection in immunosuppressed Syrian hamsters, whereas ribavirin did not show similar activity in this particular animal model (Tollefson *et al.*, 2014). Finally, in a very small number of immunocompromised patients with severe adenovirus infections including disseminated infections, CMX001 treatment resulted in a favorable response in 70% of the patients by lowering viral loads and increasing patient survival rate (Florescu *et al.*, 2012). Follow-up phase II trials of CMX001 in relatively small groups of immunocompromised patients with severe HAdV infections also appear to be quite promising, although phase III trials with larger patient populations will be necessary to determine the full extent of efficacy and safety. As is the case for most antiviral agents, it remains to be determined whether CMX001-resistant mutants of adenovirus will emerge. However, the spontaneous mutation rate of adenovirus DNA is rather low compared to that of RNA viruses. Although such resistance has not yet been observed in clinical trials with relatively small groups of patients, previous studies showed the emergence of cidofovir resistant HAdV-C5 mutants during continual passage of the virus in cell culture in the presence of this agent (Gordon *et al.*, 1996). Thus, it will be necessary to monitor for potential antiviral drug resistance in future trials with larger groups of patients.

10.3 Harnessing the Power of Immunotherapy

Due to the growing incidence of adenovirus disease in immunocompromised patients, several clinical trials have been initiated with the aim of assessing safety and efficacy of adoptively transferred immune T cells capable of recognizing HAdV-infected cells as well as cells infected with other viral pathogens. In one recent multicenter trial, prebanked HLA-matched (at least one allele per patient) peripheral blood mononuclear cells, educated to respond to HAdV *in vitro*, were used to treat transplant recipients with ongoing HAdV infections who had failed to respond to conventional antiviral therapy (Leen *et al.*, 2013). In this small phase I trial, 50 patients were selected for adoptive cell transfer using one of the 18 different cell lines that had been pre-stimulated with a HAdV-C5 vector displaying the type 35 fiber to increase transduction of immune cells. The patients receiving the cells shared at least one HLA allele with the presenting T cells against HAdV. Overall, the treatment was relatively safe with few if any adverse events related to the cell transfer. Importantly, a high percentage of the patients (~78%) showed a complete or partial response to the treatment based on the reduction in virus load and ongoing disease. Even more encouraging was that the response to the adoptive immune cell therapy appeared to be durable, with only 11% of the patients experiencing progression or recurrence of disease. It is important to note in this trial that the adoptively transferred immune T cells had also been educated against other opportunistic pathogens including Epstein–Barr virus and human cytomegalovirus. Further, the response rate to these viral pathogens was comparable to those observed with adenovirus. Although the results from this small trial are encouraging and could lead to further exploration, there are a number of hurdles that need to be overcome. One of these is that a more diverse panel of prebanked cells will need to be established in order to provide a better match for recipients. In addition, the occurrence of mild graft *vs.* host disease (GVHD) that was noted in several patients continues to be a safety concern. Finally, it remains to be determined whether this approach can be used effectively and safely during the clinical use of potent immunosuppressive regimens.

In a variation of this approach, (Papadopoulou *et al.*, 2014) stimulated peripheral blood mononuclear cells from allogenic donors (containing

CD4[+] and CD8[+] T cells) with peptide libraries corresponding to antigens from adenovirus, EBV, CMV, BK, and HHV6. These educated T cells were established within a 10–14-day period. In another small trial, peptide-stimulated T cells were used to treat 11 different recipients of allogeneic transplants, the majority of whom had developed infections with one or more of these viruses. As before, the treatment proved to be relatively safe and sustained antiviral response rates (overall 94%) were observed.

10.4 Early Stage Development of Novel Adenovirus Antivirals

10.4.1 *DNA replication inhibitors*

Despite the recent successes with the various antiviral approaches described above, additional investigations have been conducted with the goal of targeting distinct stages of the adenovirus virus life cycle. In one example, in a search for synthetic antiviral compounds from combinatorial libraries, a trisubstituted piperazin-2-one, designated '15D8,' was identified (Sanchez-Cespedes *et al.*, 2014). The synthetic molecule inhibited adenovirus infection in cells in culture at low micromolar concentrations and showed minimal cytotoxicity. 15D8 selectively inhibited adenovirus DNA replication. However, the mechanism underlying this activity has yet to be fully delineated. As these types of synthetic compounds are designed to mimic protein–protein interactions (Whitby & Boger, 2012), it is not inconceivable that 15D8 recognizes a specific component of the adenoviral DNA replication machinery such as the viral polymerase. However, further studies are needed to determine the precise mode of action of 15D8 as well as to examine its potential efficacy *in vivo*.

10.4.2 *Inhibitors of adenovirus capsid maturation*

Another antiviral approach that has been explored comprises a search for antagonists of the adenovirus cysteine protease (AVP, Adenain) (Mac Sweeney *et al.*, 2014; McGrath *et al.*, 2013). AVP is required for cleavage of multiple virus capsid proteins during maturation of virus particles, and hence the production of infectious particles (Chapter 7). Consequently, inhibitors of this critical enzyme would be likely to exert a highly detrimental effect on the virus infectious cycle. McGrath and colleagues (2003) have used the crystal structure of the AVP and its cofactor, pVIc, as well as existing docking programs, to search for small molecules available from the NCI Developmental Therapeutics Program that inhibit this adenoviral enzyme complex. They identified a few lead compounds as well as derivatives that had sub-micromolar activity against AVP or against the AVP–pVIc complex. In a similar structure guided approach, Mac Sweeney and colleagues (2014) identified tetrapeptide and pyridine-nitrile-based compounds that were capable of inhibiting AVP at low nanomolar concentrations *in vitro*. While these AVP antagonists represent potential new anti-adenoviral compounds, they have yet to be examined for their ability to block infection *in vivo*. These or related small candidate anti-AVP molecules will likely need to be of sufficient affinity to compete for the viral substrates of AVP and/or restrict encapsidation of the viral protease.

10.4.3 *Inhibitors of adenovirus attachment*

In principle, small molecules that prevent the earliest step in adenovirus host cell interactions via blockade of surface receptors might be highly effective antiviral agents. One clinical situation that may be particularly well suited for this approach is severe ocular infections caused by species D adenoviruses such as HAdV-D37. This viral pathogen recognizes a branched hexasaccharide, GD1a, that has two terminal sialic acid residues on each of its extensions (see Fig. 3.1 in Chapter 3) (Nilsson *et al.*, 2011). Structural analyses revealed the critical trivalent binding sites for the sialic acid residues on the HAdV-D37 fiber knob (see Fig. 2.5 in Chapter 2). On the basis of these findings, small synthetic and flexible scaffolds with terminal sialic acid residues in multivalent arrays were designed as potential antiviral agents (Spjut *et al.*, 2011). These multivalent synthetic compounds exhibited four orders of magnitude better inhibition of HAdV-D37 binding to human corneal cells in culture than monovalent sialic acid, indicating their potential value as antiviral agents.

In a related study, synthetic sialic acid residues were conjugated to lipid-based carriers in order to present an artificial cell-surface-like moiety

(Aplander *et al.*, 2011). These so-called 'molecular wipes' exhibited inhibition of virus binding at low micomolar concentrations and perhaps nanomolar antiviral activity in culture. More interestingly, they had a very distinct mode of antiviral activity; these lipid-based carriers displaying sialic acid residues had the ability to aggregate virus particles and thereby reduce virus interactions with susceptible target host cells.

With regard to sialic acid conjugates as potential antiviral agents, it is worth noting that carbohydrate-based or glycomimetic compounds have already entered into clinical trials. Nonetheless, the use of these compounds has been hindered by their poor pharmacokinetic profiles (Ernst & Magnani, 2009). Topical administration of sialic acid derivatives directly onto the surface of the eye for adenovirus ocular infections could very well overcome this limitation. However, it will also be necessary to monitor how well these compounds can be tolerated in the eye and how effective they are in defeating virus infection.

10.4.4 *Agents that interfere with post-attachment cell entry*

Following high-affinity binding to cells by specific receptors, the virus is taken up into cells by clathrin-mediated endocytosis (Mercer & Greber, 2013). This internalization event is mediated largely by the interactions of the RGD (Arg–Gly–Asp) motif in the penton base protein with cell-surface α_v integrins ($\alpha_v\beta_3$, $\alpha_v\beta_5$, $\alpha_v\beta_1$) (Bai *et al.*, 1993; Li *et al.*, 2001; Wickham *et al.*, 1993). The role of these integrins in adenovirus infection was subsequently confirmed using a pepidomimetic antagonist based on the virus penton base RGD sequence, which inhibited adenovirus infection of integrin-expressing 293 cells (Hippenmeyer *et al.*, 2002) and also interfered with cell adhesion (Carron *et al.*, 1998; Engleman *et al.*, 1997). These integrin antagonists only partially inhibit adenovirus infection, mainly because of the ability of the virus to use alternative receptors for cell entry. However, protein epitope mimetics (Robinson, 2011) might prove to be useful additions to the armamentarium of HAdV antivirals, especially if modern synthetic chemistry approaches can be used to improve their efficacy (Nagarajan *et al.*, 2007).

In addition to integrin antagonists as antiviral agents, other adenovirus entry inhibitors have been discovered in recent investigations. High-throughput screening of a synthetic small-molecule library identified five-membered heterocylic compounds that exhibit antiviral activity against HAdV-C5 that had been bound to clotting factor X to increase liver cell tropism (Duffy *et al.*, 2013). These compounds showed minimal toxicity and were also effective when administered *in vivo* in a murine model of liver tropism. Although these particular synthetic molecules did not restrict virus attachment, they appeared to block a subsequent step in infection, although the exact step has not yet been defined.

While steady progress has been made on well-established synthetic chemical methods of antiviral therapy, a few new unconventional approaches toward anti-HAdV drug development have been explored recently. One such approach selected a cyclic D,L-α-peptide with antiviral activity from a combinatorial library (Horne *et al.*, 2005). This type of self-assembling molecule has the capacity to form nanotubes that can readily penetrate cellular membranes, perhaps due to the amphipathic properties of these molecules (Fernandez-Lopez *et al.*, 2001; Ghadiri *et al.* 1994). In cell culture assays, the cyclic D,L-α-peptide blocked adenovirus-mediated gene delivery to multiple cell types and was effective at low micromolar concentrations against HAdV-C5 vectors displaying fibers from distinct HAdV species. The target of the cyclic D,L-α-peptide was a crucial cellular process rather than components of the virus itself; the peptide prevented the development of the low pH environment of the cell endosome. In principle, the lack of pH reduction would have the effect of preventing capsid disassembly in the endosome, a process that is needed to release membrane lytic protein VI from the interior of the capsid. In support of this mechanism, the cyclic D,L-α-peptide was shown to prevent endosome rupture by adenovirus (Horne *et al.*, 2005). Although further studies are needed to determine whether this compound is effective and well tolerated *in vivo*, these early experiments in cells in culture suggest that these self-assembling nanopeptides may represent a novel approach to the design and discovery of new synthetic compounds with broad antiviral activities.

10.4.5 *RNA interference (RNAi)*

It is well established that the HAdV genomes include one or two genes that enocode virus-associated (VA) RNAs. In addition to other functions

(Chapter 5), these molecules regulate the virus replication cycle in host cells by interfering with the host RNAi processing pathway (Carnero *et al.*, 2011). Over the past few years, further knowledge of RNA interference has provided opportunities to use the RNAi pathway as a target for adenovirus replication. In one study, small interfering (si) RNAi directed against early (E1A) or late (hexon, IVa2) RNA was used to target adenovirus replication in tissue culture cells (Eckstein *et al.*, 2010). Each of the siRNAs was capable of downregulating adenovirus gene expression. However, the siRNA used in this study against E1A was not as effective in blocking adenovirus replication as those directed against the late genes. In addition, only combinations of siRNAs targeting both early and late genes showed the ability to block adenovirus-mediated cytotoxicity. Effective inhibition of viral RNA synthesis as well as replication, as measured by genome copies in infected cells in culture, has also been achieved using siRNA directed against the viral DNA polymerase (Kneidinger *et al.*, 2012). More recent studies have shown that combining siRNA directed against adenoviral early or late genes with cidofovir (Ibrisimovic *et al.*, 2013) or with soluble recombinant Coxsackie virus and adenovirus receptor (CAR)–Fc and cidofovir (Pozzuto *et al.*, 2015) has a more profound effect on reducing production of virions than did siRNA alone. Thus, if efficient methods of delivering adenovirus-directed siRNAs *in vivo* could be established, their use in combination with more conventional antiviral agents such as cidofovir might prove very useful.

10.5 What Next?

The growing use of immunosuppressive regimens in solid organ and stem cell transplantation has been accompanied by a significant rise in the number of serious adenovirus infections in these patients. This clinical situation underscores the challenge of identifying optimal antiviral agents to combat adenovirus infections.

Currently, the most promising antiviral treatment modalities that have advanced to the clinic are adoptive T cell transfer and a new antiviral agent, CMX001. These antiviral treatments appear to have the highest success rate as well as acceptable safety profiles. Nonetheless, properly controlled clinical trials with larger patient numbers will be needed to determine whether these two antiviral methods can be used effectively and safely. Thus, the continuing development of novel antivirals, not only as alternatives to these current best treatment options but also as complementary approaches for standard of care, seems appropriate at this point in time. Moreover, the use of some newly developed compounds for topical administration to treat adenovirus ocular infections could mitigate some of the safety concerns that accompany systemic administration.

Among the latest antivirals to be examined are those that target the viral protease. New lead antiviral compounds that recognize the viral cysteine protease, AVP, a molecule that is essential for virus maturation, have now been identified and appear to show good inhibitory activity against the enzyme *in vitro*. It is possible that the use of modern medicinal chemistry techniques could drive the discovery of second-generation molecules with greater potency and selectivity as well as enhanced bioavailability and pharmacokinetics.

Essential reactions in the virus infectious cycle that have not yet exploited in any meaningful way in development of antivirals are capsid assembly or disassembly. This dearth is in large part due to the fact that fundamental information about these steps for this large and structurally elaborate virus is still lacking. However, clues to the process of capsid disassembly have been gleaned from a combination of structural and molecular virology studies that examined adenovirus neutralization by α-defensins (Smith *et al.*, 2010). These small innate immune proteins bind to critical regions of the most adenovirus types and prevent capsid disassembly and release of membrane lytic protein VI that is needed for endosome escape (Smith & Nemerow, 2008; see also Chapter 8). It is possible that a peptidiomimetic form of defensins or other synthetic molecules with similar activity might prove effective as antiviral agents for treating localized (e.g. ocular) or disseminated adenovirus infections.

Therapeutic Applications: History and Use of Replication-Defective Vectors

11

11.1 Human-Adenovirus-Mediated Gene Transfer: A Historical Perspective

11.1.1 *The development and use of vectors for respiratory diseases*

Following their much-publicized introduction into the clinic in 1993 (Crystal, 2014), adenoviral vectors have been administered to hundreds of patients and currently represent ~25% of all viral vector in clinical trials (Ginn *et al.*, 2013). The vast majority of these clinical studies employ local delivery of these vectors to achieve transient gene expression in efforts to treat cardiovascular diseases or cancer, or as vaccine vehicles for counteracting a variety of infectious diseases (Appaiahgari & Vrati, 2015; Bradshaw & Baker, 2013; Khare *et al.*, 2011; Kaminsky *et al.*, 2013; Loskog, 2015; Majhen *et al.*, 2014; Mennechet *et al.*, 2015). A description of the different types of adenovirus vectors employed for various clinical applications can be found in these or other equally good reviews, and will be discussed later in this chapter. The goal of this section is to provide an overview of the developments that led up to the more prevalent use of viral vectors, as well as some of the problems, solutions, and modest successes culminating from investigations in human subjects.

Several advances in the late 1980s and early 1990s laid the foundation for the use of adenovirus vectors in human clinical trials. One was the development of E1- and E3-deleted human adenovirus (HAdV)-C5 vectors that could be propagated in human embryonic kidney (HEK)-293 cells (Bett *et al.*, 1993; Bett *et al.*, 1994). These cells contain a portion of the left end of the viral genome including E1, thereby complementing E1-deleted viruses, while the E3 region is not required for HAdV reproduction in cells in culture. E1/E3-deleted HAdV-C5 vectors can accommodate the insertion of foreign gene sequences of ~7 kb into the E1 region for subsequent expression of cDNAs encoding potentially therapeutic proteins under the control of strong, non-selective, or tissue-specific promoters (Morsy *et al.*, 1993; Putzer *et al.*, 1997; Rosenfeld *et al.*, 1991). An important advantage for clinical applications is that replication defective HAdV-C5 vectors can be produced at nearly 10^{13} particles/ml under good manufacturing practices (GMP).

A second advance that fostered the use of HAdV-mediated gene transfer was the identification and cloning of the cystic fibrosis transmembrane conductance regulator (CFTR) gene (Riordan *et al.*, 1989; Rommens *et al.*, 1989). Patients with mutations in this chloride-transporting protein (Kerem & Kerem, 1995) frequently have life-threatening respiratory infections with bacterial pathogens such as *Pseudomonas aeruginosa*. As bronchial epithelial cells are capable of being transduced with HAdV-C5-encoded transgenes (Rosenfeld *et al.*, 1991; Rosenfeld *et al.*, 1992), this advance provided the impetus for clinical trials with an E1-, E3-deleted adenovirus encoding a wild-type version of the CFTR gene. In the first such trial in 1993, four cystic fibrosis patients were given HAdV-C5 encoding CTFR via nasal and bronchial routes using moderate doses of virus (up to 2×10^9 pfu) (Crystal *et al.*, 1994). This therapeutic regimen was shown to be safe, although a transient inflammatory response in pulmonary tissues was observed. However, only 5–14% of the cells recovered from biopsies of nasal or bronchial epithelial cells showed evidence of CFTR transduction following vector administration. This poor transduction was likely due to both the heavy mucous layer that overlays target epithelial cells in CF patients and the presence of the receptor for HAdV-C5 (i.e. Coxsackie virus and adenovirus receptor, CAR) on the basal lateral surfaces of these cells rather than on the apical (accessible) surface (Pickles *et al.*, 1998; Walters *et al.*, 2002). Similar clinical findings were obtained in subsequent CFTR trials, in which the viral vectors were shown to be relatively safe with no dramatic improvements in disease (Zabner *et al.*, 1993; Zuckerman *et al.*, 1999).

Based on studies in an animal model showing that attaining 5–10% of normal CFTR expression in airway epithelial cells would be sufficient to alleviate CF disease (Dorin *et al.*, 1996), further clinical efforts were undertaken, using repeat administration of HAdV-CTFR to overcome the physical and cell biological barriers to gene transfer (Harvey *et al.*, 1999). In these studies, the vector was administered to each CF patient multiple times over a nine-month period. Although 5% of normal CFTR production was achieved at the highest doses of virus (2×10^9 pfu), expression was transient, lasting less than 30 days. Repeated administration of HAdV-CFTR led to decreased expression of the transgene as a result of host immune responses to virus encoded gene products (Hartman *et al.*, 2008; Worgall *et al.*, 1997; Yang *et al.*, 1994; Y. Yang *et al.* 1996).

11.1.2 *The liver-directed OTC gene therapy trial*

A phase I clinical trial conducted at the University of Pennsylvania in 1999 (Raper *et al.*, 2002; Raper *et al.*, 2003) had a profound effect on the field of viral vector gene therapy, and provided important and painful lessons for the systemic administration of adenovirus vectors (Wilson, 2009). This trial was designed to evaluate the safety and tolerability of a replication-defective HAdV-C5 vector administered to patients with a deficiency in an X-linked gene that encodes ornithine transcarbamylase (OTC). OTC is a mitochondrial matrix enzyme that functions in the urea cycle, and its deficiency leads to accumulation of glutamine and ammonia in the blood. Patients with the most severe form of OTC deficiency exhibit hyperammonia in their bloodstream, which can lead to coma and death. Standards of care include hemodialysis or the intravenous delivery of sodium benzoate, arginine and sodium phenylacetate. These treatments are only partially effective in ameliorating the disease, and the general prognosis of OTC patients is poor as a result of progressively worsening cognitive abilities. Thus, in the late 1990s, liver-directed HAdV-mediated gene therapy was considered an appropriate, and ideally, a preferred experimental option for treating OTC deficiency. Despite the known limitations of HAdV-C5 vectors, such as their inability to support long-term transgene expression and their capacity to induce host immune responses, the investigators hoped that the rapid onset of expression of OTC gene carried by adenovirus would counteract the hyperammonemic crises in OTC-deficient patients. Furthermore, OTC is normally produced in the liver, an organ that is easily accessible to adenovirus.

In preparation for human investigations, preclinical studies were performed in which a first-generation (E1-deleted HAdV-C5) vector was administered to the liver of rhesus monkeys via the portal vein. At high vector doses (1×10^{13} particles/kg), severe liver damage and clotting disorders were observed (Nunes *et al.*, 1999). Later-generation vectors containing deletions of E1 and E2 or E1 and E4 had higher safety profiles at substantially lower doses (Raper *et al.*, 1998), although some toxicity was still noted in these animals. The more fully deleted gene vector that lacked the E4 gene was ultimately selected for a phase I safety trial. This trial included 19 OTC-deficient human subjects, 3–4 in each cohort, who received a single dose of HAdV-C5-OTC (from 2×10^9 to 6×10^{11} particles/kg) administered systemically via intra-arterial infusion into the right lobe of the liver. Liver biopsies, blood, and urine were collected from each patient over the next five months to evaluate toxicity, immune responses, and efficacy (as determined by metabolic correction). In the first 18 patients, there were unexpectedly excessive immune responses to the virus, and substantial liver toxicity that was not proportional to the vector dose (Raper *et al.*, 2002). In fact, there was a steep curve for toxicity that was not predicted from animal (mouse) studies (Ye *et al.*, 1996; Ye *et al.*, 1997). Even more disappointing was the fact that no significant transgene (OTC) expression was detected. Tragically, the 19th patient in the trial, an 18-year-old male who was injected with the highest number of vector particles (~10^{13} total), experienced a fulminant reaction to the virus characterized by thrombocytopenia, liver toxicity, and excessive production of pro-inflammatory cytokines that ultimately led to progressive organ failure and death 98 hours post-vector administration (Raper *et al.*, 2003). The 18th patient in this trial, preceding the fatal case, received the same quantity of the vector and also exhibited substantial signs of inflammation, but these subsided and the patient recovered from the procedure. The reasons for the dramatically different outcomes for these two human subjects are not clear, although differences in genetic predispositions for innate immune responses or in the status of pre-existing immunity to HAdV-C5 are possibilities. The OTC trial was subsequently put on hold and eventually discontinued. The systemic administration of adenovirus was judged to be too risky by the NIH Recombinant DNA Advisory Committee (RAC) and U.S. Food and Drug Administration (FDA), and nearly every subsequent trial with adenovirus has employed such vectors for local delivery at doses considerably lower than those used in the OTC trial. This early HAdV gene delivery trial underscores the limitations of adenovirus for certain applications, as well as of reliance on animal models to predict human immune responses to viral vectors in general.

11.2 Circumventing the Limitations of First-Generation Adenovirus Vectors

11.2.1 *Gutted HAdV vectors*

In attempts to overcome the limitations of HAdV-C5-mediated gene delivery, several investigators have generated vectors that have the majority of their genomes deleted to minimize immune responses to the vector. These vectors are variously termed: high-capacity, helper-dependent, third-generation, gutted, or gutless (Brunetti-Pierri & Ng, 2011; Kochanek, 1999; Kreppel, 2015; Segura *et al.*, 2008). Gutted HAdV-C5 vectors are usually propagated in the presence of a first-generation E1-deleted HAdV-C5 "helper" virus containing LoxP sites flanking its genome. A Cre-recombinase-producing 293 cell line is then used to propagate the gutted vector while selectively removing the helper virus genome (Palmer & Ng, 2003). The gutted vector contains only the *cis*-acting sequences (inverted terminal repeats [ITRs] and the packaging signal, Ψ) which are needed for replication and packaging of the viral vector genome. Consequently, approximately 36 kbp of foreign sequences can be accommodated in the gutted vector although a minimum of 28–30 kbp is needed to produce a stable vector particle (Kennedy & Parks, 2009; Parks *et al.*, 1999).

Compared to standard E1-deleted HAdV-C5 vectors, gutted HAdV-C5 based vectors have a very large cloning capacity (~36 KB) and are able to support the long-term expression of transgenes in tissues that are not highly proliferating (Brunetti-Pierri *et al.*, 2005; Brunetti-Pierri *et al.*, 2012; Ehrhardt *et al.*, 2003; Schiedner *et al.*, 1998). Gutless HAdV vectors are non-integrating, a property that eliminates the problem of insertional mutagenesis because the viral vector nucleic acid sequences remain episomal (Kreppel & Kochanek, 2004). Thus far there have been no reports of gutless HAdV vectors being used in human clinical trials. However, preclinical studies using these vectors in animal models, including non-human primates, have shown some efficacy and long-term transgene expression (Brunetti-Pierri *et al.*, 2013; Z. Jiang *et al.*, 2004; Schiedner *et al.*, 2002; Toietta *et al.*, 2005). Remaining hurdles for human applications appear to be the technical challenges and expense of producing large quantities of HAdV-C5-based gutless vectors, the pre-existing antibodies to HAdV-C5 in the majority of patients, and competition from other vectors such as AAV that provoke less substantial immune responses than standard E1-deleted HAdV-C5 vectors. It will be of interest to see if further improvements in gutless HAdV vector technology (e.g. use of different virus serotypes) enable them to move forward in the clinic.

11.2.2 *Cloaking and retargeting HAdV vectors*

An alternative approach to reduce the host immune response to adenovirus vectors is to envelope (cloak) the virus particle with many copies of a relatively non-immunogenic compound to render it invisible to immune mediators such as antibodies and complement and clotting factors. One of the most common of these cloaking compounds is polyethylene glycol (PEG) (Barry *et al.*, 2010). Cloaking is typically achieved using an N-hydroxy-succinidimide group on PEG molecules that can react covalently with free amines in the outer capsid proteins of virus particles. The size of the PEG molecules can range from 5 kDa to larger-sized molecules (Hofherr *et al.*, 2008). Coating virus particles with these compounds has successfully blunted virus interactions with antibodies, T cells, platelets, and other innate immune effector molecules (Chillon *et al.*, 1998; Croyle *et al.*, 2001; Hofherr *et al.*, 2007; Mok *et al.*, 2005; O'Riordan *et al.*,

1999). The PEG moiety can also be modified on its distal end with cell targeting motifs that facilitate HAdV-mediated gene transfer to selected target tissues (Eto *et al.*, 2005; Jung *et al.*, 2007; Lanciotti *et al.*, 2003; Oh *et al.*, 2006).

Basic biology investigations using PEG-coated HAdV have also been reported. In one example, covalent modifications of the seven hypervariable region loops of the hexon, provided insights into interactions of virus particles with cell receptors. In particular, modifications of hexons hypervariable regions (HVRs) 1, 2, 5, and 7 with PEG prevented virus uptake into liver Kupffer cells via the scavenger receptor SRA-II (Khare *et al.*, 2012).

Despite the usefulness of PEGylated HAdV vectors, there are several disadvantages to this approach that have yet to be fully addressed. The first is that PEG-associated linker moieties appear to be antigenic and thus re-administration of PEGylated vectors necessitates the use of alternative linker chemistries (Croyle *et al.*, 2002). Secondly, covalent attachment of PEG molecules onto the surface of adenovirus particles has a tendency to reduce infectivity, perhaps due to over-conjugation of critical regions of the virus (Fasbender *et al.*, 1997). Other synthetic polymers that appear to have increased half-lives in blood and lower toxicity such as poly-[N-(hydroxypropyl) methacrylamide] (pHPMA) have also been explored (Green *et al.*, 2004; Kreppel & Kochanek, 2008). Further attempts to improve virus shielding have been made by coating adenovirus particles with hybrid molecules comprising PEG linked to homopolypeptides such as poly-L-lysine (Jiang *et al.*, 2013). In this case the hybrid molecule attaches non-covalently through charge interactions with the virus particle via the poly-L-lysine moiety. This approach has the added value of being less damaging to the virus vector while still allowing effective cloaking.

Another distinct advantage of adenovirus gene transfer vectors is their ability to be manipulated to expand their tropism or to potentiate their delivery to selected cell types. As described below, such retargeting is most often accomplished by genetically appending antibodies, peptides, or other molecules to different outer capsid proteins (Banerjee *et al.*, 2011; Beatty & Curiel, 2012; Coughlan, 2014; Majhen *et al.*, 2012; Poulin *et al.*, 2010).

11.3 Vectors for Cardiovascular Diseases

One of the more successful uses of adenovirus as a gene transfer vector is for treating cardiovascular diseases (Bradshaw & Baker, 2013; Kaminsky *et al.*, 2013). Coronary artery disease (CAD) manifests itself in the narrowing of the arteries, thus restricting blood flow to the heart and peripheral sites, and is a major cause of morbidity and mortality worldwide. Standards of care for CAD include implantation of stents in concert with balloon angioplasty to open blocked vessels or the more invasive bypass surgery. However, these procedures have limitations (i.e. mostly effective for large but not smaller vessels), and each carries its own risk and safety concerns. Consequently, preclinical and clinical CAD trials using adenovirus vectors have been conducted over the past decade to investigate the safety and efficacy of this approach.

One of the main objectives of CAD gene transfer is to mimic a process that sometimes occurs naturally in some patients, namely the growth of new arterial branches from preexisting unblocked vessels, a mechanism that is generally referred to as angiogenesis. New blood vessel growth can be stimulated by a variety of naturally occurring proteins, including vascular endothelial growth factor A (VEGF-A) (Ferrara *et al.*, 2003), VEGF-B, -C, -D, -E, and -F (Ferrara, 2004; Patil *et al.*, 2012), fibroblast growth factor 1 (FGF-1) (Iwakura *et al.*, 2000; Uriel *et al.*, 2006), angiopoietins (Brindle *et al.*, 2006), hepatic growth factor (HGF) (Sala & Crepaldi, 2011; Yuan & Cantley, 2008), and stromal-derived factor 1α (SDF-1α) (Salcedo *et al.*, 1999). Given the diversity of proteins that govern angiogenesis, a major question is whether an HAdV vector carrying a single angiogenic gene or whether one encoding several angiogenic molecules is optimal for promoting blood vessel growth *in vivo*. For mostly practical reasons, the majority of adenovirus vectors used for CAD encode one potential pro-angiogenic protein. The short-term expression of transgenes directed by HAdVs is considered to be advantageous, because sustained production of angiogenic proteins, as supported by other vectors such as AAV or lentiviral vectors, might be less safe. The preferred route of HAdV administration is via intramyocardial injection, which allows efficient cardiac delivery with limited systemic spread (Magovern *et al.*, 1996).

Following extensive preclinical studies in small animal models as well as pigs (reviewed in Kaminsky *et al.*, 2013), several HAdV vectors delivering angiogenic genes have been investigated in clinical trials. Many of the dozen such clinical trials have established good safety profiles as well as encouraging signs of efficacy. Despite these positive outcomes, no approved angiogenic gene therapy has been licensed commercially. It therefore seems likely that additional improvements in vector design and/or transgenes will need to be made before CAD vectors move forward clinically in a meaningful way. One possible future application for adenovirus vectors might be an *ex vivo* approach such as the delivery of the tissue inhibitor of metalloproteinase 3 (TIMP3), to restrict restenosis of vein grafts (Bradshaw & Baker, 2013; Zhu *et al.*, 2013).

11.4 New Endeavors with Non-HAdVs and Adenovirus-Like Nanoparticles

Applications of HAdV for gene transfer are still limited by preexisting immunity to HAdV-C5, but new opportunities to eliminate this potential problem have arisen via the identification and genetic modifications of non-human adenoviruses (Alonso-Padilla *et al.*, 2015). One of the main examples of this approach is canine adenovirus type 2 (CAdV-2), which has the ability to transduce human neuronal cells and therefore could be a useful vehicle to introducing therapeutic genes for neurological disorders (Kremer, 2004; Kremer & Nemerow, 2015; Soudais *et al.*, 2001; Soudais *et al.*, 2004). Currently, clinical grade quantities of CAdV-2 based vectors are being produced using GMP methods in anticipation for use in human subject trials (Fernandes *et al.*, 2015; Silva *et al.*, 2014). As described in the next section, other non-HAdVs such as the closely related simian adenoviruses (SAdVs) that have low seroprevalence in humans (Quinn *et al.*, 2013; Xiang *et al.*, 2006; Zhang *et al.*, 2013) are making their way into the clinic as vaccine vectors. Other non-HAdV types that are being explored as vectors to avoid pre-existing immunity in humans include fowl adenoviruses, which can possess ~10 kbp larger viral genomes than HAdVs (Cherenova *et al.*, 2004; Michou *et al.*, 1999), and ovine (sheep) adenovirus vectors (Both, 2004; Hofmann *et al.*, 1999).

Thus, the presence of adenoviruses in nearly every vertebrate class may enable investigators to take advantage of nature's bounty to improve vector targeting and immune evasion.

Another approach to adenovirus gene transfer relates to the re-design of naturally occurring nanoparticles with key elements of adenovirus to facilitate gene transfer *in vivo*. In principle, naturally occurring nanoparticles should be less immunogenic than virus particles. One example of naturally occurring nanoparticles are so-called vaults, which are cytoplasmic ribonucleoproteins with a hollow, barrel-like structure with overall dimensions of $72.5 \times 41 \times 41$ nm^3 and containing a large internal cavity (Anderson *et al.*, 2007; Mikyas *et al.*, 2004; Suprenant, 2002). Vault nanoparticles can be made via self-assembly of the major vault protein (Stephen *et al.*, 2001) and such recombinant particles have been investigated as potential drug delivery vehicles (Casanas *et al.*, 2012). Vault particles containing a cell targeting motif as well as a 20-residue segment corresponding to the membrane lytic portion of HAdV-C5 protein VI (Chapter 4) were shown to promote delivery of a DNA plasmid or ribotoxin into cells in culture (Han *et al.*, 2011). However, it remains to be seen whether these particles can be tailored to carry nucleic acid sequences *in vivo*. The major hurdle that remains to be overcome is the optimal design of a nanoparticle that functions like adenovirus with regard to efficient host cell interactions, not an easy proposition as viruses have the tremendous advantage of co-evolving with host cells for millions of years.

11.5 What Next?

HAdV genomes were among the first viruses to be modified for the express purpose of therapeutic gene transfer, in part because adenoviruses were quite well understood, the virus particles can be purified readily in large quantities and they can transduce a considerable variety of cell types. Moreover, detailed structural knowledge of HAdV has enabled re-engineering of some its outer capsid proteins (e.g. fiber and hexon) to facilitate cell targeting. The recognition of the limitations imposed by both pre-existing antibodies to HAdV-C5 and vector-induced innate and adaptive immune responses has stimulated the implementation of numerous strategies to improve adenovirus vectors for gene transfer and delivery of vaccine antigens. Given the delicate balance needed to achieve efficacy while avoiding vector toxicity, especially for systemic delivery, it seems that standard HAdV vectors will continue to be employed only for local delivery in the near term. However, it remains possible that gutted HAdV vectors could gain increasing use if methods to improve their production and stability can be achieved. Finally, it would not be too surprising if some of the key components of adenovirus including its attachment ligands (fiber and penton base) or membrane disrupting molecules (protein VI) continue to be co-opted to facilitate naked DNA or nanoparticle–DNA complexes uptake into host cells.

Therapeutic Applications: Vaccines and Oncolytic Vectors   12

12.1 Introduction

Adenoviruses have been considered not only for gene transfer applications like those discussed in Chapter 11, but also as components of vaccines. As with gene transfer, pre-existing immunity to human adenovirus (HAdV)-C5 limited the utility of vaccine based on first-generation vectors, but several vectors based on rare human types or adenoviruses of other animals appear promising. These vectors, like those for gene transfer, are designed to be replication defective. In contrast, oncolytic adenovirus, a class that includes the first adenovirus derivative approved for clinical use, replicate, but do so preferentially in tumor cells. These viruses can be modified in very many different ways to enhance selectivity for replication in, and killing of tumor cells. In this chapter, we describe the strategies adopted in the quests for effective adenovirus vaccine vectors and oncolytic agents.

12.2 Vaccine Vectors

The properties of HAdVs that support efficient delivery and expression of transgenes described in the previous chapter make them excellent candidates for vaccine vectors, especially because the deletions typical of first generation vectors remove coding sequences for viral proteins (E1A, E1B, and E3) that counter innate and adaptive immune responses (Chapter 9).

Indeed, early preclinical studies in this field established that such first-generation HAdV-C5 vectors elicited strong B and T cell responses to transgene products in mice (He *et al.*, 2000; Xiang *et al.*, 1996). Because of the strong innate and adaptive immune responses induced even by wild-type HAdV-C5 in such animal models and humans (Chapters 8 and 9), these viruses act as effective and natural adjuvants. This section will highlight the development, characterization, and future challenges of HAdV-based vaccine vectors.

12.2.1 *HAdV-C5 vaccine vectors*

For the reasons described in the context of gene transfer applications, initial efforts to develop adenovirus vaccines relied on HAdV-C5. Despite some negative features, such candidate vectors have been developed as vaccines for a number of pathogenic bacteria, viruses and protozoa, and tested in pre-clinical or clinical trials (Table 12.1).

One limitation of HAdV-C5 vectors is the high prevalence in human populations worldwide of neutralizing antibodies (Chapter 9), with up to 98% of individuals seropositive in some parts of Africa (Dudareva *et al.*, 2009) where specific vaccines are sorely needed. As might be anticipated, such pre-existing immunity to HAdV-C5 reduces adaptive responses to the products of the transgenes engineered into the vector genome (e.g. Barouch *et al.*, 2004; Bruder *et al.*, 2012; Casimiro *et al.*, 2003; Lemckert *et al.*, 2005; McCoy *et al.*, 2007; Roberts *et al.*, 2006; Small *et al.*, 2014; Sumida *et al.*, 2004; Yang *et al.*, 2003). This limitation can be at least partially overcome by strategies described in the previous section and Chapter 9, such as replacing the seven hypervariable loops of the HAdV-C5 hexon with those of types rare in humans, for example, HAdV-C43 or -D48 (Bruder *et al.*, 2012; Roberts *et al.*, 2006). Effective vaccination by non-replicating vectors frequently requires multiple inoculations (prime-boost). However, robust adaptive immune responses to the vector that can severely limit the efficacy of multiple immunizations are elicited by even a single dose of HAdV-C5 (e.g. Schirmbeck *et al.*, 2008; Sullivan *et al.*, 2003; Tatsis *et al.*, 2009). Despite this problem, multiple inoculations with HAdV-C5 vectors carrying HIV-1 coding sequences were administered in large phase IIb trials.

Table 12.1. Examples of adenovirus vaccine vectors.

Vector	Target	Antigen(s)[1]	Host	Delivery	Response(s)	Reference(s)
HAdV-B35	HIV-1	Env (CladeA) Gag-RT-IN-Nef, Env	Humans	i.m.	Antibody and CD8$^+$ T cells in HAdV-C5 +ve and −ve individuals	Kopycinski *et al.*, 2014 Fuchs *et al.*, 2015
	Mycobacterium tuberculosis	Ag85A, Ag85B	Mice Rhesus macaques,	i.n or i.m., i.d. + i.m. boosts	T cells protection CD8+ T cells, protection	Radosevic *et al.*, 2007
		B10.4	Humans	i.m., BCG-boost after 3-6 months	CD4$^+$ and CD8$^+$ T cells	Hoft *et al.*, 2012; Rahman *et al.* 2012
HAdV-C5[2]	Dengue Virus type 2	E	Mice	i.p. + i.d. boost	Neutralizing antibodies, T cells	Khanam *et al.*, 2009
	Types 1–4	E	Mice	i.p.	Neutralizing antibodies, Th1 T cells	Jaiswal *et al.*, 2003
	Hepatitis B virus	C, S	Mice	Tail vein injection of vector transduced dendritic cells	Thl, IFNγ	Jia *et al.*, 2015
	Hepatitis C virus	Core, E1, E2	Mice	i.p.	Antibodies to all 3 HCV proteins	Makimura *et al.*, 1996
		NS4	Mice	i.m. or i.p.	Pro-inflammatory cytokines, Antibodies, CTLs	Singh *et al.*, 2014
	Influenza A virus H5N1	HA	Humans	Oral	Neutralizing antibodies, CTLs, IFNγ	Liebowitz *et al.*, 2015; Peters *et al.*, 2013

(Continued)

Table 12.1. *(Continued)*

Vector	Target	Antigen(s)[1]	Host	Delivery	Response(s)	Reference(s)
Adenoviruses of other animals						
ChAdV3[3]	Hepatitis C virus (Type 1B)	NS	Humans	i.m.	CD4+, CD8+ T cells, cross reaction to other HCV genotypes	Barnes *et al.* 2012
				i.m + vaccinia virus vector boost	High CD4+, CD8+ T cells; sustained effector and memory responses	Swadling *et al.* 2014
ChAdV68 (aka SiAdV24)	Influenza A virus, H5N1	NP	Mice	i.m.	T cells, some protection	Roy *et al.* 2007
	Rabies virus	GP	Mice	Inhalation	Antibodies, protection	Zhou *et al.* 2006
			Rhesus macaques, cyonomologous monkeys	i.m. + boost after 5 months	Neutralizing antibodies, protection	Xiang *et al.* 2014
CAdV2	Rabies	GP	Dogs, cats	s.c.	Neutralizing antibodies, protection	Hu *et al.* 2006; Hu *et al.* 2007
			Pigs	i.m.	Antibodies, protection	Liu *et al.* 2008

1. Immune responses listed are those to target antigens.
2. HAdV-C5 vaccines for HIV-1 are described in the text.
3. ChAdV63 vaccines for Plasmodium falciparum are described in the text.

i.d. = intradermally; i.m. = intramuscularly; i.n.= intranasally; i.p. = intraperitoneally; s.c. = subcutaneously

12.2.1.1 *The HIV-1 vaccine STEP and Phambili trials — an unanticipated outcome*

The candidate vaccine tested in these trials comprised E1-deleted HAdV-C5 vectors, one for each of the HIV-1 Gag, Pol, and Nef proteins, developed by Merck in the wake of the failure of an Env-specific vaccine to protect against HIV-1 infection, or to reduce viral load in trial patients who became HIV-1 infected (see Sekaly, 2008; Watkins *et al.*, 2008). The multiple vector design represented a shift in focus toward the induction of cytotoxic T lymphocytes (CTL) responses, which were known to confer protection against HIV-1 in humans and animals models, with the goal of reducing viral load in infected individuals and hence secondary transmissions. Replication deficient HAdV-C5 derivatives were the vectors of choice as they elicit potent T cell responses (Chapter 9). Indeed, such vectors were shown to induce $CD8^+$ T cell responses against HIV-1 in humans and animals, and lead to some protection against infection in some primate models (Barouch & Nabel, 2005; Casimiro *et al.*, 2003; Gomez-Roman *et al.*, 2006; Shiver *et al.*, 2002). The particular HIV-1 antigens were chosen on the basis of their conservation across HIV-1 clades and large size, which was expected to facilitate generation of multiple T cell epitopes from each. This inference was confirmed by large-scale analysis of T cell HIV-1 epitopes in HIV-1-positive subjects from different geographical locations and infected with different HIV-1 clades (Shiver & Emini, 2004).

Prior to the phase IIb trials, the candidate vaccine was tested extensively in phase I and IIa protocols that included more than 130 volunteers, and was demonstrated to be safe. It was also found to induce T cell responses to the HIV-1 antigens, but these were quite weak and directed against only three epitopes in the HIV-1 proteins (Priddy *et al.*, 2008). The phase IIb STEP trial enrolled 3,000 healthy people, mainly homosexual men, in North and South America, where clade B HIV-1 is predominant. It comprised three randomized inoculations (at 0, 1, and 6 months) of the vaccine or placebo, and measured rates of HIV-1 infection. Viral load was assessed in participants who became infected at three months after diagnosis of HIV-1 (Buchbinder *et al.*, 2008). A second trial, the Phambili trail (*phambili* = 'forward' in Zulu), with the same design was initiated in sexually active, HIV-1 seronegative volunteers at five sites in South Africa, where clade C HIV-1 is prevalent, and pre-existing immunity to HAdV-C5 is exceptionally high (Gray *et al.*, 2011).

Unexpectedly, however, the STEP trial had to be halted after an interim analysis revealed that, although eliciting HIV-1 specific T cell responses in 75% of recipients, the vaccine not only failed to prevent HIV-1 infection or reduce viral load, but was also associated with increased rates of HIV-1 infection in HAdV-C5-seropositive men and uncircumcised men (Buchbinder *et al.*, 2008). As a result of this lack of beneficial effect and the potential for increasing HIV-1 infections rates, the Phambili trial was stopped after only 801 of the scheduled 3,000 participants had been enrolled (Gray *et al.*, 2011). The increased risk of HIV-1 infection associated with prior HAdV-C5 infection in the STEP trial was confirmed subsequently, when data from all participants were analyzed, but shown to decrease with time from vaccination (Duerr *et al.*, 2012). For example, HAdV-C5 seropositive, uncircumcised men who were vaccinated had a >fourfold higher risk of HIV-1 infection than matched placebo recipients during the 18 months following vaccination, but the risk declined thereafter. The higher risk remained significant even when potentially confounding variables, such as unprotected and insertive anal intercourse with male partners of unknown HIV-1 status, were taken into account (Koblin *et al.*, 2012). Initial data from the Phambili trail also indicated no beneficial effect of the vaccine, and a somewhat increased risk of infection (Gray *et al.*, 2011). Follow-up studies, which included patients initially HIV-1-negative who dropped out, confirmed these conclusions and detected a small increased risk for vaccinated men, but not women (Gray *et al.*, 2014; Moodie *et al.*, 2015). Differences in the trial populations that might account for the difference in association of the magnitude of increased risk for HIV-1 infection in the two trials include sexual preference and a particularly high proportion of Phambili participants who were HAdV-C5 seropositive at the time of enrollment (Moodie *et al.*, 2015).

Various parameters might have contributed to the failure of the vaccine to prevent HIV-1 infection or reduce viral load, including pre-existing neutralizing HAdV antibodies and $CD8^+$ T cells specific for the vector and the 3 dose immunization protocol (see Sekaly, 2008; Watkins *et al.*, 2008).

However, the limited repertoire of HIV-specific T cell epitopes recognized in response to the vaccine indicates that the HIV-1 transgenes were not efficient T cell immunogens. Furthermore, subsequent studies indicated that immunological responses in rhesus monkeys, which were used in preclinical assessment of vaccine efficacy, do not necessarily correlate with those observed in clinical trials in humans (Bett *et al.*, 2010). Perhaps of even greater import for the future of HAdV-C5 as a vaccine vector is the present lack of a clear understanding of the reason for the increased rate of HIV-1 infection in specific patient groups.

In an early study to address this question, the effects on dendritic cells of HAdV-C5 vector alone or of the vector pre-incubated with neutralizing antibodies to form immune complexes were compared. The immune complexes were observed to stimulate maturation of dendritic cells, as well as of $CD8^+$ T cells, more effectively than the vector alone, and to enhance HIV-1 infection in co-cultures of dendritic and T cells (Perreau *et al.*, 2008). It was therefore proposed that this immune-enhancement mechanism could increase the likelihood of HIV-1 infection in vaccinated individuals with pre-existing antibodies against the HAdV-C5 vector. Another attractive hypothesis is based on the reproduction of HIV-1 in activated $CD4^+$ T cells, and posited that in individuals with pre-existing immunity to HAdV-C5, exposure to the vaccine vector results in activation and homing to mucosal surfaces (sites of HIV-1 entry) of HAdV-C5 memory $CD4^+$ T cells (Benlahrech *et al.*, 2009). Such an expansion of adenovirus-specific $CD4^+$ memory T cells was demonstrated when lymphocytes were co-cultured with dendritic cells exposed to HAdV-C5 (Benlahrech *et al.*, 2009). On the other hand, analysis of data from 40 subjects enrolled in a safety trial of the Merck vaccine detected no significant risk differences in those seropositive or seronegative for HAdV-C5 at the start of the trial (Hutnick *et al.*, 2009; O'Brien *et al.*, 2009), but the small numbers of patients would be expected to limit the power of this study. In trial subjects who became HIV-1 infected, no significant differences in $CD4^+$ T cell counts were evident in the groups who received the vaccine or placebo (Fitzgerald *et al.*, 2011). Similarly, no increase in mucosal trafficking of $CD4^+$ T cells was observed when rhesus monkeys were inoculated with an E1/E3- deleted HAdV-C5 vector carrying the HIV-1 Gag, Pol, and Nef genes and pre-infected with HAdV-C5 (Masek-Hammerman *et al.*, 2010).

Such observations argue that the actual number of adenovirus-specific CD4$^+$ T cells did not contribute to increased risk for HIV-1 infection, and rather that their activity or maturation state may be the crucial parameter. Indeed, a detailed analysis of the immune correlates of responses to the vaccine or placebo in trial subjects revealed that increases in secretion of IFNγ by peripheral blood mononuclear cells collected four weeks into the trial upon exposure to medium alone were associated with increased risk for HIV-1 infection, and that the difference in this parameter between vaccine and placebo groups was greater in HAdV-C5 seropositive than in seronegative men (Huang *et al.* 2014). It was therefore concluded that the proportion of cells producing IFN-γ without stimulation by exogenous antigens (most likely memory CD4$^+$ T cells), activated by unrelated infections prior to the start of the trial, could be a predictor of risk for HIV-1 infection (Y. Huang *et al.*, 2014). On the other hand, examination of the relationship between preexisting immunity to HAdV-C5 and the profile of cytokines produced by T cells in STEP trial participants not infected by HIV-1 suggested that anti-HAdV-C5 immunity might modify the T cell response following vaccination (Pine *et al.*, 2011). Consistent with this view, a systems-based approach to assess immune responses established that innate responses to HIV-1 were attenuated in participants with pre-existing neutralizing antibodies to HAdV-C5 (Zak *et al.*, 2012). These observations suggest that lack of appropriate activation of immune systems might contribute to the unanticipated increase in rates of HIV-1 infection associated with prior infection of trial recipients with HAdV-C5. The non-exclusive explanations for which there is some experimental evidence emphasize the complexity of the interplay among immune components and viruses when individuals are exposed to repeated infection by the same or different viruses.

Despite the results from the STEP trial, infection with HAdV-C5 does not appear to be intrinsically associated with increased risk for HIV-1 infection, for no such correlation was identified in participants in other studies, including 1,570 adults enrolled in efficiency trials of an HIV-1 subunit vaccine (Curlin *et al.*, 2011; Stephenson *et al.*, 2012). Nor was prior infection with other HAdV types predictive of risk for HIV-1 infection (Stephenson *et al.*, 2012). While the mechanism(s) underlying the increased risk for HIV-1 infection in HAdV-C5 seropositive men in the

Merck vaccine trials remain to be established unequivocally, this result and the lack of efficacy, associated with poor T cell immunogenicity, focused attention on alternative vaccine regimens and vaccine vectors.

12.2.1.2 *Alternative vaccination protocols using HAdV-C5*

One approach to limiting undesirable consequences of both pre-existing immunity to the vector and that induced following the initial inoculation is to use different vectors to administer sequential doses of the same antigen(s) (i.e. heterologous prime-boost). Provided that the second vector is sufficiently unrelated to the first, its efficacy is not limited by immune responses induced by the priming vector and it can deliver a second dose of antigen that results in increased antigen-specific B- and T cell responses (Liu *et al.*, 2009). This approach has been evaluated for various pairs of vectors derived from HAdV-C5 and other human types that are relatively rare in human populations or adenoviruses of other species (Barratt-Boyes *et al.*, 2006; Geisbert *et al.*, 2011; Lemckert *et al.*, 2005; Rodriguez *et al.*, 2009; Schuldt *et al.*, 2012; Thorner *et al.*, 2006), as well as with pairs of simian adenoviruses (SAdVs) (Santra *et al.*, 2009). In all cases, more robust antibody and T cell responses to the antigen of interest were observed in animal models than achieved with a second dose of the initial vector. Furthermore, boosting with a heterologous HAdV vector achieved complete protection against Ebola virus challenge in macaques, whereas the single initial dose conferred only partial protection (Geisbert *et al.*, 2011). As might be anticipated, pairs of HAdVs that are less closely related in the hypervariable hexon loops against which neutralizing antibodies are usually directed were more effective immunogens than pairs that elicited even low concentrations of cross-reacting antibodies (Thorner *et al.*, 2006). Comparison of all possible pairs of HAdV types would be a costly and lengthy exercise, so the optimal pair, or pairs, has not been defined, and in any case, might be specific for particular application or subject populations.

A simpler approach is the combination of a DNA plasmid engineered to express the antigen of interest as the priming agent followed by a HAdV-C5 vector boost: as priming is with DNA only, immune responses to an initial viral vector are avoided. This strategy was first reported using a plasmid expression vector for the Ebola virus glycoprotein (GP) prior to boosting with an HAdV-C5 vector carrying the GP gene and shown to overcome the humoral responses in mice previously exposed to E1-deleted HAdV-C5 (Yang *et al.*, 2003). Such DNA priming prior to administration of an HAdV-C5 vaccine vector was also reported to increase CTL responses against the hepatitis C virus core protein in mice (Matsui *et al.*, 2003), and to HIV-1 Gag in rhesus monkeys (Santra *et al.*, 2005). It also stimulated the appearance of neutralizing antibodies to HIV-1 proteins in mice, and increased the breadth of the antibody response (Mascola *et al.*, 2005). This vaccination protocol was subsequently shown to increase substantially the generation of HIV-1 Env-specific antibodies and CD8$^+$ T cells in humans (Churchyard *et al.*, 2011; Koup *et al.*, 2010), and to be far more effective in humans in eliciting T cell responses (CD4$^+$ and CD8$^+$), including long-term memory CD8$^+$ T cells, than both priming and boosting with an HAdV-C5 vector (De Rosa *et al.*, 2011). Despite such encouraging findings, the efficiency trial of an HIV-1 vaccine comprising DNA priming and HAdV-C5 boost for delivery of clade B Gag, Pol, and Nef and clades A, B, and C Env, which was designed to evaluate HIV-1 infection rates and viral load, was halted early for lack of efficacy (Hammer *et al.*, 2013). The reasons for this failure are not yet clear.

12.2.2 *Other HAdV vaccine vectors*

HAdV types that are relatively rare in human populations are good candidates for vaccine vectors, but they must, like HAdV-C5, support efficient expression of transgenes in differentiated cells like skeletal muscle, a typical site of vaccine inoculation, and induce robust immune responses to the antigen(s). As seroprevalence studies established that certain species B and D HAdVs are rare in various human populations, including those in geographic areas where the risk of HIV-1 infections is high (Kostense *et al.*, 2004; Nwanegbo *et al.*, 2004; Vogels *et al.*, 2003), replication-deficient vectors based on HAdV-B35, -B11, and -D49 were developed (Barouch *et al.*, 2004; Holterman *et al.*, 2004; Lemckert *et al.*, 2006; Stone *et al.*, 2005). As discussed previously, these vectors are immunogenic in mice or monkeys with pre-existing immunity to HAdV-C5. Several are in various phases of development and testing as vaccines against a number of infectious agents (Table 12.1). However, in side-by-side comparisons, these vectors proved to be less immunogenic than HAdV-C5 in mice or monkeys with no pre-existing immunity to HAdV-C5 (Barouch *et al.*, 2004; Lemckert *et al.*, 2005; Nanda *et al.*, 2005; Roberts *et al.*, 2006; Shiver & Emini, 2004; Thorner *et al.*, 2006). Subsequently, vaccine vectors have been engineered for other HAdV types with low seroprevalence, including HAdV-B26 (Abbink *et al.*, 2007), HAdV-C6 (Capone *et al.*, 2006), and HAdV-D48 (Kahl *et al.*, 2010), and demonstrated to be immunogenic in animal models. In a comparison of the immunogenicity of species B and D vectors that derived from HAdV-B26 was the most effective following delivery of a single antigen dose (Abbink *et al.*, 2007). In another study, species D HAdVs were reported to be superior to an HAdV-B35 vector in inducing T cell responses to the transgene product (Kahl *et al.*, 2010). All rare HAdVs have not been compared under the same conditions in a single study, leaving open the question of which is the "best" immunogen. Furthermore, as discussed previously, data collected in mice and monkeys do not necessarily predict responses in humans. Consequently, development of these vaccine vectors remains largely empirical. Several are being tested, alone or in combination prime-boost regimens, for example as vaccines for filoviruses (Zahn *et al.*, 2012) or influenza A viruses (Gurwith *et al.*, 2013; Weaver, 2014; see also Table 12.1).

12.2.3 *Adenoviruses of other animals as vaccine vectors*

Non-human mastadenoviruses as well as aviadenoviruses are of considerable interest as vaccine vectors to protect agriculturally important animals. For example, bovine, canine, and porcine adenovirus vectors have been developed for vaccination of farm animals, and fowl adenoviruses for delivery of antigens to chickens and turkeys (e.g. (Ayalew *et al.*, 2015; Corredor & Nagy, 2011; Hammond *et al.*, 2001; Tordo *et al.*, 2008; Tuboly & Nagy, 2001; Zakhartchouk *et al.*, 1999). As discussed in chapter 11, canine adenovirus 2 is of particular interest because of its ability to transduce neurons efficiently (Chapter 3), and lack of cross-reactivity with neutralizing antibodies to HAdVs (see Bru *et al.*, 2010). However, efforts to engineer animal adenovirus vectors for use in humans has focused on adenoviruses that infect our closest, non-human relatives, great apes such as chimpanzees.

A large number of SAdVs have been isolated from gorillas, chimpanzees (*Pan troglodytes)* and bonobos (*Pan paniscus*, pygmy chimpanzees) (Roy *et al.*, 2004a; Roy *et al.*, 2004b), including many from the feces of chimps and bonobos resident in animal facilities and zoos in the US (Colloca *et al.*, 2012). These SAdVs are generally closely related to HAdVs, and many have been classified on the basis of sequence analysis to the human species B, C, and D (Colloca *et al.*, 2012; Pantó *et al.*, 2015). Functional similarity is underscored by the efficient propagation of E1-deleted SAdVs in human cell lines that contain and express integrated copies of the HAdV-C5 E1A and E1B genes (Colloca *et al.*, 2012). Despite this close relationship, SAdV seroprevalence is low in humans in all parts of the world that have been examined (Chen *et al.*, 2010; Colloca *et al.*, 2012; Dudareva *et al.*, 2009; Ersching *et al.*, 2010; Xiang *et al.*, 2006). Most importantly, in terms of vaccine development, the ability of SAdV vectors to induce T cell responses in mice is not impaired by prior infection with HAdV-C5 or a different SAdV, and many SAdVs elicit robust and long lasting antibody and T cell responses in mice, guinea pigs and macaques (Colloca *et al.*, 2012; Kobinger *et al.*, 2006; Xiang *et al.*, 2002). Although not compared side-by-side, application of the same methods to measure the generation of T cell responses in mice and macaques infected with HAdV or SAdV vectors indicated that the latter are more effective immunogens than the less common HAdV types, such as B35 and D26, and that the best (notably ChAd3 and ChAd36) are as effective as HAdV-C5 (Colloca *et al.*, 2012).

For such reasons, SAdV vector vaccines have been developed for a variety of infectious agents, including influenza A virus, HIV-1, and protozoan parasites, and tested in animal models (Table 12.1). One such vector, based on one of the first SAdVs to be isolated, encoding the conserved influenza A virus proteins NP and M1 has been tested in dose-escalation studies in humans with boosting by a vaccinia virus vector encoding the same influenza virus antigens (Antrobus *et al.*, 2014). This vaccine was well tolerated at doses up to 2.5×10^{10} virus particles and induced strong influenza virus-specific T cell responses, which increased ~threefold in study participants who received the boosting vaccine. Initial clinical results with an SAdV vector for expression of the NS gene of hepatitis C virus (genotype 1) were also encouraging: this vaccine vector elicited T cell responses against this NS proteins, but that could also recognize NS proteins of heterologous HCV genotypes and, following boosting, the immune responses were sustained for at least a year (Barnes *et al.*, 2012). However, most progress has been made with vaccines against the filovirus Ebola virus and the protozoan parasite *Plasmodium falciparum*, which is responsible for most cases of malaria in humans.

12.2.3.1 *SAdV Ebola virus vaccines*

Outbreaks of Ebola virus are relatively rare, but are frequently associated with fatal hemorrhagic fever. Until the most recent outbreak that spread rapidly to several countries in West Africa and has been estimated to have killed more than 11,000 people (de La Vega *et al.*, 2015), infections tended to be local and limited in scope. The high fatality rate and lack of specific treatments prompted development of vaccines against Ebola virus, including several using adenoviral vectors. Early studies demonstrated that HAdV-C5 vectors for the Zaire Ebola virus GP protected all cynomologous macaques tested against a lethal dose of the virus when administered in a DNA prime-HAdV-C5 boost regimen, as did two doses of the HAdV-C5 vaccine (Sullivan *et al.*, 2003). An HAdV-C5 vector delivering a codon-optimized GP coding sequence was also fully protective when given as a single dose (Choi *et al.*, 2015). Although the initial HAdV-C5 vaccine proved safe and immunogenic in humans (Ledgerwood *et al.*, 2010), it suffers from the limitations imposed by pre-existing immunity to HAdV-C5, which, as noted previously, is especially common in Africa where all Ebola virus outbreaks have occurred. Indeed, in a phase I trial of a mixture of HAdV-C5 vaccine vectors for GPs of the Sudan and Zaire strains of the virus, only 1/22 subjects developed neutralizing antibodies to the viral GPs (Ledgerwood *et al.*, 2010).

Several SAdV vaccines vectors based on viruses isolated from chimpanzees generated strong antibody and T cell responses to Ebola virus GP in mice, guinea pigs, and macaques (Kobinger *et al.*, 2006; Stanley *et al.*, 2014). However, a ChAd3 vaccine vector encoding the Zaire Ebola virus GP protected all macaques against Ebola virus challenge at a dose at which a ChAd63 based vaccine protected only 1 of the 4 animals challenged (Stanley *et al.*, 2014). Clinical trials planned for the SAdV vectors were accelerated with the continuing escalation of the West African Ebola virus epidemic, with three phase I trials completed in late 2015. In the first, 20 volunteers from the Washington D.C. area received 1×10^{10} or 2×10^{11} virus particles of a ChAdV3 vector encoding both the Zaire and the Sudan Ebola virus GPs. No safety concerns were noted, although two of the high-dose recipients developed transient fever (Ledgerwood *et al.*, 2014). GP-specific antibodies were produced in all participants and CD4[+] T cell responses in all members of the high-dose cohort, with CD8[+] T cell responses in 7/10 of this group. A second trial comprised a dose-escalation study of a ChAd3 vector for the Zaire Ebola virus GP in 60 volunteers in the U.K. All doses were judged to be safe and all recipients in the high-dose group exhibited Ebola virus-specific T cell responses, and the majority developed antibodies to the antigen (Rampling *et al.*, 2015). A third, larger phase I study examined this same vaccine vector in 91 participants in Mali and 20 in the Baltimore area of the U.S. in a single-blind rand-omized-dose protocol (Tapia *et al.*, 2016). In addition, 52 Malian partici-pants were boosted with a vaccinia virus vector encoding the Zaire and Sudan Ebola virus GPs previously reported to be safe and effective as a boost in human volunteers (Swadling *et al.*, 2014), or with a saline pla-cebo. As in the other trials, some recipients of the higher dose of the SadV vaccine vector developed transient fever, and 56% of those who received the vaccinia virus vector boost reported injection site pain, but there were no significant safety concerns. Again, neutralizing antibodies were induced in most recipients, and increased with SAdV vaccine dose and to a greater degree upon heterologous boost, but T cell responses were limited to 31% of the participants after the priming dose (Tapia *et al.*, 2016). Boosting increased both the prevalence and magnitude of CD4[+] and CD8[+] T cell responses (Tapia *et al.*, 2016). In view of these results, a phase IIb/III trial that includes this ChAd3 vaccine vector in one of its three arms, the others being placebo or a vesicular stomatitis virus vaccine vector, was initiated in Liberia in February 2015. This trial is ongoing, and no reports have been published, but analysis of data from more than 600 participants has con-firmed the safety of the adenoviral vector, opening the way for the phase III portion of the trial (NIAID, 2015).

12.2.3.2 *SAdV malaria vaccines*

Malaria afflicts millions of people largely in tropical and sub-tropical regions of the world, and is a leading cause of illness and death, particularly in young children. It is caused by four species of the protozoan parasite *Plasmodium*, most commonly *P. falciparum* and *P. vivaz*, which are transmitted by female mosquitoes of the genus *Anopheles* and have a complicated life cycle that includes multiple stages in both the insect and the human host (see Crompton *et al.*, 2014; Miller *et al.*, 2013). The high toll on human health and the increasing resistance to frontline drugs (Fairhurst, 2015; Sinha *et al.*, 2014) has made the development of a preventative vaccine a high priority.

Proof-of-principle of the potential of adenoviruses as vaccine vectors for malaria was obtained using *P. berghei*, which infects mice and other rodents, but not humans. SAdV vectors that delivered coding sequences for a set of CD4$^+$ and CD8$^+$ T cell epitopes from early parasite stages of the parasite fused to the thrombospondin-related adhesion protein (ME-TRAP) elicited strong malaria-specific CD8$^+$ T cell responses and protected 67–92% (depending on the vector) of the mice tested against *P. berghei* infection (Reyes-Sandoval *et al.*, 2008). As with adenoviral Ebola virus vaccine candidates, HAdV-C5 was engineered for development of malaria vaccines, for example, one designed to target multiple stages of the *P. falciparum* life cycle. Although safe and immunogenic, such vaccines did not induce protection in human volunteers (e.g. Tamminga *et al.*, 2013). A prime-boost regimen appeared to be more effective, as protection was achieved in 4/15 volunteers (Chuang *et al.*, 2013; Sedegah *et al.*, 2014).

Development of a SAdV (CHAd63) vector for the ME-TRAP antigen was also pursued. Following demonstrations that this vaccine vector generated protective CD8$^+$ T cell responses in mice, particularly when combined with a vaccinia virus vector boost (Reyes-Sandoval *et al.*, 2010), it was shown to exhibit similar immunogenicity in non-human primates (Capone *et al.*, 2010) and was safe in humans (O'Hara *et al.*, 2012). Consequently, phase I/IIa controlled human malaria infection (CHMI) trials were initiated. Although ChAd63-ME-TRAP alone was not protective, addition of a vaccinia virus boost resulted in complete protection of 3/14 volunteers and a delay in time to parasitemia in 5/14, with large reductions in the numbers of parasites emerging from the liver (Ewer *et al.*, 2013). Similar results were obtained in a second trial of this kind (Hodgson *et al.*, 2015). The protective responses correlated with the frequency of activated, malaria antigen-specific CD8$^+$ T cells, suggesting that achieving activation of a sufficient number of these CTLs is critical for protection against malaria. Subsequently, this vaccination strategy was tested in phase I/IIb trials in regions of Kenya and The Gambia in which malaria is endemic (Kimani *et al.*, 2014; Ogwang *et al.*, 2015). Similar immune responses to those seen in CHMI trials were observed, again skewed to CD8$^+$ T cells, albeit of somewhat lower magnitude. Furthermore, a study of 121 male volunteers in Kenya randomly assigned to vaccination with ChAd63-ME-TRAP plus vaccinia virus vectors or with a control rabies vaccine, revealed that the former reduced the risk of infection by 67% during an eight-week period (Ogwang *et al.*, 2015). Clearly, much larger phase III studies in which participants are monitored for considerably longer periods are required to assess this promising vaccine more rigorously. Nevertheless, it is noteworthy that that the degree of protection achieved in this phase IIb trial is significantly greater than that reported in a phase III trial of the current leading malaria vaccine candidate (see Ewer *et al.*, 2015).

Such advantages offered by the ChAdV63 vector as ease of manipulation, ready propagation and efficient expression of transgenes have also been exploited to identify new antigens produced in various *Plasmodium* forms (e.g. pre-erythrocyte and blood stage forms) that can induce immune responses that are protective (Goodman *et al.*, 2011) or that block development of the parasite in the mosquito (see Ewer *et al.*, 2015).

12.3 Replication-Competent Oncolytic Adenoviruses

12.3.1 *Contemporary proof-of-principle*

HAdVs were first considered for cancer therapy, specifically treatment of cervical carcinoma, some 60 years ago when they were shown to mediate extensive tumor necrosis (Georgiades *et al.*, 1959; Huebner *et al.*, 1956). However, the antitumor effects were not sustained and did not improve overall survival. Consequently, enthusiasm for this approach waned, until reignited in 1996 by the report by Bischoff and colleagues that an E1B 55 kDa protein-null mutant of HAdV-C5 replicated in, and killed, tumor cells selectively. In these experiments, the cytotoxicity of this mutant virus, designated ONYX-015 (a.k.a. dl1520) (Barker & Berk, 1987), correlated with the absence of the tumor suppressor p53, consistent with the role of the E1B 55-kDa protein in targeting p53 for proteasomal degradation (Chapter 6) (Bischoff *et al.*, 1996). However, studies by several other investigators soon established that the cytotoxicity of E1B 55 kDa protein-null mutant viruses bears no relationship to the presence or absence of p53, or whether this protein is wild type or altered, in human tumor cell lines and primary human cells (Goodrum & Ornelles, 1997; Goodrum & Ornelles, 1998; Hall *et al.*, 1998; Harada & Berk, 1999; Rothmann *et al.*, 1998; Turnell *et al.*, 1999). Subsequently, it was reported that impaired reproduction of ONYX-015 and other like mutant viruses in normal human cells correlates with the loss of selective export of viral late mRNAs from the nucleus that results from these mutations (Gonzalez *et al.*, 2006; O'Shea *et al.*, 2005). However, the difference in the degree of dependence of efficient viral late mRNA export and late protein synthesis in normal and tumor cells is quite small (Gonzalez & Flint, 2002; Gonzalez *et al.*, 2006), suggesting that other parameters may be important. Furthermore, the molecular basis of the selective export of viral late mRNA remains incompletely understood (Chapter 5), a limitation that precludes rational choice of target tumors and patients in clinical applications.

ONYX-015 was soon tested in the clinic, and shown to be safe and well tolerated, even at doses of 2×10^{12} virus particles, in phase I and II trials in patients with recurrent head and neck tumors (Ganly *et al.*, 2000; McCormick, 2003; Reid *et al.*, 2002). In this and other types of tumors, the efficacy of ONYX-015 is limited, with local tumor regression rates of only 0–14% reported. More encouraging results were obtained when the virus was used in combination with chemotherapy, such as treatment with cisplatin or 5-flourouracil (Dobbelstein, 2004; Hecht *et al.*, 2003; Heise *et al.*, 2000b; Khuri *et al.*, 2000; Lu *et al.*, 2004; Nemunaitis *et al.*, 2003). Although a phase III trial in the U.S. was halted for financial and business reasons (see Larson *et al.*, 2015), a very similar E1B 55 kDa mutant virus, H101, was tested in such trials in China, where a high response rate was achieved (Xia *et al.*, 2004). This virus was approved in 2005 in that country for treatment of late-stage nasopharyngeal carcinoma in combination with chemotherapy, the first oncolytic virus to enter the clinic. Much clinical experience with H101 (under the brand name Oncorine) has accrued in China, and this virus has also been examined in preclinical studies for other applications (Lei *et al.*, 2015; Yu & Fang, 2007). However, the results of this experience have not been widely published outside of China, and work with oncolytic adenoviruses elsewhere has focused on increasing tumor cell specificity and cytotoxicity.

12.3.2 *Increasing tumor cell specificity*

Since the original work of Bischoff and colleagues (1996), a great many conditionally replicating adenoviruses (CRAds) designed for the destruction of various types of cancer cells have been developed and tested in pre-clinical settings. Despite considerable variation in specific molecular properties, three principal strategies have been devised to increase tumor cell selectivity.

12.3.2.1 *Modification of viral gene products*

The ability of the viral E1A proteins to bind to, and reverse transcriptional repression by, retinoblastoma (RB) family proteins is necessary for efficient reproduction of HAdV-C5 in normal human cells (Chapter 6). As the genomes of many cancer cells carry deletions or other alterations in RB or RB pathway proteins (Sherr & McCormick, 2002), precise alterations in the E1A gene that block this function but not others were considered likely to confer tumor cell-specific reproduction. Indeed, deletions in the CR2 RB-binding site (Fig. 5.6) greatly impaired replication of HAdV-C5 in normal primary human cells such as lung fibroblast and small airway epithelial cells, but the mutant viruses (Δ24 and dl922-947) reproduced efficiently (and more efficiently than ONYX-015) in tumor cell lines in culture and in tumor xenografts in mice (Fueyo *et al.*, 2000; Heise *et al.*, 2000). Nevertheless, subsequent studies in which the effects of several E1A sequence deletions on induction of S phase, virus reproduction and cytotoxicity were compared indicated that deletion of only the RB-binding region did not inhibit S-phase induction and replication in all normal cell types examined (Sauthoff *et al.*, 2004). When combined with additional deletions, for example, of the E1A N-terminal sequence, such RB deletions resulted in tumor cell selectivity without loss of cytotoxicity (Sauthoff *et al.*, 2004). Similarly, a mutant HAdV-C5 in which the E1A gene contains both a CR2 deletion and mutations that prevent binding of the viral proteins to p300/CBP (Fig. 5.6) attenuated replication in normal human astrocytes or normal human brain tissues, but enhanced killing, as compared to the single mutant viruses, of infected glioma cell lines (Ulasov *et al.*, 2008). Combination of E1A mutations that ablate RB binding with deletion of the coding sequences of the E1B 19 kDa (Oberg *et al.*, 2010; Yamada *et al.*, 2007) or 55-kDa (Kim *et al.*, 2007) proteins have also been reported to reduce reproduction in normal human cells and increase killing of tumor cells in culture and xenografts.

Other modifications of the HAdV-C5 genome have been examined in more specific contexts. For example, deletion of E1B 19 kDa protein coding sequence, which blocks apoptosis (Chapter 6), in the background of an E1B 55 kDa deletion resulted in greater cytoxocity because of induction of apoptosis in combination with radiation in tumor cells in culture and in xenografts (Kim *et al.*, 2009). In a more targeted application, deletion of the virus-associated (VA) RNA I gene was shown to impair HAdV-C5 reproduction in primary human cells and Epstein–Barr virus (EBV) negative tumor cells, but the mutant virus replicated efficiently in and killed EBV-positive tumor cells (Wang *et al.*, 2005). Such specificity was ascribed to the ability of the small EBV-encoded RNA1 to complement the lack of VA RNA I. The small EBV RNA is synthesized in the majority of EBV-associated tumors and hence could confer selectivity for replication of this type of HAdV-C5 mutant virus.

12.3.2.2 *Engineering tumor cell-specific expression of viral genes and transgenes*

A popular method of enhancing the tumor cell selectivity of oncolytic viruses has been to place viral genes necessary for efficient progression through the infectious cycle, typically the E1A gene, and/or transgenes (see next section), under the control of promoters or promoter elements that are active only in cancer cells. Depending on the choice of promoter sequences, this approach can allow HAdV-C5 replication and/or transgene expression in specific or multiple types of tumor cells.

The promoter of the human gene for the reverse transcriptase subunit of telomerase (TERT) has been exploited in several CRAds, as it is active in most tumor and immortal cells, but inactive in somatic cells (see Hahn & Meyerson, 2001; Kim, 1997). Initial studies indicated that placing the viral E1A gene under the control of a modified, more active TERT promoter led to increased viral reproduction and cytopathic effect (CPE) only in tumor cells (Kim *et al.*, 2003; Irving *et al.*, 2004; Wirth *et al.*, 2003). Similar results were obtained when the E1A gene was expressed from the human E2F1 promoter, which is de-repressed in tumor cells with defects in the RB pathway (Jakubczak *et al.*, 2003; Rojas *et al.*, 2009). Inclusion of E2F binding sites in the E1A promoter for the Δ24 E1A gene described above has also been reported to reduce CRAd toxicity in mice, while stimulating systemic antitumor activity (Rojas *et al.*, 2010). Furthermore, inclusion of E2F binding sites in the modified TERT promoter conferred selectivity for efficient replication only in RB-deficient and TERT-positive tumor cells (Lei *et al.*, 2009). Controlling expression of the viral E1A and E1B genes via the E2F1 and TERT promoters, respectively, appeared to be even more effective, as no replication of this CRAd was observed in normal tissues obtained in clinical biopsies, yet the virus exhibited good anti-tumor activity in several xenograft models (G. Li *et al.*, 2005). Control of expression of both E1Aand E1B coding sequences (separated by an internal ribosome entry site to allow synthesis of E1B proteins) via the TERT promoter also leads to high tumor cell selectivity (Kawashima *et al.*, 2004). The CRAds that rely on the tumor cell specificity of the TERT promoter alone, or in viruses that deliver toxic transgenes (see next section), have been reported to achieve effective destruction of various types of tumor cells in culture or in xenografts (e.g. Li *et al.*, 2010; G. Li *et al.*, 2011; X. Li *et al.*, 2012). Other promoters active in many tumors that have been incorporated in the genomes of CRAds include those of the human survivin gene (Zhu *et al.*, 2006) and hypoxia responsive promoters or promoter elements (Guse *et al.*, 2009; Hernandez-Alcoceba *et al.*, 2002; Kwon *et al.*, 2010; Zhang *et al.*, 2006).

Control of expression of viral genes or potentially beneficial transgenes from tissue-specific promoters or those of genes expressed only in specific tumor types has also been of widespread interest. Examples of such promoters and their applications are listed in Table 12.2. Variations on this approach to restrict virus replication in normal cells include use of different tumor-specific promoters to drive expression of the viral E1A and E4 genes (Hoffmann & Wildner, 2006), placing the E1A gene with mutations that prevent binding to RB under the control of such promoters (Bauerschmitz *et al.*, 2006) and including tumor-specific translational in addition to transcriptional control (Stoff-Khalili *et al.*, 2008).

More recently, the cell-type-specific production of certain microRNAs (miRNAs) has been considered as a means to promote tumor-cell-specific replication of CRAds. This approach was stimulated by the finding that inclusion of sequences complementary to the seed sequence of the liver-specific miR-122 in the 3' untranslated region of the HAdV-C5 E1A gene reduced accumulation of E1A mRNA and proteins, and impaired virus replication in cells of hepatic origin (Ylosmaki *et al.*, 2008). Rendering E1A mRNAs sensitive to miR122 also decreased expression of this viral gene following intravenous delivery of the modified virus to mice, and reduced toxicity substantially, but had no effect on replication in cancer cells in which miR122 is not made (Cawood *et al.*, 2009). Introduction of target sequences for the seed sequences of a variety of miRNAs synthe-sized in specific cell types has improved selectivity for tumor-cell-specific replication and cytotoxicity, compared to the corresponding normal cells (Callegari *et al.*, 2013; Jin *et al.*, 2011; Yao *et al.*, 2014).

12.3.2.3 *Tumor-cell-specific delivery of CRAds*

An important complement to engineering the viral genome for tumor-cell-specific replication of expression of transgenes is to improve the efficiency and specificity with which CRAds enter tumor cells. Retargeting of oncolytic derivatives of HAdV-C5 employing the fiber knob chimeras or the insertion of RGD motifs into this fiber domain described in the context of gene transfer vectors (Chapter 11) was rap-idly adopted, following the recognition that the levels of the Coxsackie virus and adenovirus receptor (CAR) are variable, and often low, on the surfaces of human cancer cells, whereas $\alpha_v\beta_3$ integrins are usually well expressed (see Coughlan, 2014; Noureddini & Curiel, 2005; Seguin *et al.*, 2015). Both types of fiber modification increased transduction of human tumor cells, including those from ovarian (Kanerva *et al.*, 2002; Rocconi *et al.*, 2007; Z.B. Zhu *et al.*, 2008), head and neck (Reddy *et al.*, 2006; Stoff-Khalili *et al.*, 2007), pancreatic (Hammer *et al.*, 2015),

Table 12.2. Examples of oncolytic HAdVs in clinical trials.

Tumor	Vector Modification(s)	Transgene	Combination	Phase/Protocol	Outcomes	Reference(s)
Prostate	PSA promoter control of E1A	—	—	I, dose escalation	DLTs at $\geq 6 \times 10^{12}$ v.p.; some decrease [PSA] at highest doses	Small *et al.*, 2006)
	Mutations that prevent E1B 55 kDa protein synthesis	CD–TK fusion	5–FC, GCV	I, dose escalation	Increased [PSA] doubling time	Freytag *et al.*, 2007a
	Mutations that prevent E1B 55 kDa protein synthesis	yCD–mutTK	1MRT	I, dose escalation	Some effect in intermediate risk patients	Freytag *et al.*, 2007b
	Mutations that prevent E1B 55 kDa protein synthesis	yCD–mutTK	1MRT	II, compared to 1MRT alone	Decrease in positive biopsy at 2 years	Freytag *et al.*, 2014
Ovarian	Δ24E1A; HAdV-B3 Fiber knob	—	—	I, dose escalation	No PR or CR; Stable disease in some patients	Kim *et al.*, 2013
Ovarian, other gynecological tumors	Δ24 E1A; RGD in H1 fiber loop	—	—	I, dose escalation	DLT not reached; no PR or CR	Kimball *et al.*, 2010
Bladder	E2F1 promoter control of E1A	GM-CSF	—	I, escalation single or multiple doses	48.6% CR overall; 63% CR in multidose cohorts	Burke *et al.*, 2012

(Continued)

Table 12.2. (*Continued*)

Tumor	Vector Modification(s)	Transgene	Combination	Phase/Protocol	Outcomes	Reference(s)
Sarcoma (bone)	Δ24 E1A; HAdV-B3 fiber knob	GM-CSF	Cyclophosphamide (if no contra-indications)	I, single or three doses	2/15 minor response; 6/15 stable disease	Bramante *et al.*, 2014
Advanced solid tumors	Mutations that prevent E1B 55 kDa protein synthesis	HSP-70	—	I, dose escalation	DLTs at $\geq 1.5 \times 10^{12}$ v.p.; 11.1% PR + CR	Li *et al.*, 2009
	hTERT promoter control of E1A-1RES-E1B; all E3 sequences present	—	—	I, dose escalation	1/12 PR, 7/12 stable disease	Nemunaitis *et al.*, 2010
	E2F1 promoter control of E1A; addition Kozak consensus sequence to E1A	—	—	I, dose escalation	Evidence of antitumor activity in 9/17 patients	Nokisalmi *et al.*, 2010
	Δ24 E1A	GM-CSF	Cyclophosphamide	I, single dose and + ↑ [cyclophosphamide]	Combination increased progression-free survival	Cerullo *et al.*, 2011
	Δ24; RGD in H1 fiber loop; all E3 sequences present	GM-CSF	—	I, dose escalation	Stable disease in 3/6 patients; transient induction antitumor T-cell responses	Pesonen *et al.*, 2012

Note: 5-FC = 5-fluorocytosine; CD = cytosine deaminase; CR = complete response; DLT = dose-limiting toxicity; E2F1 = E2 factor 1; GCV = ganciclovir; GM-CSF = granulocyte macrophage-colony stimulating factor; HSP70 - heat shock protein 70; IRES = internal ribosome entry site; PR = partial response; PSA = prostate specific antigen; TERT = telomerase reverse transcriptase; TK = thymidine kinase; v.p. = virus particles.

colorectal (Lavilla-Alonso *et al.*, 2010), and biliary (Wakayama *et al.*, 2007) cancer, and melanoma (Rivera *et al.*, 2004; Z.B. Zhu *et al.*, 2008) and glioma (Hoffmann *et al.*, 2007; Nandi *et al.*, 2009). Attachment of CRAds to receptors not known to be bound by any HAdV has also been achieved by incorporation of mutations that eliminate the normal fiber knob-cell receptor interaction and the introduction of ligands for alternative receptors, including the EGF and FGF receptors (Blackwell *et al.*, 1999; Gupta *et al.*, 2006; Piao *et al.*, 2009; Uusi-Kerttula *et al.*, 2015; van der Poel *et al.*, 2002), the epithelial-cell-specific integrin $\alpha_v\beta_6$ (Coughlan *et al.*, 2009) and the vesicular stomatitis virus receptor phosphatidyl serine (Yoon *et al.*, 2015). Incorporation of appropriate ligands into the hypervariable regions of the hexon can also enhance transduction of the targeted cell types (Lucas *et al.*, 2015). In addition, HAdV-C5 tropism can be altered and expanded by inclusion in the H1 loop of the fiber knob of protein transduction domains, relatively small sequences that mediate efficient cell entry, such as those of the HIV-1 Tat protein or fish (herring) protamine (Kurachi *et al.*, 2007b; Youn *et al.*, 2008), or synthetic sequences identified experimentally (Puig-Saus *et al.*, 2014; Yamamoto *et al.*, 2014). Replacement of the entire fiber knob by larger retargeting modules, for example, a single-chain T cell receptor specific for the melanoma associated cancer testis antigen MAGE A1, is also possible (Sebestyen *et al.*, 2007).

The constraints imposed by the need for assembly of virus particles that support initiation of the infectious cycle limit the genetic manipulation of structural proteins to alter tropism. A more flexible alternative is chemical modification of virus particles, in the case of HAdVs, typically covalent addition of polyethylene glycol (PEG), but also polyethylenimine (PEI), which has been reported to stimulate uptake of CRAds and killing of CAR-negative cells (see Choi *et al.*, 2015). More recently, the power of this approach has been improved by replacement of Met residues in HAdV-C5 structural proteins with the analog azidohomoalanine (AHA), a reactive amino acid that allows subsequent and specific modification of virus particles using "click" chemistry (Banerjee *et al.*, 2010). Decoration of HAdV-C5 particles with folate using this method led to large increase (~20-fold) in infection of murine breast cancer cells (Banerjee *et al.*, 2011), as did conjugation of an oncolytic HAdV-C5 to a compound that releases taxol upon endocytosis and a folate targeting motif (Banerjee *et al.*, 2011).

More exotic delivery methods are also being explored. These include "magnofection," the direction of migration of oncolytic HAdV-C5 particles modified by covalent association with iron oxide nanoparticles (Choi *et al.*, 2015; Scherer *et al.*, 2002) and delivery within cells, such as macrophages (Muthana *et al.*, 2011), B cells (Zhang *et al.*, 2013), or dendritic cells (e.g. Brossart *et al.*, 1997; Steitz *et al.*, 2001; Tuettenberg *et al.*, 2003). For example, in a clever use of cell-mediated delivery, macrophages were transduced with a plasmid expression vector for the HAdV-C5 E1A and E1B coding sequences under the control of hypoxia-regulated promoter elements and an oncolytic CRAd with prostate-specific E1A gene expression (Muthana *et al.*, 2011). When injected into mice with human prostate tumor xenografts, the macrophages migrated to hypoxic tumors, where E1A and E1B genes were expressed from the plasmid vector, allowing reproduction of the CRAd. The virus released from the macrophages infected neighboring cells, but could replicate only in prostate tumor cells because expression of its E1A gene was controlled by prostate-specific promoter elements. This strategy led to increased destruction of primary human prostate tumors in the mice, as well as elimination of lung metastases, and was considerably more effective in both respects than systemic introduction of the same CRAd alone (Muthana *et al.*, 2011).

12.3.3 *Enhancing oncolytic activity*

Many of the CRAd designs described in the previous section improve killing of tumor cells indirectly, as a result of increased efficiency of transduction or enhanced expression of viral genes in the target tumor cells. The cytotoxicity of these vectors can also be increased by more direct strategies.

12.3.3.1 *Manipulation of viral genes*

Although CRAds are, by definition, replication competent, most do not have a wild-type genome, but rather carry deletions in the E3 region that were introduced to facilitate insertion of transgenes into the HAdV-C5 backbones used to make most CRAds, dl309 (Jones & Shenk, 1979) or AdEasy (He *et al.*, 1998). The absence of sequences that encode several E3 proteins would be expected to increase the sensitivity of infected cells to exogenous signals that induce apoptosis (Chapter 6). However, the deletions also prevent synthesis of the E3 ADP, which promotes lysis of infected cells and release of virus particles (Chapter 7). Modification of the genomes of CRAds to ensure overproduction of the ADP substantially enhances cytotoxicity in various types of tumor cells (Doronin *et al.*, 2000; Toth *et al.*, 2004). Such viruses have been demonstrated to be safe in permissive animal models (Chapter 9).

An interesting, and potentially powerful complement to such rational design to increase cytotoxicity is unbiased selection following random chemical mutagenesis of HAdV-C5. Selection for cytotoxicity was achieved by serial passages (up to 19) of stocks harvested as soon as the first signs of CPE were evident in a colorectal cancer cell line (HT129) that is quite resistant to lysis by wild-type HAdV-C5. Subsequent plaque assay on the same cells identified two mutant viruses that formed large plaques (Yan *et al.*, 2003). These mutants lysed HT129 cells up to 100 times more efficiently than did the parental virus, with no change in the induction of CPE in normal human cells. Sequencing of the mutant virus genomes identified four common mutations. One of these led to a truncation of the C-terminal 20 amino acids from the protein encoded by a fourth 5' exon found in some major late mRNAs, the i-leader (Fig. 5.2), and was shown to be essential for increased oncolytic activity (Yan *et al.*, 2003). Remarkably, a second independent experiment of this type using an RGD-modified HAdV-C5 and selection in cancer-associated fibroblasts also identified truncation of the i-leader protein, in this case of the 17 C-terminal amino acids, as increasing cytotoxicity (Puig-Saus *et al.*, 2012). These observations illustrate the power of unbiased methods to identify viral gene products that modulate cell killing. Despite its potential value, only one other application using this approach has been reported, and led to selection of a mutant virus carrying an insertion that truncated the E3 gp19K protein in its endoplasmic reticulum (ER) retention domain (Chapter 8). As a result, the protein localized to the plasma membrane, with concomitant increases in the release of virus particles and local and systemic antitumor activity in animals (Gros *et al.*, 2008).

12.3.3.2 *Arming conditionally replicating HAdVs*

An important theme in the development of oncolytic CRAds has been combining modifications that increase tumor cell selectivity with insertion into the viral genome of transgenes specifying proteins or small RNAs that foster tumor cell killing. A very large number of such manipulations have been reported, and many described in detailed reviews (Alemany, 2012; Cody & Douglas, 2009), including recent reviews of CRAds for immunotherapy (de Gruijl & van de Ven, 2012; Jiang *et al.*, 2015) and application in combination with drugs (Hallden *et al.*, 2012). The principal strategies adopted for increasing tumor killing by HAdV-C5 derivatives are described in this section.

12.3.3.2.1 Enhancing cell killing

Insertion into the viral genome of various pro-apoptotic genes, including those that encode the tumor suppressor p53, a p53 variant resistant to degradation, and activators of the caspase cascade like SMAC (second mitochondria-derived activator of caspase) results in induction of apoptosis, as expected, and increased destruction of tumor cells in xenograft models (see Cody & Douglas, 2009). Overexpression of the tumor suppressor cell adhesion molecule 1 (a.k.a. tumor suppressor lung cancer 1) and p73 (in conjunction with an shRNA [small hairpin RNA] targeting HDAC1) was also reported to improve killing of infected tumor cells in which the tumor suppressor genes are lost or perturbed, hepatocellular carcinoma (He *et al.*, 2012) and metastatic melanoma (Schipper *et al.*, 2014), respectively. Similarly, overexpression of the pro-apoptotic gene TNF-related apoptosis-inducing ligand (TRAIL) allowed efficient killing of glioblastoma and lung cancer-like stem cells infected by CRAds (X. Li *et al.*, 2012; Yang *et al.*, 2015), and elimination of liver metastases of colon cancer in mice after systemic delivery of the CRAd (Sova *et al.*, 2004). As expected, increased tumor cell killing and inhibition of tumor growth in xenograft models can also be achieved when antisense DNA or sequences that encode shRNAs designed to inhibit synthesis of proteins that promote cell proliferation are built into the genomes of oncolytic CRAds. This approach has been applied using antisense cDNA against protein kinases that regulate cell cycle progression and cell division such as polo like kinase 1 (Zhou *et al.*, 2005) and shRNAs targeting MYC-N, which is amplified in aggressive neuroblastomas (Li *et al.*, 2013), as well as anti-apoptotic proteins (e.g. Liu *et al.*, 2012).

An alternative method of cell killing is to introduce so-called 'suicide' genes, which code for enzymes that catalyze the conversion of non-toxic prodrugs to toxic compounds. The toxic molecules produced in the infected cells that express such suicide genes diffuse to, and kill, neighboring cells (bystander effect), a major advantage when the HAdV-C5 vectors do not infect all cells in a tumor. In addition, there is potential to control toxicity, by addition or removal of the prodrug. Suicide genes/prodrug combinations that have that have been tested with CRAds include herpes simplex virus thymidine kinase (TK) / gancyclovir, cytosine deaminase (CD) from bacteria or yeast (which has greater catalytic activity)/5-flourocytosine and TK-CD fusions with both prodrugs (Cody & Douglas, 2009). These systems increase potency of CRAds against tumor cells in culture and animal models, but the effects can be dependent on the viral backbone and prodrug dose. For example, exposure to 5-flourocytosine was reported to enhance tumor cell killing following low, but not by high, multiplicity of infection (MOI) with a CRAd carrying a CD gene (Liu & Deisseroth, 2006), presumably because the drug inhibits viral nucleic acid synthesis. When introduced via CRAds, the TK–CD fusion gene was more effective than either single gene alone, and enhanced the sensitivity of tumor cells to irradiation sufficiently to induce complete remission of cervical carcinoma xenografts in animal models (Rogulski *et al.*, 2000). This oncolytic derivative of HAdV-C5 was one of the first to enter clinical trials (see next section).

Other suicide genes that have been examined in pre-clinical systems include carboxypeptidase G2, nitroreductase, carboxylesterase, and, more recently, *Drosphila* deoxyribonucleoside kinase, which has broader substrate specificity and greater catalytic activity than herpesviral TK (Cody & Douglas, 2009; Zhu *et al.*, 2010). Exposing lung cancer cell xeongrafts infected with a CRAd armed with the *E. coli* nitroreductase gene to a more potent prodrug was particularly effective, with 62.5% of the mice tumor free on day 120 (Singleton *et al.*, 2007).

12.3.3.2.2 Enhancing immune responses

A hallmark of cancer is the establishment of a local immunosuppressive environment via release of inhibitory cytokines made in tumor cells and recruitment of regulatory cells that limit immune responses (de Gruijl & van de Ven, 2012). Consequently, increasing attention has been paid to developing oncolytic CRAds that include in their genomes coding sequences for proteins that circumvent immune suppression by tumor cells. As described in detail elsewhere (de Gruijl & van de Ven, 2012; Jiang *et al.*, 2015), this strategy has been investigated with intratumoral expression of transgenes that encode specific stimulators of immune responses, such as IFNα, TNFα, granulocyte macrophage-colony stimulating factor (GM-CSF) and tumor necrosis factor ligand superfamily 9 (a.k.a. 4-1BB ligand), chemokines that attract dendritic and T cells or proteins that block the action of immune checkpoint inhibitors, notably an anti-T-lymphocyte associated protein 4 monoclonal antibody and a soluble, inhibitory form of programmed death ligand 1 (Shin *et al.*, 2013). In general, including such immunomodulatory genes in the genomes of CRAds increased tumor cell destruction in xenograft models, and correlated with increased antitumor immune responses. Furthermore, CRAds that carry GM-CSF genes are now in clinical trials (see next section).

12.3.3.2.3 Manipulating the tumor microenvironment

Several CRAds have been engineered for delivery of genes for proteins that inhibit angiogenesis to solid tumors, which depend on the development of new blood vessels for continued expansion. Such anti-angiogenic genes include those that encode murine or human endostatin (Fang *et al.*, 2010; G. Li *et al.*, 2005), soluble VEGF receptors (decoys for VEGF) (Thorne *et al.*, 2006; Zhang *et al.*, 2005), and a VEGF inhibitor (Xiao *et al.*, 2010). They are expressed well in the various tumor cells tested, inhibit angiogenesis and improve oncolytic activity (see Cody & Douglas, 2009). Angiogenesis has also been inhibited more indirectly, for example, by arming CRAds with genes that encode tissue inhibitor of metalloproteinase 3 (TIMP3), which degrades proteins of the extracellular matrix (ECM) and other components of the tumor microenvironment and in so doing induces release of proteins that stimulate angiogenesis and tumor growth. In initial experiments, CRAd delivery of this gene was observed to increase oncolytic activity and suppress proliferation of tumor cells in culture, but did not improve activity against glioblastoma in xenograft models (Lamfers *et al.*, 2005). However, more effective inhibition of angiogenesis and tumor growth has been reported subsequently with CRAds that deliver TIMP2 alone or in combination with chemotherapy and radiation therapy (McNally *et al.*, 2009; Yang *et al.*, 2011). Delivery of shRNAs that inhibit production of cell-membrane-associated proteins of the ECM (Ulasov *et al.*, 2015), VEGF (Yoo *et al.*, 2007), or proteins that stimulate destruction of the ECM, which supports migration of cells into the tumor, e.g. for angiogenesis, has also been investigated. The latter approach can promote oncolytic activity when the efficiency of CRAd transduction is low (Cody & Douglas, 2009).

12.3.4 *Clinical trials of oncolytic CRAds*

The strategies to improve tumor-cell-selective reproduction and cytotoxicity of HAdV-C5 vectors considered in the previous sections have been integrated in many different ways to generate a legion of CRAds that have, for the most part, exhibited encouraging properties in preclinical systems, typically xenografts of human tumor cells in mice. To date, however, a relatively small number have followed ONYX-015 and H101 into clinical trials in humans (Table 12.2).

Several CRAds engineered solely for reproduction are specific tumor types, such as prostate cancer (DeWeese *et al.*, 2001; Small *et al.*, 2006), ovarian, and other gynocological tumors (Kim *et al.*, 2013; Kimball *et al.*, 2010), or in tumor cells more generally (Nemunaitis *et al.*, 2010; Nokisalmi *et al.*, 2010) have been examined in phase I trials. They are generally safe and well tolerated, and some instances of positive responses have been reported in some studies (DeWeese *et al.*, 2001; Nemunaitis *et al.*, 2010; Nokisalmi *et al.*, 2010). Two armed CRAds have been investigated in the clinic in more detail.

The first such CRAd to be tested clinically carries mutations that prevent synthesis of the E1B 55 kDa protein, a deletion in the E3 region and coding sequences for a CD/HSV TK fusion protein under the control of the strong, constitutive HCVMV IE enhancer and promoter in an E1B intron (Freytag *et al.*, 1998). It was well tolerated, with no dose-limiting toxicities following injection into prostate tumors followed by administration of 5-fluorocytosine and ganciclovir for one or two weeks, and there was evidence of tumor destruction at the infection site (Freytag *et al.*, 2002; Freytag *et al.*, 2003). In a five-year follow-up study of the phase I trial patients, measurement of prostate antigen-specific doubling time (a surrogate endpoint) indicated that the treatment delayed the time to initiation of salvage androgen suppression therapy by an average of two years (Freytag *et al.*, 2007b). A second-generation vector of this kind included the E3 ADP gene under the control of the HCMV IE enhancer and promoter and coding sequences for CD and TK enzymes with greater catalytic activity (Barton *et al.*, 2006). This CRAd was also well tolerated in combination with intensity-modulated radiotherapy (IMRT), to which tumor cells were sensitized by infection (Freytag *et al.*, 2002), with some benefit to low-risk patients (Freytag *et al.*, 2007a). A phase II trial of this second generation CRAd plus IMRT compared to IMRT alone reported a 60% relative reduction positive biopsy at two years in the CRAd-treated patients (Freytag *et al.*, 2014).

Phase I trials in which CRAds that deliver GM-CSF and that have been modified to increase tumor cell selectivity using various approaches have also been conducted in patients with recurrent head and neck cancer (Chang *et al.*, 2009) invasive bladder cancer (Burke *et al.* 2012), and various advanced solid tumors refractory to standard treatments (Koski *et al.*, 2010; Pesonen & Diaconu, 2012). GM-CSF promotes the proliferation, maturation, and migration of dendritic cells and is widely used to increase antitumor immunity. The viruses were well tolerated, and some evidence of clinical effect detected. A subsequent case-control analysis assessed the survival of 270 patients with advanced tumors treated in the Advanced Therapy Access Program in Finland and in 186 control patients from the same hospital and discerned no overall increase in survival in the CRAd-treated cohort (Kanerva *et al.*, 2015). However, a significant improvement in median survival was observed in patients with GM-CSF-sensitive tumor types and treated with CRAds that deliver this protein. Analysis of immunological responses in tumor biopsies or fluid samples correlated poor prognosis and shorter survival time in patients treated with oncolytic CRAds with activation of innate immunity prior to the onset of treatment (Taipale *et al.*, 2015). As the patients in this program have advanced disease, some died relatively soon after treatment. The autopsies required to establish cause of death (disease progression in all cases) allowed a unique biodistribution study of oncolytic CRAds in humans. Viral DNA was detected in a wide range of tissues, including tissues that were not injected and normal tissues, indicating dissemination of CRAds via viremia (Koski *et al.*, 2015). Virus that could be propagated in cells in culture was recovered from an untreated brain metastasis of one patient, consistent with dissemination by this route to tumors distant from site of infection.

The data collected from the substantial number of patients who have now been treated by intratumoral infection of a CRAd establish that this procedure is safe and usually well tolerated. It therefore seems likely that that the number of clinical trials of these agents will continue to increase as more of the strategies for CRAd development described above complete preclinical testing.

12.4 What Next?

The cumulative experience with vectors for vaccine delivery strongly argues that rare HAdVs and adenovirus of non-human mammals are likely to be most useful, especially as HAdV-C5 vectors can be deleterious in some circumstances. Indeed, encouraging data have been collected in recent trails of chimpanzee adenovirus vectors for malaria and Ebola virus vaccines, and continued testing of these and other adenovirus vaccines seems likely to lead to approved therapeutic uses in the not too distant future.

In contrast to these vectors, HAdVs for cancer therapy are replication competent, and are purposely designed (or selected) to reproduce efficiently in, and kill, tumor cells. A very large number of molecular strategies have been adopted in attempts to limit or prevent replication in normal human cells while increasing destruction of tumor cells. One such CRAd has been approved for use in China, in combination with chemotherapy, and several others are in clinical trials. Although safe, these viruses are typically not highly efficacious, and seem likely to be more potent when used in combination with chemotherapy or immunotherapy. The great majority of CRAds described have been tested only in cells in culture or animals models, typically xenografts of human tumors in mice. As testing all or even the majority, of the large number of these agents in humans is impractical, comparison of the CRAds for a specific application in preclinical studies would be valuable, particularly if performed in a permissive animal model, or humanized mice.

Bibliography

Abbink, P., A. A. Lemckert, B. A. Ewald, D. M. Lynch, M. Denholtz, S. Smits, L. Holterman, I. Damen, R. Vogels, A. R. Thorner, K. L. O'Brien, A. Carville, K. G. Mansfield, J. Goudsmit, M. J. Havenga and D. H. Barouch (2007). "Comparative seroprevalence and immunogenicity of six rare serotype recombinant adenovirus vaccine vectors from subgroups B and D." *J Virol* **81**(9): 4654–63.

Abdelhakim, A. H., E. N. Salgado, X. Fu, M. Pasham, D. Nicastro, T. Kirchhausen and S. C. Harrison (2014). "Structural correlates of rotavirus cell entry." *PLoS Pathog* **10**(9): e1004355.

Ablack, J. N., M. Cohen, G. Thillainadesan, G. J. Fonseca, P. Pelka, J. Torchia and J. S. Mymryk (2012). "Cellular GCN5 is a novel regulator of human adenovirus E1A-conserved region 3 transactivation." *J Virol* **86**(15): 8198–209.

Ablack, J. N., P. Pelka, A. F. Yousef, A. S. Turnell, R. J. Grand and J. S. Mymryk (2010). "Comparison of E1A CR3-dependent transcriptional activation across six different human adenovirus subgroups." *J Virol* **84**(24): 12771–81.

Abrescia, N. G., J. J. Cockburn, J. M. Grimes, G. C. Sutton, J. M. Diprose, S. J. Butcher, S. D. Fuller, C. San Martin, R. M. Burnett, D. I. Stuart, D. H. Bamford and J. K. Bamford (2004). "Insights into assembly from structural analysis of bacteriophage PRD1." *Nature* **432**(7013): 68–74.

Ackrill, A. M. and G. E. Blair (1988a). "Expression of hamster MHC class I antigens in transformed cells and tumours induced by human adenoviruses." *Eur J Cancer Clin Oncol* **24**(11): 1745–50.

Ackrill, A. M. and G. E. Blair (1988b). "Regulation of major histocompatibility class I gene expression at the level of transcription in highly oncogenic adenovirus transformed rat cells." *Oncogene* **3**(4): 483–7.

Ackrill, A. M., G. R. Foster, C. D. Laxton, D. M. Flavell, G. R. Stark and I. M. Kerr (1991). "Inhibition of the cellular response to interferons by products of the adenovirus type 5 E1A oncogene." *Nucleic Acids Res* **19**(16): 4387–93.

Adam, S., T. Nakagawa, M. S. Swanson, T. K. Woodruff and G. Dreyfuss (1986). "mRNA polyadenylate-binding protein: gene isolation and identification of a ribonucleoprotein consensus sequence." *Mol. Cell. Biol.* **6**: 2932–2943.

Ahi, Y. S., S. V. Vemula, A. O. Hassan, G. Costakes, C. Stauffacher and S. K. Mittal (2015). "Adenoviral L4 33K forms ring-like oligomers and stimulates ATPase activity of IVa2: implications in viral genome packaging." *Front Microbiol* **6**: 318.

Ahi, Y. S., S. V. Vemula and S. K. Mittal (2013). "Adenoviral E2 IVa2 protein interacts with L4 33K protein and E2 DNA-binding protein." *J Gen Virol* **94**(Pt 6): 1325–34.

Aissi-Rothe, L., V. Decot, V. Venard, H. Jeulin, A. Salmon, L. Clement, A. Kennel, C. Mathieu, J. H. Dalle, G. Rauser, C. Cambouris, M. de Carvalho, J. F. Stoltz, P. Bordigoni and D. Bensoussan (2010). "Rapid generation of full clinical-grade human antiadenovirus cytotoxic T cells for adoptive immunotherapy." *J Immunother* **33**(4): 414–24.

Alba, R., A. C. Bradshaw, N. Mestre-Frances, J. M. Verdier, D. Henaff and A. H. Baker (2012). "Coagulation factor X mediates adenovirus type 5 liver gene transfer in non-human primates (Microcebus murinus)." *Gene Ther* **19**(1): 109–13.

Albinsson, B. and A. H. Kidd (1999). "Adenovirus type 41 lacks an RGD alpha(v)-integrin binding motif on the penton base and undergoes delayed uptake in A549 cells." *Virus Res* **64**(2): 125–36.

Aldhamen, Y. A., S. S. Seregin and A. Amalfitano (2011). "Immune recognition of gene transfer vectors: focus on adenovirus as a paradigm." *Front Immunol* **2**: 40.

Alemany, R. (2012). "Chapter four — Design of improved oncolytic adenoviruses." *Adv Cancer Res* **115**: 93–114.

Alemany, R., K. Suzuki and D. T. Curiel (2000). "Blood clearance rates of adenovirus type 5 in mice." *J Gen Virol* **81**(Pt 11): 2605–9.

Ali, H., G. LeRoy, G. Bridge and S. J. Flint (2007). "The adenovirus L4 33-kilodalton protein binds to intragenic sequences of the major late promoter required for late phase-specific stimulation of transcription." *J Virol* **81**(3): 1327–38.

Allen, B. L. and D. J. Taatjes (2015). "The Mediator complex: a central integrator of transcription." *Nat Rev Mol Cell Biol* **16**(3): 155–66.

Alonso-Padilla, J., T. Papp, G. L. Kajan, M. Benko, M. Havenga, A. Lemckert, B. Harrach and A. H. Baker (2015). "Development of Novel Adenoviral Vectors to Overcome Challenges Observed With HAdV-5-based Constructs." *Mol Ther*.

Amstutz, B., M. Gastaldelli, S. Kalin, N. Imelli, K. Boucke, E. Wandeler, J. Mercer, S. Hemmi and U. F. Greber (2008). "Subversion of CtBP1-controlled macropinocytosis by human adenovirus serotype 3." *EMBO J* **27**(7): 956–69.

Anderson, C. W. (1990). "The proteinase polypeptide of adenovirus serotype 2 virions." *Virology* **177**(1): 259–72.

Anderson, C. W., P. R. Baum and R. F. Gesteland (1973). "Processing of adenovirus 2-induced proteins." *J Virol* **12**(2): 241–52.

Anderson, D. H., V. A. Kickhoefer, S. A. Sievers, L. H. Rome and D. Eisenberg (2007). "Draft crystal structure of the vault shell at 9-A resolution." *PLoS Biol* **5**(11): e318.

Anderson, K. P. and E. H. Fenni (1987). "Adenovirus early region 1A modulation of interferon antiviral activity." *J Virol* **61**(3): 787–95.

Anderson, K. P., E. A. Wong and D. F. Klessig (1985). "Microinjection of mRNA enhances translational efficiency of human adenovirus fiber message in monkey cells." *Mol Cell Biol* **5**(10): 2870–3.

Andersson, M. G., P. C. Haasnoot, N. Xu, S. Berenjian, B. Berkhout and G. Akusjarvi (2005). "Suppression of RNA interference by adenovirus virus-associated RNA." *J Virol* **79**(15): 9556–65.

Antrobus, R. D., L. Coughlan, T. K. Berthoud, M. D. Dicks, A. V. Hill, T. Lambe and S. C. Gilbert (2014). "Clinical assessment of a novel recombinant simian adenovirus ChAdOx1 as a vectored vaccine expressing conserved Influenza A antigens." *Mol Ther* **22**(3): 668–74.

Aoki, K., M. Benko, A. J. Davison, M. Echavarria, D. D. Erdman, B. Harrach, A. E. Kajon, D. Schnurr, G. Wadell and Members of the Adenovirus Research (2011). "Toward an integrated human adenovirus designation system that utilizes molecular and serological data and serves both clinical and fundamental virology." *J Virol* **85**(11): 5703–4.

Aparicio, O., E. Carnero, X. Abad, N. Razquin, E. Guruceaga, V. Segura and P. Fortes (2010). "Adenovirus VA RNA-derived miRNAs target cellular genes involved in cell growth, gene expression and DNA repair." *Nucleic Acids Res* **38**(3): 750–63.

Aparicio, O., N. Razquin, M. Zaratiegui, I. Narvaiza and P. Fortes (2006). "Adenovirus virus-associated RNA is processed to functional interfering RNAs involved in virus production." *J Virol* **80**(3): 1376–84.

Aplander, K., M. Marttila, S. Manner, N. Arnberg, O. Sterner and U. Ellervik (2011). "Molecular wipes: application to epidemic keratoconjuctivitis." *J Med Chem* **54**(19): 6670–5.

Appaiahgari, M. B. and S. Vrati (2015). "Adenoviruses as gene/vaccine delivery vectors: promises and pitfalls." *Expert Opin Biol Ther* **15**(3): 337–51.

Appella, E. and C. W. Anderson (2001). "Post-translational modifications and activation of p53 by genotoxic stresses." *Eur J Biochem* **268**(10): 2764–72.

Appledorn, D. M., A. McBride, S. Seregin, J. M. Scott, N. Schuldt, A. Kiang, S. Godbehere and A. Amalfitano (2008). "Complex interactions with several arms of the complement system dictate innate and humoral immunity to adenoviral vectors." *Gene Ther* **15**(24): 1606–17.

Appledorn, D. M., S. Patial, A. McBride, S. Godbehere, N. Van Rooijen, N. Parameswaran and A. Amalfitano (2008). "Adenovirus vector-induced innate inflammatory mediators, MAPK signaling, as well as adaptive immune responses are dependent upon both TLR2 and TLR9 in vivo." *J Immunol* **181**(3): 2134–44.

Araujo, F. D., T. H. Stracker, C. T. Carson, D. V. Lee and M. D. Weitzman (2005). "Adenovirus type 5 E4orf3 protein targets the Mre11 complex to cytoplasmic aggresomes." *J Virol* **79**(17): 11382–91.

Ariga, H., H. Klein, A. J. Levine and M. S. Horwitz (1980). "A cleavage product of the adenovirus DNA binding protein is active in DNA replication in vitro." *Virology* **101**(1): 307–10.

Armentero, M. T., M. Horwitz and N. Mermod (1994). "Targeting of DNA polymerase to the adenovirus origin of DNA replication by interaction with nuclear factor I." *Proc Natl Acad Sci U S A* **91**(24): 11537–41.

Arnberg, N. (2012). "Adenovirus receptors: implications for targeting of viral vectors." *Trends Pharmacol Sci* **33**(8): 442–8.

Arnberg, N., K. Edlund, A. H. Kidd and G. Wadell (2000). "Adenovirus type 37 uses sialic acid as a cellular receptor." *J Virol* **74**(1): 42–8.

Arnberg, N., A. H. Kidd, K. Edlund, J. Nilsson, P. Pring-Akerblom and G. Wadell (2002). "Adenovirus type 37 binds to cell surface sialic acid through a charge-dependent interaction." *Virology* **302**(1): 33–43.

Arnberg, N., A. H. Kidd, K. Edlund, F. Olfat and G. Wadell (2000). "Initial interactions of subgenus D adenoviruses with A549 cellular receptors: sialic acid versus alpha(v) integrins." *J Virol* **74**(16): 7691–3.

Arnberg, N., Y. Mei and G. Wadell (1997). "Fiber genes of adenoviruses with tropism for the eye and the genital tract." *Virology* **227**(1): 239–44.

Arnberg, N., P. Pring-Akerblom and G. Wadell (2002). "Adenovirus type 37 uses sialic acid as a cellular receptor on Chang C cells." *J Virol* **76**(17): 8834–41.

Arulsundaram, V. D., P. Webb, A. F. Yousef, P. Pelka, G. J. Fonseca, J. D. Baxter, P. G. Walfish and J. S. Mymryk (2014). "The adenovirus 55 residue E1A protein is a transcriptional activator and binds the unliganded thyroid hormone receptor." *J Gen Virol* **95**(Pt 1): 142–52.

Aspegren, A., C. Rabino and E. Bridge (1998). "Organization of splicing factors in adenovirus-infected cells reflects changes in gene expression during the early to late phase transition." *Exp Cell Res* **245**(1): 203–13.

Avvakumov, N., R. Wheeler, J. C. D'Halluin and J. S. Mymryk (2002). "Comparative sequence analysis of the largest E1A proteins of human and simian adenoviruses." *J Virol* **76**(16): 7968–75.

Axelrod, N. (1978). "Phosphoproteins of adenovirus 2." *Virology* **87**: 366–383.

Ayabe, T., D. P. Satchell, C. L. Wilson, W. C. Parks, M. E. Selsted and A. J. Ouellette (2000). "Secretion of microbicidal alpha-defensins by intestinal Paneth cells in response to bacteria." *Nat Immunol* **1**(2): 113–8.

Ayalew, L. E., P. Kumar, A. Gaba, N. Makadiya and S. K. Tikoo (2015). "Bovine adenovirus-3 as a vaccine delivery vehicle." *Vaccine* **33**(4): 493–9.

Babich, A., L. T. Feldman, J. R. Nevins, J. E. Darnell, Jr. and C. Weinberger (1983). "Effect of adenovirus on metabolism of specific host mRNAs: transport control and specific translational discrimination." *Mol Cell Biol* **3**(7): 1212–21.

Backstrom, E., K. B. Kaufmann, X. Lan and G. Akusjarvi (2010). "Adenovirus L4-22K stimulates major late transcription by a mechanism requiring the intragenic late-specific transcription factor-binding site." *Virus Res* **151**(2): 220–8.

Baehrecke, E. H. (2005). "Autophagy: dual roles in life and death?" *Nat Rev Mol Cell Biol* **6**(6): 505–10.

Bagchi, S., P. Raychaudhuri and J. R. Nevins (1990). "Adenovirus E1A proteins can dissociate heteromeric complexes involving the E2F transcription factor: a novel mechanism for E1A trans-activation." *Cell* **62**: 659–669.

Bai, M., B. Harfe and P. Freimuth (1993). "Mutations that alter an Arg-Gly-Asp (RGD) sequence in the adenovirus type 2 penton base protein abolish its cell-rounding activity and delay virus reproduction in flat cells." *J Virol* **67**(9): 5198–205.

Bailey, A. and V. Mautner (1994). "Phylogenetic relationships among adenovirus serotypes." *Virology* **205**(2): 438–52.

Baker, A., K. J. Rohleder, L. A. Hanakahi and G. Ketner (2007). "Adenovirus E4 34k and E1b 55k oncoproteins target host DNA ligase IV for proteasomal degradation." *J Virol* **81**(13): 7034–40.

Baker, A. H., S. A. Nicklin and D. M. Shayakhmetov (2013). "FX and host defense evasion tactics by adenovirus." *Mol Ther* **21**(6): 1109–11.

Banerjee, P. S., P. Ostapchuk, P. Hearing and I. Carrico (2010). "Chemoselective attachment of small molecule effector functionality to human adenoviruses facilitates gene delivery to cancer cells." *J Am Chem Soc* **132**(39): 13615–7.

Banerjee, P. S., P. Ostapchuk, P. Hearing and I. S. Carrico (2011). "Unnatural amino acid incorporation onto adenoviral (Ad) coat proteins facilitates chemoselective modification and retargeting of Ad type 5 vectors." *J Virol* **85**(15): 7546–54.

Banerjee, S., A. Bartesaghi, A. Merk, P. Rao, S. L. Bulfer, Y. Yan, N. Green, B. Mroczkowski, R. J. Neitz, P. Wipf, V. Falconieri, R. J. Deshaies, J. L. Milne, D. Huryn, M. Arkin and S. Subramaniam (2016). "2.3 A resolution cryo-EM structure of human p97 and mechanism of allosteric inhibition." *Science* **351**(6275): 871–5.

Barbeau, D., R. Charbonneau, S. G. Whalen, S. T. Bayley and P. E. Branton (1994). "Functional interactions within adenovirus E1A protein complexes." *Oncogene* **9**(2): 359–73.

Barker, D. D. and A. J. Berk (1987). "Adenovirus proteins from both E1B reading frames are required for transformation of rodent cells by viral infection and DNA transfection." *Virology* **156**(1): 107–21.

Barlan, A. U., P. Danthi and C. M. Wiethoff (2011). "Lysosomal localization and mechanism of membrane penetration influence nonenveloped virus activation of the NLRP3 inflammasome." *Virology* **412**(2): 306–14.

Barlan, A. U., T. M. Griffin, K. A. McGuire and C. M. Wiethoff (2011). "Adenovirus membrane penetration activates the NLRP3 inflammasome." *J Virol* **85**(1): 146–55.

Barnes, E., A. Folgori, S. Capone, L. Swadling, S. Aston, A. Kurioka, J. Meyer, R. Huddart, K. Smith, R. Townsend, A. Brown, R. Antrobus, V. Ammendola,

M. Naddeo, G. O'Hara, C. Willberg, A. Harrison, F. Grazioli, M. L. Esposito, L. Siani, C. Traboni, Y. Oo, D. Adams, A. Hill, S. Colloca, A. Nicosia, R. Cortese and P. Klenerman (2012). "Novel adenovirus-based vaccines induce broad and sustained T cell responses to HCV in man." *Sci Transl Med* **4**(115): 115ra1. doi: 10.1126/scitranslmed.3003155.

Barouch, D. H., S. V. Kik, G. J. Weverling, R. Dilan, S. L. King, L. F. Maxfield, S. Clark, D. Ng'ang'a, K. L. Brandariz, P. Abbink, F. Sinangil, G. de Bruyn, G. E. Gray, S. Roux, L. G. Bekker, A. Dilraj, H. Kibuuka, M. L. Robb, N. L. Michael, O. Anzala, P. N. Amornkul, J. Gilmour, J. Hural, S. P. Buchbinder, M. S. Seaman, R. Dolin, L. R. Baden, A. Carville, K. G. Mansfield, M. G. Pau and J. Goudsmit (2011). "International seroepidemiology of adenovirus serotypes 5, 26, 35, and 48 in pediatric and adult populations." *Vaccine* **29**(32): 5203–9.

Barouch, D. H. and G. J. Nabel (2005). "Adenovirus vector-based vaccines for human immunodeficiency virus type 1." *Hum Gene Ther* **16**(2): 149–56.

Barouch, D. H., M. G. Pau, J. H. Custers, W. Koudstaal, S. Kostense, M. J. Havenga, D. M. Truitt, S. M. Sumida, M. G. Kishko, J. C. Arthur, B. Korioth-Schmitz, M. H. Newberg, D. A. Gorgone, M. A. Lifton, D. L. Panicali, G. J. Nabel, N. L. Letvin and J. Goudsmit (2004). "Immunogenicity of recombinant adenovirus serotype 35 vaccine in the presence of pre-existing anti-Ad5 immunity." *J Immunol* **172**(10): 6290–7.

Barratt-Boyes, S. M., A. C. Soloff, W. Gao, E. Nwanegbo, X. Liu, P. A. Rajakumar, K. N. Brown, P. D. Robbins, M. Murphey-Corb, R. D. Day and A. Gambotto (2006). "Broad cellular immunity with robust memory responses to simian immunodeficiency virus following serial vaccination with adenovirus 5- and 35-based vectors." *J Gen Virol* **87**(Pt 1): 139–49.

Barry, M. A., E. A. Weaver and S. E. Hofherr (2010). "Rescue, amplification, purification, and PEGylation of replication defective first-generation adenoviral vectors." *Methods Mol Biol* **651**: 227–39.

Bartel, D. P. (2009). "MicroRNAs: target recognition and regulatory functions." *Cell* **136**(2): 215–33.

Barton, K. N., D. Paielli, Y. Zhang, S. Koul, S. L. Brown, M. Lu, J. Seely, J. H. Kim and S. O. Freytag (2006). "Second-generation replication-competent oncolytic adenovirus armed with improved suicide genes and ADP gene demonstrates greater efficacy without increased toxicity." *Mol Ther* **13**(2): 347–56.

Bastian, A. and H. Schafer (2001). "Human alpha-defensin 1 (HNP-1) inhibits adenoviral infection in vitro." *Regul Pept* **101**(1–3): 157–61.

Baud, V. and M. Karin (2001). "Signal transduction by tumor necrosis factor and its relatives." *Trends Cell Biol* **11**(9): 372–7.

Bauerschmitz, G. J., K. Guse, A. Kanerva, A. Menzel, I. Herrmann, R. A. Desmond, M. Yamamoto, D. M. Nettelbeck, T. Hakkarainen, P. Dall, D. T. Curiel and A. Hemminki (2006). "Triple-targeted oncolytic adenoviruses featuring the cox2 promoter, E1A transcomplementation, and serotype chimerism for enhanced selectivity for ovarian cancer cells." *Mol Ther* **14**(2): 164–74.

Beatty, M. S. and D. T. Curiel (2012). "Chapter two--Adenovirus strategies for tissue-specific targeting." *Adv Cancer Res* **115**: 39–67.

Bednenko, J., G. Cingolani and L. Gerace (2003). "Nucleocytoplasmic transport: navigating the channel." *Traffic* **4**(3): 127–35.

Begin, M. and J. Weber (1975). "Genetic analysis of adenovirus type 2. I. Isolation and genetic characterization of temperature-sensitive mutants." *J Virol* **15**(1): 1–7.

Bellutti, F., M. Kauer, D. Kneidinger, T. Lion and R. Klein (2015). "Identification of RISC-associated adenoviral microRNAs, a subset of their direct targets, and global changes in the targetome upon lytic adenovirus 5 infection." *J Virol* **89**(3): 1608–27.

Beltz, G. A. and S. J. Flint (1979). "Inhibition of HeLa cell protein synthesis during adenovirus infection: restriction of cellular messenger RNA sequences to the nucleus." *J. Mol. Biol.* **131**: 353–373.

Benedict, C. A., P. S. Norris, T. I. Prigozy, J. L. Bodmer, J. A. Mahr, C. T. Garnett, F. Martinon, J. Tschopp, L. R. Gooding and C. F. Ware (2001). "Three adenovirus E3 proteins cooperate to evade apoptosis by tumor necrosis factor-related apoptosis-inducing ligand receptor-1 and -2." *J Biol Chem* **276**(5): 3270–8.

Benevento, M., S. Di Palma, J. Snijder, C. L. Moyer, V. S. Reddy, G. R. Nemerow and A. J. Heck (2014). "Adenovirus composition, proteolysis, and disassembly studied by in-depth qualitative and quantitative proteomics." *J Biol Chem* **289**(16): 11421–30.

Benko, M. and B. Harrach (2003). "Molecular evolution of adenoviruses." *Curr Top Microbiol Immunol* **272**: 3–35.

Benlahrech, A., J. Harris, A. Meiser, T. Papagatsias, J. Hornig, P. Hayes, A. Lieber, T. Athanasopoulos, V. Bachy, E. Csomor, R. Daniels, K. Fisher, F. Gotch, L. Seymour, K. Logan, R. Barbagallo, L. Klavinskis, G. Dickson and S. Patterson (2009). "Adenovirus vector vaccination induces expansion of memory CD4 T cells with a mucosal homing phenotype that are readily susceptible to HIV-1." *Proc Natl Acad Sci U S A* **106**(47): 19940–5.

Bennett, E. M., J. R. Bennink, J. W. Yewdell and F. M. Brodsky (1999). "Cutting edge: adenovirus E19 has two mechanisms for affecting class I MHC expression." *J Immunol* **162**(9): 5049–52.

Benson, R. E., E. B. Gottlin, D. J. Christensen and P. T. Hamilton (2003). "Intracellular expression of peptide fusions for demonstration of protein essentiality in bacteria." *Antimicrob Agents Chemother* **47**(9): 2875–81.

Benson, S. D., J. K. Bamford, D. H. Bamford and R. M. Burnett (1999). "Viral evolution revealed by bacteriophage PRD1 and human adenovirus coat protein structures." *Cell* **98**(6): 825–33.

Bergelson, J. M., J. A. Cunningham, G. Droguett, E. A. Kurt-Jones, A. Krithivas, J. S. Hong, M. S. Horwitz, R. L. Crowell and R. W. Finberg (1997). "Isolation of a common receptor for Coxsackie B viruses and adenoviruses 2 and 5." *Science* **275**(5304): 1320–3.

Bergelson, J. M., A. Krithivas, L. Celi, G. Droguett, M. S. Horwitz, T. Wickham, R. L. Crowell and R. W. Finberg (1998). "The murine CAR homolog is a receptor for coxsackie B viruses and adenoviruses." *J Virol* **72**(1): 415–9.

Berget, S. M., C. Moore and P. A. Sharp (1977). "Spliced segments at the 5' terminus of adenovirus 2 late mRNA." *Proc Natl Acad Sci U S A* **74**(8): 3171–5.

Berhane, S., C. Areste, J. N. Ablack, G. B. Ryan, D. J. Blackbourn, J. S. Mymryk, A. S. Turnell, J. C. Steele and R. J. Grand (2011). "Adenovirus E1A interacts directly with, and regulates the level of expression of, the immunoproteasome component MECL1." *Virology* **421**(2): 149–58.

Berk, A. J. (2005). "Recent lessons in gene expression, cell cycle control, and cell biology from adenovirus." *Oncogene* **24**(52): 7673–85.

Berk, A. J. (2013). *Adenoviridae.* Fields Virology. D. M. Knipe and P. M. Howley. Philadelphia, PA, Wolters Kluwer/Lippincott Williams and Wilkins.

Berk, A. J., F. Lee, T. Harrison, J. F. Williams and P. A. Sharp (1979). "A pre-early adenovirus 5 gene product regulates synthesis of early viral messenger RNAs." *Cell* **17**: 935–944.

Berk, A. J. and P. A. Sharp (1977). "Sizing and Mapping of early adenovirus mRNAs by gel electrophoresis of S1 endonuclease digested hybrids." *Cell* **12**: 721–732.

Berman, A. J., S. Kamtekar, J. L. Goodman, J. M. Lazaro, M. de Vega, L. Blanco, M. Salas and T. A. Steitz (2007). "Structures of phi29 DNA polymerase complexed with substrate: the mechanism of translocation in B-family polymerases." *EMBO J* **26**(14): 3494–505.

Bernards, R., A. Houweling, P. I. Schrier, J. L. Bos and A. J. Van der Eb (1982). "Characterization of cells transformed by Ad5/Ad12 hybrid early region I plasmids." *Virology* **120**(2): 422–32.

Berscheminski, J., P. Groitl, T. Dobner, P. Wimmer and S. Schreiner (2013). "The adenoviral oncogene E1A-13S interacts with a specific isoform of the tumor suppressor PML to enhance viral transcription." *J Virol* **87**(2): 965–77.

Berscheminski, J., P. Wimmer, J. Brun, W. H. Ip, P. Groitl, T. Horlacher, E. Jaffray, R. T. Hay, T. Dobner and S. Schreiner (2014). "Sp100 isoform-specific regulation of human adenovirus 5 gene expression." *J Virol* **88**(11): 6076–92.

Bett, A. J., S. A. Dubey, D. V. Mehrotra, L. Guan, R. Long, K. Anderson, K. Collins, C. Gaunt, R. Fernandez, S. Cole, S. Meschino, A. Tang, X. Sun, S. Gurunathan, J. Tartaglia, M. N. Robertson, J. W. Shiver and D. R. Casimiro (2010). "Comparison of T cell immune responses induced by vectored HIV vaccines in non-human primates and humans." *Vaccine* **28**(50): 7881–9.

Bett, A. J., W. Haddara, L. Prevec and F. L. Graham (1994). "An efficient and flexible system for construction of adenovirus vectors with insertions or deletions in early regions 1 and 3." *Proc Natl Acad Sci U S A* **91**(19): 8802–6.

Bett, A. J., L. Prevec and F. L. Graham (1993). "Packaging capacity and stability of human adenovirus type 5 vectors." *J Virol* **67**(10): 5911–21.

Bewley, M. C., K. Springer, Y. B. Zhang, P. Freimuth and J. M. Flanagan (1999). "Structural analysis of the mechanism of adenovirus binding to its human cellular receptor, CAR." *Science* **286**(5444): 1579–83.

Beyer, A. L., A. H. Bouton, L. D. Hodge and O. L. Miller, Jr. (1981). "Visualization of the major late R strand transcription unit of adenovirus serotype 2." *J Mol Biol* **147**(2): 269–95.

Beyer, I., H. Cao, J. Persson, H. Song, M. Richter, Q. Feng, R. Yumul, R. van Rensburg, Z. Li, R. Berenson, D. Carter, S. Roffler, C. Drescher and A. Lieber (2012). "Coadministration of epithelial junction opener JO-1 improves the efficacy and safety of chemotherapeutic drugs." *Clin Cancer Res* **18**(12): 3340–51.

Beyer, I., H. Cao, J. Persson, H. Wang, Y. Liu, R. Yumul, Z. Li, D. Woodle, R. Manger, M. Gough, D. Rocha, J. Bogue, A. Baldessari, R. Berenson, D. Carter and A. Lieber (2013). "Transient removal of CD46 is safe and increases B-cell depletion by rituximab in CD46 transgenic mice and macaques." *Mol Ther* **21**(2): 291–9.

Beyer, I., R. van Rensburg, R. Strauss, Z. Li, H. Wang, J. Persson, R. Yumul, Q. Feng, H. Song, J. Bartek, P. Fender and A. Lieber (2011). "Epithelial junction opener JO-1 improves monoclonal antibody therapy of cancer." *Cancer Res* **71**(22): 7080–90.

Bhat, R. A. and B. Thimmappaya (1984). "Adenovirus mutants with DNA sequence perturbations in the intragenic promoter of VA$_\text{I}$ RNA gene allow the enhanced transcription of VA$_\text{II}$ RNA gene in HeLa cells." *Nucleic Acids Res.* **12**: 7377–7388.

Bhattacharya, S., R. Eckner, S. Grossman, E. Oldread, Z. Arany, A. D'Andrea and D. M. Livingston (1996). "Cooperation of Stat2 and p300/CBP in signalling induced by interferon-alpha." *Nature* **383**(6598): 344–7.

Biasiotto, R. and G. Akusjarvi (2015). "Regulation of human adenovirus alternative RNA splicing by the adenoviral L4-33K and L4-22K proteins." *Int J Mol Sci* **16**(2): 2893–912.

Bidgood, S. R., J. C. Tam, W. A. McEwan, D. L. Mallery and L. C. James (2014). "Translocalized IgA mediates neutralization and stimulates innate immunity inside infected cells." *Proc Natl Acad Sci U S A* **111**(37): 13463–8.

Binger, M. H. and S. J. Flint (1984). "Accumulation of early and intermediate mRNA species during subgroup C adenovirus productive infections." *Virology* **136**: 387–403.

Biron, C. A., K. B. Nguyen, G. C. Pien, L. P. Cousens and T. P. Salazar-Mather (1999). "Natural killer cells in antiviral defense: function and regulation by innate cytokines." *Annu Rev Immunol* **17**: 189–220.

Bischoff, J. R., D. H. Kirn, A. Williams, C. Heise, S. Horn, M. Muna, L. Ng, J. A. Nye, A. Sampson-Johannes, A. Fattaey and F. McCormick (1996). "An adenovirus mutant that replicates selectively in p53-deficient human tumor cells." *Science* **274**(5286): 373–6.

Bishop, J. M. (1985). "Proto-oncogenes: Clues to the puzzle of purpose." *Nature* **316**(6028): 483–4.

Blackford, A. N., R. K. Bruton, O. Dirlik, G. S. Stewart, A. M. Taylor, T. Dobner, R. J. Grand and A. S. Turnell (2008). "A role for E1B-AP5 in ATR signaling pathways during adenovirus infection." *J Virol* **82**(15): 7640–52.

Blackwell, J. L., C. R. Miller, J. T. Douglas, H. Li, P. N. Reynolds, W. R. Carroll, G. E. Peters, T. V. Strong and D. T. Curiel (1999). "Retargeting to EGFR enhances adenovirus infection efficiency of squamous cell carcinoma." *Arch Otolaryngol Head Neck Surg* **125**(8): 856–63.

Blainey, P. C., V. Graziano, A. J. Perez-Berna, W. J. McGrath, S. J. Flint, C. San Martin, X. S. Xie and W. F. Mangel (2013). "Regulation of a viral proteinase by a peptide and DNA in one-dimensional space: IV. viral proteinase slides along DNA to locate and process its substrates." *J Biol Chem* **288**(3): 2092–102.

Blanchette, P., K. Kindsmüller, P. Groitl, F. Dallaire, T. Speiseder, P. E. Branton and T. Dobner (2008). "Control of mRNA export by adenovirus E4orf6 and E1B55K proteins during productive infection requires E4orf6 ubiquitin ligase activity." *J Virol* **82**(6): 2642–51.

Blusch, J. H., F. Deryckere, M. Windheim, Z. Ruzsics, N. Arnberg, T. Adrian and H. G. Burgert (2002). "The novel early region 3 protein E3/49K is specifically expressed by adenoviruses of subgenus D: implications for epidemic keratoconjunctivitis and adenovirus evolution." *Virology* **296**(1): 94–106.

Bosher, J., E. C. Robinson and R. T. Hay (1990). "Interactions between the adenovirus type 2 DNA polymerase and the DNA binding domain of nuclear factor I." *New Biol* **2**(12): 1083–90.

Both, G. W. (2004). "Ovine atadenovirus: a review of its biology, biosafety profile and application as a gene delivery vector." *Immunol Cell Biol* **82**(2): 189–95.

Botting, C. H. and R. T. Hay (1999). "Characterisation of the adenovirus preterminal protein and its interaction with the POU homeodomain of NFIII (Oct-1)." *Nucleic Acids Res* **27**(13): 2799–805.

Boudin, M. L., M. Moncany, J. C. D'Halluin and P. A. Boulanger (1979). "Isolation and characterization of adenovirus type 2 vertex capsomer (penton base)." *Virology* **92**(1): 125–38.

Boulakia, C. A., G. Chen, F. W. Ng, J. G. Teodoro, P. E. Branton, D. W. Nicholson, G. G. Poirier and G. C. Shore (1996). "Bcl-2 and adenovirus E1B 19 kDa protein prevent E1A-induced processing of CPP32 and cleavage of poly(ADP-ribose) polymerase." *Oncogene* **12**(3): 529–35.

Boulanger, P. A. and B. Hennache (1973). "Adenovirus uncoating: an additional evidence for the involvement of cell surface in capsid labilization." *FEBS Lett* **35**(1): 15–8.

Boyd, J. M., T. Subramanian, U. Schaeper, M. La Regina, S. Bayley and G. Chinnadurai (1993). "A region in the C-terminus of adenovirus 2/5 E1a protein is required for association with a cellular phosphoprotein and important for the negative modulation of T24-ras mediated transformation, tumorigenesis and metastasis." *Embo J* **12**(2): 469–78.

Boyer, J., K. Rohleder and G. Ketner (1999). "Adenovirus E4 34k and E4 11k inhibit double strand break repair and are physically associated with the cellular DNA-dependent protein kinase." *Virology* **263**(2): 307–12.

Boyer, J. L. and G. Ketner (2000). "Genetic analysis of a potential zinc-binding domain of the adenovirus E4 34k protein." *J. Biol. Chem.* **275**(20): 14969–14978.

Boyer, T. G., M. E. Martin, E. Lees, R. P. Ricciardi and A. J. Berk (1999). "Mammalian Srb/Mediator complex is targeted by adenovirus E1A protein." *Nature* **399**: 276–279.

Bradley, R. R., L. F. Maxfield, D. M. Lynch, M. J. Iampietro, E. N. Borducchi and D. H. Barouch (2012). "Adenovirus serotype 5-specific neutralizing antibodies target multiple hexon hypervariable regions." *J Virol* **86**(2): 1267–72.

Bradshaw, A. C. and A. H. Baker (2013). "Gene therapy for cardiovascular disease: perspectives and potential." *Vascul Pharmacol* **58**(3): 174–81.

Brady, H. A., A. Scaria and W. S. Wold (1992). "Map of cis-acting sequences that determine alternative pre-mRNA processing in the E3 complex transcription unit of adenovirus." *J Virol* **66**(10): 5914–23.

Braithwaite, A., C. Nelson, A. Skulimowski, J. McGovern, D. Pigott and J. Jenkins (1990). "Transactivation of the p53 oncogene by E1a gene products." *Virology* **177**(2): 595–605.

Braithwaite, A. W., B. F. Cheetham, P. Li, C. R. Parish, L. K. Waldron-Stevens and A. Bellett (1983). "Adenovirus-induced alterations of the cell growth cycle: a requirement for expression of E1A but not of E1B." *J Virol* **45**: 192–199.

Bramante, S., A. Koski, A. Kipar, I. Diaconu, I. Liikanen, O. Hemminki, L. Vassilev, S. Parviainen, V. Cerullo, S. K. Pesonen, M. Oksanen, R. Heiskanen, N. Rouvinen-Lagerstrom, M. Merisalo-Soikkeli, T. Hakonen, T. Joensuu, A. Kanerva, S. Pesonen and A. Hemminki (2014). "Serotype chimeric oncolytic adenovirus coding for GM-CSF for treatment of sarcoma in rodents and humans." *Int J Cancer* **135**(3): 720–30.

Branton, P. E., S. T. Bayley and F. L. Graham (1985). "Transformation by human adenoviruses." *Biochim Biophys Acta* **780**(1): 67–94.

Breathnach, R., C. Benoist, K. O'Hare, F. Gannon and P. Chambon (1978). "Ovalbumin gene: evidence for a leader sequence in mRNA and DNA sequences at the exon-intron boundaries." *Proc Natl Acad Sci U S A* **75**(10): 4853–7.

Brehm, A., E. Miska, J. Reid, A. Bannister and T. Kouzarides (1999). "The cell cycle-regulating transcription factors E2F-RB." *Br J Cancer* **80 Suppl 1**: 38–41.

Bremner, K. H., J. Scherer, J. Yi, M. Vershinin, S. P. Gross and R. B. Vallee (2009). "Adenovirus transport via direct interaction of cytoplasmic dynein with the viral capsid hexon subunit." *Cell Host Microbe* **6**(6): 523–35.

Brenkman, A. B., E. C. Breure and P. C. van der Vliet (2002). "Molecular architecture of adenovirus DNA polymerase and location of the protein primer." *J Virol* **76**(16): 8200–7.

Bridge, E. and G. Ketner (1990). "Interaction of adenoviral E4 and E1b products in late gene expression." *Virology* **174**: 345–353.

Bridge, E. and U. Pettersson (1996). "Nuclear organization of adenovirus RNA biogenesis." *Exp. Cell Res.* **229**(2): 233–239.

Bridge, E., D. X. Xia, M. Carmo-Fonseca, B. Cardinali, A. I. Lamond and U. Pettersson (1995). "Dynamic organization of splicing factors in adenovirus-infected cells." *J Virol* **69**(1): 281–90.

Bridges, R. G., S. Y. Sohn, J. Wright, K. N. Leppard and P. Hearing (2016). "The Adenovirus E4-ORF3 Protein Stimulates SUMOylation of General Transcription Factor TFII-I to Direct Proteasomal Degradation." *MBio* **7**(1) e02184-15. doi: 10.1128/mBio.02184-15.

Brindle, N. P., P. Saharinen and K. Alitalo (2006). "Signaling and functions of angiopoietin-1 in vascular protection." *Circ Res* **98**(8): 1014–23.

Brossart, P., A. W. Goldrath, E. A. Butz, S. Martin and M. J. Bevan (1997). "Virus-mediated delivery of antigenic epitopes into dendritic cells as a means to induce CTL." *J Immunol* **158**(7): 3270–6.

Brough, D. E., G. Droguett, M. S. Horwitz and D. F. Klessig (1993). "Multiple functions of the adenovirus DNA-binding protein are required for efficient viral DNA synthesis." *Virology* **196**(1): 269–81.

Brown, D. T., M. Westphal, B. T. Burlingham, U. Winterhoff and W. Doerfler (1975). "Structure and composition of the adenovirus type 2 core." *J Virol* **16**(2): 366–87.

Brown, M. T., K. M. McBride, M. L. Baniecki, N. C. Reich, G. Marriott and W. F. Mangel (2002). "Actin can act as a cofactor for a viral proteinase in the cleavage of the cytoskeleton." *J Biol Chem* **277**(48): 46298–303.

Bru, T., S. Salinas and E. J. Kremer (2010). "An update on canine adenovirus type 2 and its vectors." *Viruses* **2**(9): 2134–53.

Bruder, J. T. and P. Hearing (1989). "Nuclear Factor Ef-1a Binds to the Adenovirus E1a Core Enhancer Element and to Other Transcriptional Control Regions." *Mol Cell Biol* **9**(11): 5143–5153.

Bruder, J. T. and P. Hearing (1991). "Cooperative binding of EF-1A to the E1A enhancer region mediates synergistic effects on E1A transcription during adenovirus infection." *J. Virol.* **65**: 5084–5087.

Bruder, J. T., T. Jie, D. L. McVey and I. Kovesdi (1997). "Expression of gp19K increases the persistence of transgene expression from an adenovirus vector in the mouse lung and liver." *J Virol* **71**(10): 7623–8.

Bruder, J. T., E. Semenova, P. Chen, K. Limbach, N. B. Patterson, M. E. Stefaniak, S. Konovalova, C. Thomas, M. Hamilton, C. R. King, T. L. Richie and D. L. Doolan (2012). "Modification of Ad5 hexon hypervariable regions circumvents pre-existing Ad5 neutralizing antibodies and induces protective immune responses." *PLoS One* **7**(4): e33920.

Brunetti-Pierri, N., A. Liou, P. Patel, D. Palmer, N. Grove, M. Finegold, P. Piccolo, E. Donnachie, K. Rice, A. Beaudet, C. Mullins and P. Ng (2012). "Balloon catheter delivery of helper-dependent adenoviral vector results in sustained, therapeutic hFIX expression in rhesus macaques." *Mol Ther* **20**(10): 1863–70.

Brunetti-Pierri, N. and P. Ng (2011). "Helper-dependent adenoviral vectors for liver-directed gene therapy." *Hum Mol Genet* **20**(R1): R7–13.

Brunetti-Pierri, N., T. Ng, D. Iannitti, W. Cioffi, G. Stapleton, M. Law, J. Breinholt, D. Palmer, N. Grove, K. Rice, C. Bauer, M. Finegold, A. Beaudet, C. Mullins and P. Ng (2013). "Transgene expression up to 7 years in nonhuman primates following hepatic transduction with helper-dependent adenoviral vectors." *Hum Gene Ther* **24**(8): 761–5.

Brunetti-Pierri, N., D. J. Palmer, V. Mane, M. Finegold, A. L. Beaudet and P. Ng (2005). "Increased hepatic transduction with reduced systemic dissemination and proinflammatory cytokines following hydrodynamic injection of helper-dependent adenoviral vectors." *Mol Ther* **12**(1): 99–106.

Bruton, R. K., P. Pelka, K. L. Mapp, G. J. Fonseca, J. Torchia, A. S. Turnell, J. S. Mymryk and R. J. Grand (2008). "Identification of a second CtBP binding site in adenovirus type 5 E1A conserved region 3." *J Virol* **82**(17): 8476–86.

Buchbinder, S. P., D. V. Mehrotra, A. Duerr, D. W. Fitzgerald, R. Mogg, D. Li, P. B. Gilbert, J. R. Lama, M. Marmor, C. Del Rio, M. J. McElrath, D. R. Casimiro, K. M. Gottesdiener, J. A. Chodakewitz, L. Corey, M. N. Robertson; Step Study Protocol Team (2008). "Efficacy assessment of a cell-mediated immunity HIV-1 vaccine (the Step Study): a double-blind, randomised, placebo-controlled, test-of-concept trial." *Lancet* **372**(9653): 1881–93.

Buck, C. B., P. M. Day, C. D. Thompson, J. Lubkowski, W. Lu, D. R. Lowy and J. T. Schiller (2006). "Human alpha-defensins block papillomavirus infection." *Proc Natl Acad Sci U S A* **103**(5): 1516–21.

Burckhardt, C. J., M. Suomalainen, P. Schoenenberger, K. Boucke, S. Hemmi and U. F. Greber (2011). "Drifting motions of the adenovirus receptor CAR and immobile integrins initiate virus uncoating and membrane lytic protein exposure." *Cell Host Microbe* **10**(2): 105–17.

Burgert, H. G. and J. H. Blusch (2000). "Immunomodulatory functions encoded by the E3 transcription unit of adenoviruses." *Virus Genes* **21**(1–2): 13–25.

Burgert, H. G. and S. Kvist (1985). "An adenovirus type 2 glycoprotein blocks cell surface expression of human histocompatibility class I antigens." *Cell* **41**(3): 987–97.

Burgert, H. G. and S. Kvist (1987). "The E3/19K protein of adenovirus type 2 binds to the domains of histocompatibility antigens required for CTL recognition." *EMBO J* **6**(7): 2019–26.

Burgert, H. G., Z. Ruzsics, S. Obermeier, A. Hilgendorf, M. Windheim and A. Elsing (2002). "Subversion of host defense mechanisms by adenoviruses." *Curr Top Microbiol Immunol* **269**: 273–318.

Burke, J. M., D. L. Lamm, M. V. Meng, J. J. Nemunaitis, J. J. Stephenson, J. C. Arseneau, J. Aimi, S. Lerner, A. W. Yeung, T. Kazarian, D. J. Maslyar and J. M. McKiernan (2012). "A first in human phase 1 study of CG0070, a GM-CSF expressing oncolytic adenovirus, for the treatment of nonmuscle invasive bladder cancer." *J Urol* **188**(6): 2391–7.

Burmeister, W. P., D. Guilligay, S. Cusack, G. Wadell and N. Arnberg (2004). "Crystal structure of species D adenovirus fiber knobs and their sialic acid binding sites." *J Virol* **78**(14): 7727–36.

Cai, J., J. Yang and D. P. Jones (1998). "Mitochondrial control of apoptosis: the role of cytochrome c." *Biochim Biophys Acta* **1366**(1–2): 139–49.

Cai, X., Y. H. Chiu and Z. J. Chen (2014). "The cGAS-cGAMP-STING pathway of cytosolic DNA sensing and signaling." *Mol Cell* **54**(2): 289–96.

Callegari, E., B. K. Elamin, L. D'Abundo, S. Falzoni, G. Donvito, F. Moshiri, M. Milazzo, G. Altavilla, L. Giacomelli, F. Fornari, A. Hemminki, F. Di Virgilio, L. Gramantieri, M. Negrini and S. Sabbioni (2013). "Anti-tumor activity of a miR-199-dependent oncolytic adenovirus." *PLoS One* **8**(9): e73964.

Cammas, F., K. Khetchoumian, P. Chambon and R. Losson (2012). "TRIM involvement in transcriptional regulation." *Adv Exp Med Biol* **770**: 59–76.

Cantin, G. T., J. L. Stevens and A. J. Berk (2003). "Activation domain-mediator interactions promote transcription preinitiation complex assembly on promoter DNA." *Proc Natl Acad Sci U S A* **100**(21): 12003–8.

Cao, C., X. Dong, X. Wu, B. Wen, G. Ji, L. Cheng and H. Liu (2012). "Conserved fiber-penton base interaction revealed by nearly atomic resolution cryo-electron microscopy of the structure of adenovirus provides insight into receptor interaction." *J Virol* **86**(22): 12322–9.

Capone, S., A. Meola, B. B. Ercole, A. Vitelli, M. Pezzanera, L. Ruggeri, M. E. Davies, R. Tafi, C. Santini, A. Luzzago, T. M. Fu, A. Bett, S. Colloca, R. Cortese, A. Nicosia and A. Folgori (2006). "A novel adenovirus type 6 (Ad6)-based hepatitis C virus vector that overcomes preexisting anti-ad5 immunity and induces potent and broad cellular immune responses in rhesus macaques." *J Virol* **80**(4): 1688–99.

Capone, S., A. Reyes-Sandoval, M. Naddeo, L. Siani, V. Ammendola, C. S. Rollier, A. Nicosia, S. Colloca, R. Cortese, A. Folgori and A. V. Hill (2010). "Immune responses against a liver-stage malaria antigen induced by simian adenoviral vector AdCh63 and MVA prime-boost immunisation in non-human primates." *Vaccine* **29**(2): 256–65.

Caraballo, R., M. Saleeb, J. Bauer, A. M. Liaci, N. Chandra, R. J. Storm, L. Frangsmyr, W. Qian, T. Stehle, N. Arnberg and M. Elofsson (2015). "Triazole linker-based trivalent sialic acid inhibitors of adenovirus type 37 infection of human corneal epithelial cells." *Org Biomol Chem* **13**(35): 9194–205.

Caravokyri, C. and K. N. Leppard (1996). "Human adenovirus type 5 variants with sequence alterations flanking the E2A gene: effects on E2 expression and DNA replication." *Virus Genes* **12**(1): 65–75.

Carcamo, J., E. Maldonado, P. Cortes, M. H. Ahn, I. Ha, Y. Kasai, J. Flint and D. Reinberg (1990). "A TATA-like sequence located downstream of the transcription initiation site is required for expression of an RNA polymerase II transcribed gene." *Genes Dev* **4**(9): 1611–22.

Cardone, J., G. Le Friec, P. Vantourout, A. Roberts, A. Fuchs, I. Jackson, T. Suddason, G. Lord, J. P. Atkinson, A. Cope, A. Hayday and C. Kemper (2010). "Complement regulator CD46 temporally regulates cytokine production by conventional and unconventional T cells." *Nat Immunol* **11**(9): 862–71.

Carlisle, R. C., Y. Di, A. M. Cerny, A. F. Sonnen, R. B. Sim, N. K. Green, V. Subr, K. Ulbrich, R. J. Gilbert, K. D. Fisher, R. W. Finberg and L. W. Seymour (2009). "Human erythrocytes bind and inactivate type 5 adenovirus by presenting Coxsackie virus-adenovirus receptor and complement receptor 1." *Blood* **113**(9): 1909–18.

Carmody, R. J., K. Maguschak and Y. H. Chen (2006). "A novel mechanism of nuclear factor-kappaB regulation by adenoviral protein 14.7K." *Immunology* **117**(2): 188–95.

Carnero, E., J. D. Sutherland and P. Fortes (2011). "Adenovirus and miRNAs." *Biochim Biophys Acta* **1809**(11–12): 660–7.

Carrasco, L. (1994). "Entry of animal viruses and macromolecules into cells." *FEBS Lett* **350**(2–3): 151–4.

Carron, C. P., D. M. Meyer, J. A. Pegg, V. W. Engleman, M. A. Nickols, S. L. Settle, W. F. Westlin, P. G. Ruminski and G. A. Nickols (1998). "A peptidomimetic antagonist of the integrin alpha(v)beta3 inhibits Leydig cell tumor growth and the development of hypercalcemia of malignancy." *Cancer Res* **58**(9): 1930–5.

Carson, C. T., N. I. Orazio, D. V. Lee, J. Suh, S. Bekker-Jensen, F. D. Araujo, S. S. Lakdawala, C. E. Lilley, J. Bartek, J. Lukas and M. D. Weitzman (2009). "Mislocalization of the MRN complex prevents ATR signaling during adenovirus infection." *EMBO J* **28**(6): 652–62.

Carson, C. T., R. A. Schwartz, T. H. Stracker, C. E. Lilley, D. V. Lee and M. D. Weitzman (2003). "The Mre11 complex is required for ATM activation and the G2/M checkpoint." *EMBO J* **22**(24): 6610–20.

Carthew, R. W., L. A. Chodosh and P. A. Sharp (1985). "An RNA polymerase II transcription factor binds to an upstream element in the adenovirus major late promoter." *Cell* **43**(2 Pt 1): 439–48.

Carvalho, T., J. S. Seeler, K. Ohman, P. Jordan, U. Pettersson, G. Akusjarvi, M. Carmo-Fonseca and A. Dejean (1995). "Targeting of adenovirus E1A and E4-ORF3 proteins to nuclear matrix-associated PML bodies." *J Cell Biol* **131**(1): 45–56.

Casanas, A., P. Guerra, I. Fita and N. Verdaguer (2012). "Vault particles: a new generation of delivery nanodevices." *Curr Opin Biotechnol* **23**(6): 972–7.

Casasnovas, J. M., M. Larvie and T. Stehle (1999). "Crystal structure of two CD46 domains reveals an extended measles virus-binding surface." *EMBO J* **18**(11): 2911–22.

Casimiro, D. R., L. Chen, T. M. Fu, R. K. Evans, M. J. Caulfield, M. E. Davies, A. Tang, M. Chen, L. Huang, V. Harris, D. C. Freed, K. A. Wilson, S. Dubey, D. M. Zhu, D. Nawrocki, H. Mach, R. Troutman, L. Isopi, D. Williams, W. Hurni, Z. Xu, J. G. Smith, S. Wang, X. Liu, L. Guan, R. Long, W. Trigona, G. J. Heidecker, H. C. Perry, N. Persaud, T. J. Toner, Q. Su, X. Liang, R. Youil,

M. Chastain, A. J. Bett, D. B. Volkin, E. A. Emini and J. W. Shiver (2003). "Comparative immunogenicity in rhesus monkeys of DNA plasmid, recombinant vaccinia virus, and replication-defective adenovirus vectors expressing a human immunodeficiency virus type 1 gag gene." *J Virol* **77**(11): 6305–13.

Cassany, A., J. Ragues, T. Guan, D. Begu, H. Wodrich, M. Kann, G. R. Nemerow and L. Gerace (2015). "Nuclear import of adenovirus DNA involves direct interaction of hexon with an N-terminal domain of the nucleoporin Nup214." *J Virol* **89**(3): 1719–30.

Cattaneo, R. (2004). "Four viruses, two bacteria, and one receptor: membrane cofactor protein (CD46) as pathogens' magnet." *J Virol* **78**(9): 4385–8.

Cawood, R., H. H. Chen, F. Carroll, M. Bazan-Peregrino, N. van Rooijen and L. W. Seymour (2009). "Use of tissue-specific microRNA to control pathology of wild-type adenovirus without attenuation of its ability to kill cancer cells." *PLoS Pathog* **5**(5): e1000440.

Cemerski, S. and A. Shaw (2006). "Immune synapses in T-cell activation." *Curr Opin Immunol* **18**(3): 298–304.

Cepko, C. L. and P. A. Sharp (1982). "Assembly of adenovirus major capsid protein is mediated by a nonvirion protein." *Cell* **31**(2 Pt 1): 407–15.

Cepko, C. L. and P. A. Sharp (1983). "Analysis of Ad5 hexon and 100K ts mutants using conformation-specific monoclonal antibodies." *Virology* **129**(1): 137–54.

Cerullo, V., I. Diaconu, L. Kangasniemi, M. Rajecki, S. Escutenaire, A. Koski, V. Romano, N. Rouvinen, T. Tuuminen, L. Laasonen, K. Partanen, S. Kauppinen, T. Joensuu, M. Oksanen, S. L. Holm, E. Haavisto, A. Karioja-Kallio, A. Kanerva, S. Pesonen, P. T. Arstila and A. Hemminki (2011). "Immunological effects of low-dose cyclophosphamide in cancer patients treated with oncolytic adenovirus." *Mol Ther* **19**(9): 1737–46.

Cerullo, V., M. P. Seiler, V. Mane, N. Brunetti-Pierri, C. Clarke, T. K. Bertin, J. R. Rodgers and B. Lee (2007). "Toll-like receptor 9 triggers an innate immune response to helper-dependent adenoviral vectors." *Mol Ther* **15**(2): 378–85.

Chahal, J. S. and S. J. Flint (2012). "Timely synthesis of the adenovirus type 5 E1B 55-kilodalton protein is required for efficient genome replication in normal human cells." *J Virol* **86**(6): 3064–72.

Chahal, J. S., C. Gallagher, C. J. DeHart and S. J. Flint (2013). "The repression domain of the E1B 55-kilodalton protein participates in countering interferon-induced inhibition of adenovirus replication." *J Virol* **87**(8): 4432–44.

Chahal, J. S., J. Qi and S. J. Flint (2012). "The human adenovirus type 5 E1B 55 kDa protein obstructs inhibition of viral replication by type I interferon in normal human cells." *PLoS Pathog* **8**(8): e1002853.

Chakrabarti, A., B. K. Jha and R. H. Silverman (2011). "New insights into the role of RNase L in innate immunity." *J Interferon Cytokine Res* **31**(1): 49–57.

Chakrabarti, S., V. Mautner, H. Osman, K. E. Collingham, C. D. Fegan, P. E. Klapper, P. A. Moss and D. W. Milligan (2002). "Adenovirus infections following allogeneic stem cell transplantation: incidence and outcome in relation to graft manipulation, immunosuppression, and immune recovery." *Blood* **100**(5): 1619–27.

Chakupurakal, G., D. Onion, M. Cobbold, V. Mautner and P. A. Moss (2010). "Adenovirus vector-specific T cells demonstrate a unique memory phenotype with high proliferative potential and coexpression of CCR5 and integrin alpha4beta7." *AIDS* **24**(2): 205–10.

Challberg, M. D., S. V. Desiderio and T. J. Kelly (1980). "Adenovirus DNA replication in vitro: Characterization of a protein covalently linked to nascent DNA strands." *Proc Natl Acad Sci Usa* **77**: 5105–0.

Challberg, M. D. and T. J. Kelly, Jr. (1979). "Adenovirus DNA replication in vitro." *Proc Natl Acad Sci U S A* **76**(2): 655–9.

Chang, J., X. Zhao, X. Wu, Y. Guo, H. Guo, J. Cao, D. Lou, D. Yu and J. Li (2009). "A Phase I study of KH901, a conditionally replicating granulocyte-macrophage colony-stimulating factor: armed oncolytic adenovirus for the treatment of head and neck cancers." *Cancer Biol Ther* **8**(8): 676–82.

Chang, L. S. and T. Shenk (1990). "The adenovirus DNA-binding protein stimulates the rate of transcription directed by adenovirus and adeno-associated virus promoters." *J Virol* **64**: 2103–2109.

Chardonnet, Y. and S. Dales (1970). "Early events in the interaction of adenoviruses with HeLa cells. I. Penetration of type 5 and intracellular release of the DNA genome." *Virology* **40**(3): 462–77.

Chatterjee-Kishore, M., F. van Den Akker and G. R. Stark (2000). "Adenovirus E1A down-regulates LMP2 transcription by interfering with the binding of stat1 to IRF1." *J Biol Chem* **275**(27): 20406–11.

Chatterjee, P. K., M. M. Cervera and S. Penman (1984). "Formation of vesicular stomatitis virus nucleocapsid from cytoskeletal framweork-bound N protein: possible model for structure assembly." *Mol. Cell Biol.* **4**: 2231–2234.

Chatterjee, P. K., M. E. Vayda and S. J. Flint (1986). "Adenoviral protein VII packages intracellular viral DNA throughout the early phase of infection." *Embo. J.* **5**: 1633–1644.

Chatterjee, P. K., M. E. Vayda and S. J. Flint (1986). "Identification of proteins and protein domains that contact DNA within adenovirus nucleoprotein cores by ultraviolet light crosslinking of oligonucleotides ^{32}P-labelled in vivo." *J Mol Biol* **188**(1): 23–37.

Chatton, B., J. L. Bocco, M. Gaire, C. Hauss, B. Reimund, J. Goetz and C. Kedinger (1993). "Transcriptional activation by the adenovirus larger E1a product is mediated by members of the cellular transcription factor ATF family which can directly associate with E1a." *Mol Cell Biol* **13**(1): 561–70.

Chattopadhyay, S., M. Yamashita, Y. Zhang and G. C. Sen (2011). "The IRF-3/Bax-mediated apoptotic pathway, activated by viral cytoplasmic RNA and DNA, inhibits virus replication." *J Virol* **85**(8): 3708–16.

Chelius, D., A. F. Huhmer, C. H. Shieh, E. Lehmberg, J. A. Traina, T. K. Slattery and E. Pungor, Jr. (2002). "Analysis of the adenovirus type 5 proteome by liquid chromatography and tandem mass spectrometry methods." *J Proteome Res* **1**(6): 501–13.

Chen, H. and S. J. Flint (1992). "Mutational analysis of the adenovirus 2 IVa2 initiator and downstream elements." *J Biol Chem* **267**(35): 25457–65.

Chen, H., R. Vinnakota and S. J. Flint (1994). "Intragenic activating and repressing elements control transcription from the adenovirus IVa2 initiator." *Mol Cell Biol* **14**(1): 676–85.

Chen, H., Z. Q. Xiang, Y. Li, R. K. Kurupati, B. Jia, A. Bian, D. M. Zhou, N. Hutnick, S. Yuan, C. Gray, J. Serwanga, B. Auma, P. Kaleebu, X. Zhou, M. R. Betts and H. C. Ertl (2010). "Adenovirus-based vaccines: comparison of vectors from three species of adenoviridae." *J Virol* **84**(20): 10522–32.

Chen, J., N. Morral and D. A. Engel (2007). "Transcription releases protein VII from adenovirus chromatin." *Virology* **369**(2): 411–22.

Chen, M. and M. S. Horwitz (1989). "Dissection of functional domains of adenovirus DNA polymerase by linker-insertion mutagenesis." *Proc. Natl. Acad. Sci. USA* **86**: 6116–6120.

Chen, M., N. Mermod and M. S. Horwitz (1990). "Protein-protein interactions between adenovirus DNA polymerase and nuclear factor I mediate formation of the DNA replication preinitiation complex." *J Biol Chem* **265**(30): 18634–42.

Cheng, C., J. G. Gall, W. P. Kong, R. L. Sheets, P. L. Gomez, C. R. King and G. J. Nabel (2007). "Mechanism of ad5 vaccine immunity and toxicity: fiber shaft targeting of dendritic cells." *PLoS Pathog* **3**(2): e25.

Cheng, L., X. Huang, X. Li, W. Xiong, W. Sun, C. Yang, K. Zhang, Y. Wang, H. Liu, G. Ji, F. Sun, C. Zheng and P. Zhu (2014). "Cryo-EM structures of two bovine adenovirus type 3 intermediates." *Virology* **450–451**: 174–81.

Cherenova, L. V., D. Y. Logunov, E. V. Shashkova, M. M. Shmarov, L. V. Verkhovskaya, G. L. Neugodova, D. B. Kazansky, K. K. Doronin and B. S. Naroditsky (2004). "Recombinant avian adenovirus CELO expressing the human interleukin-2: characterization in vitro, in ovo and in vivo." *Virus Res* **100**(2): 257–61.

Cheresh, D. A. and J. R. Harper (1987). "Arg-Gly-Asp recognition by a cell adhesion receptor requires its 130-kDa alpha subunit." *J Biol Chem* **262**(4): 1434–7.

Chillon, M., J. H. Lee, A. Fasbender and M. J. Welsh (1998). "Adenovirus complexed with polyethylene glycol and cationic lipid is shielded from neutralizing antibodies in vitro." *Gene Ther* **5**(7): 995–1002.

Chin, Y. R. and M. S. Horwitz (2005). "Mechanism for removal of tumor necrosis factor receptor 1 from the cell surface by the adenovirus RIDalpha/beta complex." *J Virol* **79**(21): 13606–17.

Chinnadurai, G. (2002). "CtBP, an unconventional transcriptional corepressor in development and oncogenesis." *Mol Cell* **9**(2): 213–24.

Chinnadurai, G. (2004). "Modulation of oncogenic transformation by the human adenovirus E1A C-terminal region." *Curr Top Microbiol Immunol* **273**: 139–61.

Chiocca, S., A. Baker and M. Cotten (1997). "Identification of a novel antiapoptotic protein, GAM-1, encoded by the CELO adenovirus." *J Virol* **71**(4): 3168–77.

Chiou, S. K., C. C. Tseng, L. Rao and E. White (1994). "Functional complementation of the adenovirus E1B 19-kilodalton protein with Bcl-2 in the inhibition of apoptosis in infected cells." *J Virol* **68**(10): 6553–66.

Chitaev, N. A. and S. M. Troyanovsky (1997). "Direct Ca^{2+}-dependent heterophilic interaction between desmosomal cadherins, desmoglein and desmocollin, contributes to cell-cell adhesion." *J Cell Biol* **138**(1): 193–201.

Chiu, C. Y., P. Mathias, G. R. Nemerow and P. L. Stewart (1999). "Structure of adenovirus complexed with its internalization receptor, alphavbeta5 integrin." *J Virol* **73**(8): 6759–68.

Choi, J. W., J. W. Park, Y. Na, S. J. Jung, J. K. Hwang, D. Choi, K. G. Lee and C. O. Yun (2015). "Using a magnetic field to redirect an oncolytic adenovirus complexed with iron oxide augments gene therapy efficacy." *Biomaterials* **65**: 163–74.

Chow, L. T., T. R. Broker and J. Lewis (1979). "Complex splicing patterns of RNAs from the early region of adenovirus-2." *J. Mol. Biol.* **134**: 265–303.

Chow L. T., R. E. Gelinas, T. R. Broker and R. J. Roberts (1977). An amazing sequence arrangement at the 5' ends of adenovirus 2 messenger RNA. *Cell* **12**(1): 1–8.

Christensen, J. B., S. A. Byrd, A. K. Walker, J. R. Strahler, P. C. Andrews and M. J. Imperiale (2008). "Presence of the adenovirus IVa2 protein at a single vertex of the mature virion." *J Virol* **82**(18): 9086–93.

Chroboczek, J., R. W. Ruigrok and S. Cusack (1995). "Adenovirus fiber." *Curr Top Microbiol Immunol* **199** (**Pt 1**): 163–200.

Chroboczek, J., F. Viard and J. C. D'Halluin (1986). "Human adenovirus 2 temperature-sensitive mutant 112 contains three mutations in the protein IIIa gene." *Gene* **49**(1): 157–60.

Chuang, I., M. Sedegah, S. Cicatelli, M. Spring, M. Polhemus, C. Tamminga, N. Patterson, M. Guerrero, J. W. Bennett, S. McGrath, H. Ganeshan, M. Belmonte, F. Farooq, E. Abot, J. G. Banania, J. Huang, R. Newcomer, L. Rein, D. Litilit, N. O. Richie, C. Wood, J. Murphy, R. Sauerwein, C. C. Hermsen, A. J. McCoy, E. Kamau, J. Cummings, J. Komisar, A. Sutamihardja, M. Shi, J. E. Epstein, S. Maiolatesi, D. Tosh, K. Limbach, E. Angov, E. Bergmann-Leitner, J. T. Bruder, D. L. Doolan, C. R. King, D. Carucci, S. Dutta, L. Soisson, C. Diggs, M. R. Hollingdale, C. F. Ockenhouse and T. L. Richie (2013). "DNA prime/Adenovirus boost malaria vaccine encoding P. falciparum CSP and AMA1 induces sterile protection associated with cell-mediated immunity." *PLoS One* **8**(2): e55571.

Churchyard, G. J., C. Morgan, E. Adams, J. Hural, B. S. Graham, Z. Moodie, D. Grove, G. Gray, L. G. Bekker, M. J. McElrath, G. D. Tomaras, P. Goepfert, S. Kalams, L. R. Baden, M. Lally, R. Dolin, W. Blattner, A. Kalichman, J. P. Figueroa, J. Pape, M. Schechter, O. Defawe, S. C. De Rosa, D. C. Montefiori, G. J. Nabel, L. Corey, M. C. Keefer and N. H. V. T. Network (2011). "A phase IIA randomized clinical trial of a multiclade HIV-1 DNA prime followed by a multiclade rAd5 HIV-1 vaccine boost in healthy adults (HVTN204)." *PLoS One* **6**(8): e21225.

Cichon, G., S. Boeckh-Herwig, H. H. Schmidt, E. Wehnes, T. Muller, P. Pring-Akerblom and R. Burger (2001). "Complement activation by recombinant adenoviruses." *Gene Ther* **8**(23): 1794–800.

Ciesla, S. L., J. Trahan, W. B. Wan, J. R. Beadle, K. A. Aldern, G. R. Painter and K. Y. Hostetler (2003). "Esterification of cidofovir with alkoxyalkanols increases oral bioavailability and diminishes drug accumulation in kidney." *Antiviral Res* **59**(3): 163–71.

Clarke, P. A. and M. B. Mathews (1995). "Interactions between the double-stranded RNA binding motif and RNA: definition of the binding site for the interferon-induced protein kinase DAI (PKR) on adenovirus VA RNA." *RNA* **1**(1): 7–20.

Clarke, P. A., T. Pe'ery, Y. Ma and M. B. Mathews (1994). "Structural features of adenovirus 2 virus-associated RNA required for binding to the protein kinase DAI." *Nucleic Acids Res* **22**(21): 4364–74.

Cleat, P. H. and R. T. Hay (1989). "Co-operative interactions between NFI and the adenovirus DNA binding protein at the adenovirus origin of replication." *EMBO J* **8**(6): 1841–8.

Cody, J. J. and J. T. Douglas (2009). "Armed replicating adenoviruses for cancer virotherapy." *Cancer Gene Ther* **16**(6): 473–88.

Coenjaerts, F. E., J. A. van Oosterhout and P. C. van der Vliet (1994). "The Oct-1 POU domain stimulates adenovirus DNA replication by a direct interaction between the viral precursor terminal protein-DNA polymerase complex and the POU homeodomain." *EMBO J* **13**(22): 5401–9.

Cohen, C. J., J. Gaetz, T. Ohman and J. M. Bergelson (2001). "Multiple regions within the coxsackievirus and adenovirus receptor cytoplasmic domain are required for basolateral sorting." *J Biol Chem* **276**(27): 25392–8.

Cohen, C. J., J. T. Shieh, R. J. Pickles, T. Okegawa, J. T. Hsieh and J. M. Bergelson (2001). "The coxsackievirus and adenovirus receptor is a transmembrane component of the tight junction." *Proc Natl Acad Sci U S A* **98**(26): 15191–6.

Cohen, M. J., A. F. Yousef, P. Massimi, G. J. Fonseca, B. Todorovic, P. Pelka, A. S. Turnell, L. Banks and J. S. Mymryk (2013). "Dissection of the C-terminal region of E1A redefines the roles of CtBP and other cellular targets in oncogenic transformation." *J Virol* **87**(18): 10348–55.

Colloca, S., E. Barnes, A. Folgori, V. Ammendola, S. Capone, A. Cirillo, L. Siani, M. Naddeo, F. Grazioli, M. L. Esposito, M. Ambrosio, A. Sparacino, M. Bartiromo, A. Meola, K. Smith, A. Kurioka, G. A. O'Hara, K. J. Ewer, N. Anagnostou, C. Bliss, A. V. Hill, C. Traboni, P. Klenerman, R. Cortese and A. Nicosia (2012). "Vaccine vectors derived from a large collection of simian adenoviruses induce potent cellular immunity across multiple species." *Sci Transl Med* **4**(115): 115ra2.

Conaway, R. C. and J. W. Conaway (2011). "Origins and activity of the Mediator complex." *Semin Cell Dev Biol* **22**(7): 729–34.

Concino, M. F., R. F. Lee, J. B. Merryweather and R. Weinmann (1984). "The adenovirus major late promoter TATA box and initiation site are both necessary for transcription *in vitro*." *Nuc. Acids. Res.* **12**: 7423–7433.

Cook, J. L., C. H. Kirkpatrick, A. S. Rabson and A. M. Lewis, Jr. (1979). "Rejection of adenovirus 2-transformed cell tumors and immune responsiveness in Syrian hamsters." *Cancer Res* **39**(12): 4949–55.

Cook, J. L., A. M. Lewis, Jr. and C. H. Kirkpatrick (1979). "Age-related and thymus-dependent rejection of adenovirus 2-transformed cell tumors in the Syrian hamster." *Cancer Res* **39**(9): 3335–40.

Cooper, N. R. and G. R. Nemerow (1983). "Complement, viruses, and virus-infected cells." *Springer Semin Immunopathol* **6**(4): 327–47.

Corredor, J. C. and E. Nagy (2011). "Antibody response and virus shedding of chickens inoculated with left end deleted fowl adenovirus 9-based recombinant viruses." *Avian Dis* **55**(3): 443–6.

Cotten, M. and J. M. Weber (1995). "The adenovirus protease is required for virus entry into host cells." *Virology* **213**(2): 494–502.

Cotter, M. J., A. K. Zaiss and D. A. Muruve (2005). "Neutrophils interact with adenovirus vectors via Fc receptors and complement receptor 1." *J Virol* **79**(23): 14622–31.

Coughlan, L. (2014). "Genetically engineering adenoviral vectors for gene therapy." *Methods Mol Biol* **1108**: 23–40.

Coughlan, L., S. Vallath, A. Saha, M. Flak, I. A. McNeish, G. Vassaux, J. F. Marshall, I. R. Hart and G. J. Thomas (2009). "In vivo retargeting of adenovirus type 5 to alphavbeta6 integrin results in reduced hepatotoxicity and improved tumor uptake following systemic delivery." *J Virol* **83**(13): 6416–28.

Coventry, V. K. and G. L. Conn (2008). "Analysis of adenovirus VA RNAI structure and stability using compensatory base pair modifications." *Nucleic Acids Res* **36**(5): 1645–53.

Cox, J. H., J. R. Bennink and J. W. Yewdell (1991). "Retention of adenovirus E19 glycoprotein in the endoplasmic reticulum is essential to its ability to block antigen presentation." *J Exp Med* **174**(6): 1629–37.

Crawford-Miksza, L. and D. P. Schnurr (1996). "Analysis of 15 adenovirus hexon proteins reveals the location and structure of seven hypervariable regions containing serotype-specific residues." *J Virol* **70**(3): 1836–44.

Cress, W. D. and J. R. Nevins (1994). "Interacting domains of E2F1, DP1, and the adenovirus E4 protein." *J Virol* **68**(7): 4213–9.

Crimeen-Irwin, B., S. Ellis, D. Christiansen, M. J. Ludford-Menting, J. Milland, M. Lanteri, B. E. Loveland, D. Gerlier and S. M. Russell (2003). "Ligand binding determines whether CD46 is internalized by clathrin-coated pits or macropinocytosis." *J Biol Chem* **278**(47): 46927–37.

Crompton, P. D., J. Moebius, S. Portugal, M. Waisberg, G. Hart, L. S. Garver, L. H. Miller, C. Barillas-Mury and S. K. Pierce (2014). "Malaria immunity in man and mosquito: insights into unsolved mysteries of a deadly infectious disease." *Annu Rev Immunol* **32**: 157–87.

Crowther, R. A. and R. M. Franklin (1972). "The structure of the groups of nine hexons from adenovirus." *J Mol Biol* **68**(1): 181–4.

Croyle, M. A., N. Chirmule, Y. Zhang and J. M. Wilson (2001). ""Stealth" adenoviruses blunt cell-mediated and humoral immune responses against the virus and allow for significant gene expression upon readministration in the lung." *J Virol* **75**(10): 4792–801.

Croyle, M. A., N. Chirmule, Y. Zhang and J. M. Wilson (2002). "PEGylation of E1-deleted adenovirus vectors allows significant gene expression on readministration to liver." *Hum Gene Ther* **13**(15): 1887–900.

Crystal, R. G. (2014). "Adenovirus: the first effective in vivo gene delivery vector." *Hum Gene Ther* **25**(1): 3–11.

Crystal, R. G., N. G. McElvaney, M. A. Rosenfeld, C. S. Chu, A. Mastrangeli, J. G. Hay, S. L. Brody, H. A. Jaffe, N. T. Eissa and C. Danel (1994). "Administration of an adenovirus containing the human CFTR cDNA to the respiratory tract of individuals with cystic fibrosis." *Nat Genet* **8**(1): 42–51.

Cubizolle, A., N. Serratrice, N. Skander, M. A. Colle, S. Ibanes, A. Gennetier, N. Bayo-Puxan, K. Mazouni, F. Mennechet, B. Joussemet, Y. Cherel, Y. Lajat, C. Vite, F. Bernex, V. Kalatzis, M. E. Haskins and E. J. Kremer (2014). "Corrective GUSB transfer to the canine mucopolysaccharidosis VII brain." *Mol Ther* **22**(4): 762–73.

Cuconati, A., C. Mukherjee, D. Perez and E. White (2003). "DNA damage response and MCL-1 destruction initiate apoptosis in adenovirus-infected cells." *Genes Dev* **17**(23): 2922–32.

Cuconati, A. and E. White (2002). "Viral homologs of BCL-2: role of apoptosis in the regulation of virus infection." *Genes Dev* **16**(19): 2465–78.

Cuesta, R., Q. Xi and R. J. Schneider (2001). "Preferential translation of adenovirus mRNAs in infected cells." *Cold Spring Harb. Symp. Quant. Biol.* **66**: 259–267.

Cuesta, R., Q. Xi and R. J. Schneider (2004). "Structural basis for competitive inhibition of eIF4G-Mnk1 interaction by the adenovirus 100-kilodalton protein." *J Virol* **78**(14): 7707–16.

Cullen, B. R. (2004). "Derivation and function of small interfering RNAs and micro-RNAs." *Virus Res* **102**(1): 3–9.

Culp, J. S., L. C. Webster, D. J. Friedman, C. L. Smith, W. J. Huang, F. H. Wu, M. Rosenberg and R. P. Ricciardi (1988). "The 289-amino acid E1A protein of adenovirus binds zinc in a region that is important for trans-activation." *Proc Natl Acad Sci Usa* **85**: 6450–6454.

Cupelli, K., S. Muller, B. D. Persson, M. Jost, N. Arnberg and T. Stehle (2010). "Structure of adenovirus type 21 knob in complex with CD46 reveals key differences in receptor contacts among species B adenoviruses." *J Virol* **84**(7): 3189–200.

Curlin, M. E., F. Cassis-Ghavami, A. S. Magaret, G. A. Spies, A. Duerr, C. L. Celum, J. L. Sanchez, J. B. Margolick, R. Detels, M. J. McElrath and L. Corey (2011). "Serological immunity to adenovirus serotype 5 is not associated with risk of HIV infection: a case-control study." *AIDS* **25**(2): 153–8.

Cutt, J. R., T. Shenk and P. Hearing (1987). "Analysis of adenovirus early region 4-encoded polypeptides synthesized in productively infected cells." *J. Virol.* **61**: 543–552.

Cuzange, A., J. Chroboczek and B. Jacrot (1994). "The penton base of human adenovirus type 3 has the RGD motif." *Gene* **146**(2): 257–9.

D'Halluin, J. C., M. Milleville, P. A. Boulanger and G. R. Martin (1978). "Temperature-sensitive mutant of adenovirus type 2 blocked in virion assembly: accumulation of light intermediate particles." *J Virol* **26**(2): 344–56.

Dai, W., C. Fu, D. Raytcheva, J. Flanagan, H. A. Khant, X. Liu, R. H. Rochat, C. Haase-Pettingell, J. Piret, S. J. Ludtke, K. Nagayama, M. F. Schmid, J. A. King and W. Chiu (2013). "Visualizing virus assembly intermediates inside marine cyanobacteria." *Nature* **502**(7473): 707–10.

Dales, S. and Y. Chardonnet (1973). "Early events in the interaction of adenoviruses with HeLa cells. IV. Association with microtubules and the nuclear pore complex during vectorial movement of the inoculum." *Virology* **56**(2): 465–83.

Daniell, E., D. E. Groff and M. J. Fedor (1981). "Adenovirus chromatin structure at different stages of infection." *Mol Cell Biol* **1**(12): 1094–105.

Daniell, E. and T. Mullenbach (1978). "Synthesis of defective viral DNA in HeLa cells infected with adenovirus type 3." *J Virol* **26**(1): 61–70.

Davies, M. V., M. Furtado, J. W. Hershey, B. Thimmappaya and R. J. Kaufman (1989). "Complementation of adenovirus virus-associated RNA I gene deletion by expression of a mutant eukaryotic translation initiation factor." *Proc Natl Acad Sci U S A* **86**(23): 9163–7.

Davison, A. J., M. Benko and B. Harrach (2003). "Genetic content and evolution of adenoviruses." *J Gen Virol* **84**(Pt 11): 2895–908.

De Clercq, E. (2011). "The clinical potential of the acyclic (and cyclic) nucleoside phosphonates: the magic of the phosphonate bond." *Biochem Pharmacol* **82**(2): 99–109.

de Gruijl, T. D. and R. van de Ven (2012). "Chapter six--Adenovirus-based immunotherapy of cancer: promises to keep." *Adv Cancer Res* **115**: 147–220.

de Jong, R. N., M. E. Mysiak, L. A. Meijer, M. van der Linden and P. C. van der Vliet (2002). "Recruitment of the priming protein pTP and DNA binding occur by overlapping Oct-1 POU homeodomain surfaces." *EMBO J* **21**(4): 725–35.

de Jong, R. N. and P. C. van der Vliet (1999). "Mechanism of DNA replication in eukaryotic cells: cellular host factors stimulating adenovirus DNA replication." *Gene* **236**(1): 1–12.

de Jong, R. N., P. C. van der Vliet and A. B. Brenkman (2003). "Adenovirus DNA replication: protein priming, jumping back and the role of the DNA binding protein DBP." *Curr Top Microbiol Immunol* **272**: 187–211.

de La Vega, M. A., D. Stein and G. P. Kobinger (2015). "Ebolavirus Evolution: Past and Present." *PLoS Pathog* **11**(11): e1005221.

De Rosa, S. C., E. P. Thomas, J. Bui, Y. Huang, A. deCamp, C. Morgan, S. A. Kalams, G. D. Tomaras, R. Akondy, R. Ahmed, C. Y. Lau, B. S. Graham, G. J. Nabel, M. J. McElrath, A. National Institute of and H. I. V. V. T. N. Infectious Diseases (2011). "HIV-DNA priming alters T cell responses to HIV-adenovirus vaccine even when responses to DNA are undetectable." *J Immunol* **187**(6): 3391–401.

Debbas, M. and E. White (1993). "Wild-type p53 mediates apoptosis by E1A, which is inhibited by E1B." *Genes Dev* **7**(4): 546–54.

DeCaprio, J. A. (2009). "How the Rb tumor suppressor structure and function was revealed by the study of Adenovirus and SV40." *Virology* **384**(2): 274–84.

Dechecchi, M. C., P. Melotti, A. Bonizzato, M. Santacatterina, M. Chilosi and G. Cabrini (2001). "Heparan sulfate glycosaminoglycans are receptors sufficient to mediate the initial binding of adenovirus types 2 and 5." *J Virol* **75**(18): 8772–80.

Dechecchi, M. C., A. Tamanini, A. Bonizzato and G. Cabrini (2000). "Heparan sulfate glycosaminoglycans are involved in adenovirus type 5 and 2-host cell interactions." *Virology* **268**(2): 382–90.

Degli-Esposti, M. A. and M. J. Smyth (2005). "Close encounters of different kinds: dendritic cells and NK cells take centre stage." *Nat Rev Immunol* **5**(2): 112–24.

DeHart, C. J., J. S. Chahal, S. J. Flint and D. H. Perlman (2014). "Extensive post-translational modification of active and inactivated forms of endogenous p53." *Mol Cell Proteomics* **13**(1): 1–17.

DeHart, C. J., D. H. Perlman and S. J. Flint (2015). "Impact of the adenoviral E4 Orf3 protein on the activity and posttranslational modification of p53." *J Virol* **89**(6): 3209–20.

Dehghan, S., J. Seto, M. S. Jones, D. W. Dyer, J. Chodosh and D. Seto (2013). "Simian adenovirus type 35 has a recombinant genome comprising human and simian adenovirus sequences, which predicts its potential emergence as a human respiratory pathogen." *Virology* **447**(1–2): 265–73.

Dekker, J., P. N. Kanellopoulos, A. K. Loonstra, J. A. van Oosterhout, K. Leonard, P. A. Tucker and P. C. van der Vliet (1997). "Multimerization of the adenovirus DNA-binding protein is the driving force for ATP-independent DNA unwinding during strand displacement synthesis." *EMBO J* **16**(6): 1455–63.

Dempsey, A. and A. G. Bowie (2015). "Innate immune recognition of DNA: A recent history." *Virology* **479–480**: 146–52.

Dermody, T. S., E. Kirchner, K. M. Guglielmi and T. Stehle (2009). "Immunoglobulin superfamily virus receptors and the evolution of adaptive immunity." *PLoS Pathog* **5**(11): e1000481.

Dery, C. V., C. H. Herrmann and M. B. Mathews (1987). "Response of individual adenovirus promoters to the products of the E1A gene." *Oncogene* **2**(1): 15–23.

Deryckere, F. and H. G. Burgert (1996). "Early region 3 of adenovirus type 19 (subgroup D) encodes an HLA-binding protein distinct from that of subgroups B and C." *J Virol* **70**(5): 2832–41.

Desai, S. Y., R. C. Patel, G. C. Sen, P. Malhotra, G. D. Ghadge and B. Thimmapaya (1995). "Activation of interferon-inducible 2'-5' oligoadenylate synthetase by adenoviral VAI RNA." *J Biol Chem* **270**(7): 3454–61.

Desiderio, S. V. and T. J. Kelly, Jr. (1981). "Structure of the linkage between adenovirus DNA and the 55,000 molecular weight terminal protein." *J Mol Biol* **145**(2): 319–37.

Devaux, C., M. L. Caillet-Boudin, B. Jacrot and P. Boulanger (1987). "Crystallization, enzymatic cleavage, and the polarity of the adenovirus type 2 fiber." *Virology* **161**(1): 121–8.

DeWeese, T. L., H. van der Poel, S. Li, B. Mikhak, R. Drew, M. Goemann, U. Hamper, R. DeJong, N. Detorie, R. Rodriguez, T. Haulk, A. M. DeMarzo, S. Piantadosi, D. C. Yu, Y. Chen, D. R. Henderson, M. A. Carducci, W. G. Nelson and J. W. Simons (2001). "A phase I trial of CV706, a replication-competent, PSA selective oncolytic adenovirus, for the treatment of locally recurrent prostate cancer following radiation therapy." *Cancer Res* **61**(20): 7464–72.

Di Guilmi, A. M., A. Barge, P. Kitts, E. Gout and J. Chroboczek (1995). "Human adenovirus serotype 3 (Ad3) and the Ad3 fiber protein bind to a 130-kDa membrane protein on HeLa cells." *Virus Res* **38**(1): 71–81.

Di Paolo, N. C., E. A. Miao, Y. Iwakura, K. Murali-Krishna, A. Aderem, R. A. Flavell, T. Papayannopoulou and D. M. Shayakhmetov (2009). "Virus binding to a plasma membrane receptor triggers interleukin-1 alpha-mediated proinflammatory macrophage response in vivo." *Immunity* **31**(1): 110–21.

Dieckmann, A. and B. Krippl (1994). "The E1A transcriptional control region is efficiently activated in proliferating tissues of transgenic mice." *Oncogene* **9**(8): 2227–33.

Diouri, M., H. Keyvani-Amineh, K. F. Geoghegan and J. M. Weber (1996). "Cleavage efficiency by adenovirus protease is site-dependent." *J Biol Chem* **271**(51): 32511–4.

DiPaolo, N., S. Ni, A. Gaggar, R. Strauss, S. Tuve, Z. Y. Li, D. Stone, D. Shayakhmetov, N. Kiviat, P. Toure, S. Sow, B. Horvat and A. Lieber (2006). "Evaluation of adenovirus vectors containing serotype 35 fibers for vaccination." *Mol Ther* **13**(4): 756–65.

Dix, I. and K. N. Leppard (1993). "Regulated splicing of adenovirus type 5 E4 transcripts and regulated cytoplasmic accumulation of E4 mRNA." *J Virol* **67**(6): 3226–31.

Dobbelstein, M. (2004). "Replicating adenoviruses in cancer therapy." *Curr Top Microbiol Immunol* **273**: 291–334.

Dobner, T. and J. Kzhyshkowska (2001). "Nuclear export of adenovirus RNA." *Curr. Top. Microbiol. Immunol.* **259**: 25–34.

Dolph, P. J., J. T. Huang and R. J. Schneider (1990). "Translation by the adenovirus tripartite leader: elements which determine independence from cap-binding protein complex." *J Virol* **64**(6): 2669–77.

Domanov, Y. A. and P. K. Kinnunen (2006). "Antimicrobial peptides temporins B and L induce formation of tubular lipid protrusions from supported phospholipid bilayers." *Biophys J* **91**(12): 4427–39.

Dong, X., L. Z. Mi, J. Zhu, W. Wang, P. Hu, B. H. Luo and T. A. Springer (2012). "alpha(V)beta(3) integrin crystal structures and their functional implications." *Biochemistry* **51**(44): 8814–28.

Dorin, J. R., R. Farley, S. Webb, S. N. Smith, E. Farini, S. J. Delaney, B. J. Wainwright, E. W. Alton and D. J. Porteous (1996). "A demonstration using mouse models that successful gene therapy for cystic fibrosis requires only partial gene correction." *Gene Ther* **3**(9): 797–801.

Doronin, K., J. W. Flatt, N. C. Di Paolo, R. Khare, O. Kalyuzhniy, M. Acchione, J. P. Sumida, U. Ohto, T. Shimizu, S. Akashi-Takamura, K. Miyake, J. W. MacDonald, T. K. Bammler, R. P. Beyer, F. M. Farin, P. L. Stewart and D. M. Shayakhmetov (2012). "Coagulation factor X activates innate immunity to human species C adenovirus." *Science* **338**(6108): 795–8.

Doronin, K., K. Toth, M. Kuppuswamy, P. Krajcsi, A. E. Tollefson and W. S. Wold (2003). "Overexpression of the ADP (E3–11.6K) protein increases cell lysis and spread of adenovirus." *Virology* **305**(2): 378–87.

Doronin, K., K. Toth, M. Kuppuswamy, P. Ward, A. E. Tollefson and W. S. Wold (2000). "Tumor-specific, replication-competent adenovirus vectors overexpressing the adenovirus death protein." *J Virol* **74**(13): 6147–55.

Dou, S., X. Zeng, P. Cortes, H. Erdjument-Bromage, P. Tempst, T. Honjo and L. D. Vales (1994). "The recombination signal sequence-binding protein RBP-2N functions as a transcriptional repressor." *Mol. Cell. Biol.* **14**(5): 3310–3319.

Doucas, V., A. M. Ishov, A. Romo, H. Juguilon, M. D. Weitzman, R. M. Evans and G. G. Maul (1996). "Adenovirus replication is coupled with the dynamic properties of the PML nuclear structure." *Genes Dev* **10**(2): 196–207.

Dudareva, M., L. Andrews, S. C. Gilbert, P. Bejon, K. Marsh, J. Mwacharo, O. Kai, A. Nicosia and A. V. Hill (2009). "Prevalence of serum neutralizing antibodies against chimpanzee adenovirus 63 and human adenovirus 5 in Kenyan children, in the context of vaccine vector efficacy." *Vaccine* **27**(27): 3501–4.

Duerr, A., Y. Huang, S. Buchbinder, R. W. Coombs, J. Sanchez, C. del Rio, M. Casapia, S. Santiago, P. Gilbert, L. Corey and M. N. Robertson; Step/HVTN 504 Study Team (2012). "Extended follow-up confirms early vaccine-enhanced risk of HIV acquisition and demonstrates waning effect over time among participants in a randomized trial of recombinant adenovirus HIV vaccine (Step Study)." *J Infect Dis* **206**(2): 258–66.

Duffy, M. R., A. L. Parker, E. R. Kalkman, K. White, D. Kovalskyy, S. M. Kelly and A. H. Baker (2013). "Identification of novel small molecule inhibitors of adenovirus gene transfer using a high throuput screening approach." *J Control Release* **170**(1): 132–40.

Dupont, N., S. Lacas-Gervais, J. Bertout, I. Paz, B. Freche, G. T. Van Nhieu, F. G. van der Goot, P. J. Sansonetti and F. Lafont (2009). "Shigella phagocytic vacuolar membrane remnants participate in the cellular response to pathogen invasion and are regulated by autophagy." *Cell Host Microbe* **6**(2): 137–49.

Durmort, C., C. Stehlin, G. Schoehn, A. Mitraki, E. Drouet, S. Cusack and W. P. Burmeister (2001). "Structure of the fiber head of Ad3, a non-CAR-binding serotype of adenovirus." *Virology* **285**(2): 302–12.

Dyson, N. (1994). "pRB, p107 and the regulation of the E2F transcription factor." *J Cell Sci Suppl* **18**: 81–7.

Dyson, N., K. Buchkovich, P. Whyte and E. Harlow (1989). "The cellular 107K protein that binds to adenovirus E1A also associates with the large T antigens of SV40 and JC virus." *Cell* **58**(2): 249–55.

Dzananovic, E., T. R. Patel, G. Chojnowski, M. J. Boniecki, S. Deo, K. McEleney, S. E. Harding, J. M. Bujnicki and S. A. McKenna (2014). "Solution conformation of adenovirus virus associated RNA-I and its interaction with PKR." *J Struct Biol* **185**(1): 48–57.

Eager, K. B., J. Williams, D. Breiding, S. Pan, B. Knowles, E. Appella and R. P. Ricciardi (1985). "Expression of histocompatibility antigens H-2K, -D, and -L is reduced in adenovirus-12-transformed mouse cells and is restored by interferon gamma." *Proc Natl Acad Sci U S A* **82**(16): 5525–9.

Eagle, P. A. and D. F. Klessig (1992). "A zinc-binding motif located between amino acids 273 and 286 in the adenovirus DNA-binding protein is necessary for ssDNA binding." *Virology* **187**(2): 777–87.

Ebbinghaus, C., A. Al-Jaibaji, E. Operschall, A. Schoffel, I. Peter, U. F. Greber and S. Hemmi (2001). "Functional and selective targeting of adenovirus to high-affinity Fcgamma receptor I-positive cells by using a bispecific hybrid adapter." *J Virol* **75**(1): 480–9.

Echavarria, M. (2008). "Adenoviruses in immunocompromised hosts." *Clin Microbiol Rev* **21**(4): 704–15.

Eckstein, A., T. Grossl, A. Geisler, X. Wang, S. Pinkert, T. Pozzuto, C. Schwer, J. Kurreck, S. Weger, R. Vetter, W. Poller and H. Fechner (2010). "Inhibition of adenovirus infections by siRNA-mediated silencing of early and late adenoviral gene functions." *Antiviral Res* **88**(1): 86–94.

Egan, C., T. N. Jelsma, J. A. Howe, S. T. Bayley, B. Ferguson and P. E. Branton (1988). "Mapping of cellular protein-binding sites on the products of early-region 1A of human adenovirus type 5." *Mol Cell Biol* **8**(9): 3955–9.

Ehrhardt, A., H. Xu, A. M. Dillow, D. A. Bellinger, T. C. Nichols and M. A. Kay (2003). "A gene-deleted adenoviral vector results in phenotypic correction of canine hemophilia B without liver toxicity or thrombocytopenia." *Blood* **102**(7): 2403–11.

Elkon, K. B., C. C. Liu, J. G. Gall, J. Trevejo, M. W. Marino, K. A. Abrahamsen, X. Song, J. L. Zhou, L. J. Old, R. G. Crystal and E. Falck-Pedersen (1997). "Tumor necrosis factor alpha plays a central role in immune-mediated clearance of adenoviral vectors." *Proc Natl Acad Sci U S A* **94**(18): 9814–9.

Ellsworth, D., R. L. Finnen and S. J. Flint (2001). "Superimposed promoter sequences of the adenoviral E2 early RNA polymerase III and RNA polymerase II transcription units." *J. Biol. Chem.* **276**(1): 827–834.

Elsing, A. and H. G. Burgert (1998). "The adenovirus E3/10.4K-14.5K proteins down-modulate the apoptosis receptor Fas/Apo-1 by inducing its internalization." *Proc Natl Acad Sci U S A* **95**(17): 10072–7.

Endter, C., B. Hartl, T. Spruss, J. Hauber and T. Dobner (2005). "Blockage of CRM1-dependent nuclear export of the adenovirus type 5 early region 1B 55-kDa protein augments oncogenic transformation of primary rat cells." *Oncogene* **24**(1): 55–64.

Engleman, V. W., G. A. Nickols, F. P. Ross, M. A. Horton, D. W. Griggs, S. L. Settle, P. G. Ruminski and S. L. Teitelbaum (1997). "A peptidomimetic antagonist of the alpha(v)beta3 integrin inhibits bone resorption in vitro and prevents osteoporosis in vivo." *J Clin Invest* **99**(9): 2284–92.

Enomoto, T., J. H. Lichy, J. E. Ikeda and J. Herwitz (1981). "Adenovirus DNA replication in vitro: Purification of the terminal protein in a functional form." *Proc Natl Acad Sci Usa* **78**: 6779–0.

Epand, R. M. and R. F. Epand (2000). "Modulation of membrane curvature by peptides." *Biopolymers* **55**(5): 358–63.

Erkmann, J. A. and U. Kutay (2004). "Nuclear export of mRNA: from the site of transcription to the cytoplasm." *Exp Cell Res* **296**(1): 12–20.

Ernst, B. and J. L. Magnani (2009). "From carbohydrate leads to glycomimetic drugs." *Nat Rev Drug Discov* **8**(8): 661–77.

Ersching, J., M. I. Hernandez, F. S. Cezarotto, J. D. Ferreira, A. B. Martins, W. M. Switzer, Z. Xiang, H. C. Ertl, C. R. Zanetti and A. R. Pinto (2010). "Neutralizing antibodies to human and simian adenoviruses in humans and New-World monkeys." *Virology* **407**(1): 1–6.

Estmer Nilsson, C., S. Petersen-Mahrt, C. Durot, R. Shtrichman, A. R. Krainer, T. Kleinberger and G. Akusjarvi (2001). "The adenovirus E4-ORF4 splicing enhancer protein interacts with a subset of phosphorylated SR proteins." *EMBO J* **20**(4): 864–71.

Eto, Y., J. Q. Gao, F. Sekiguchi, S. Kurachi, K. Katayama, M. Maeda, K. Kawasaki, H. Mizuguchi, T. Hayakawa, Y. Tsutsumi, T. Mayumi and S. Nakagawa (2005). "PEGylated adenovirus vectors containing RGD peptides on the tip of PEG show high transduction efficiency and antibody evasion ability." *J Gene Med* **7**(5): 604–12.

Evans, J. D. and P. Hearing (2003). "Distinct roles of the Adenovirus E4 ORF3 protein in viral DNA replication and inhibition of genome concatenation." *J Virol* **77**(9): 5295–304.

Evans, J. D. and P. Hearing (2005). "Relocalization of the Mre11-Rad50-Nbs1 complex by the adenovirus E4 ORF3 protein is required for viral replication." *J Virol* **79**(10): 6207–15.

Everett, S. F. and H. S. Ginsberg (1958). "A toxin-like material separable from type 5 adenovirus particles." *Virology* **6**(3): 770–1.

Everitt, E., A. de Luca and Y. Blixt (1992). "Antibody-mediated uncoating of adenovirus in vitro." *FEMS Microbiol Lett* **77**(1–3): 21–7.

Ewer, K. J., G. A. O'Hara, C. J. Duncan, K. A. Collins, S. H. Sheehy, A. Reyes-Sandoval, A. L. Goodman, N. J. Edwards, S. C. Elias, F. D. Halstead, R. J. Longley, R. Rowland, I. D. Poulton, S. J. Draper, A. M. Blagborough, E. Berrie, S. Moyle, N. Williams, L. Siani, A. Folgori, S. Colloca, R. E. Sinden, A. M. Lawrie, R. Cortese, S. C. Gilbert, A. Nicosia and A. V. Hill (2013). "Protective CD8+ T-cell immunity to human malaria induced by chimpanzee adenovirus-MVA immunisation." *Nat Commun* **4**: 2836.

Ewer, K. J., K. Sierra-Davidson, A. M. Salman, J. J. Illingworth, S. J. Draper, S. Biswas and A. V. Hill (2015). "Progress with viral vectored malaria vaccines: A multi-stage approach involving "unnatural immunity"." *Vaccine* **33**(52): 7444–51.

Ewing, S. G., S. A. Byrd, J. B. Christensen, R. E. Tyler and M. J. Imperiale (2007). "Ternary complex formation on the adenovirus packaging sequence by the IVa2 and L4 22-kilodalton proteins." *J Virol* **81**(22): 12450–7.

Excoffon, K. J., N. D. Gansemer, M. E. Mobily, P. H. Karp, K. R. Parekh and J. Zabner (2010). "Isoform-specific regulation and localization of the coxsackie and adenovirus receptor in human airway epithelia." *PLoS One* **5**(3): e9909.

Fabian, M. R. and N. Sonenberg (2012). "The mechanics of miRNA-mediated gene silencing: a look under the hood of miRISC." *Nat Struct Mol Biol* **19**(6): 586–93.

Fabry, C. M., M. Rosa-Calatrava, J. F. Conway, C. Zubieta, S. Cusack, R. W. Ruigrok and G. Schoehn (2005). "A quasi-atomic model of human adenovirus type 5 capsid." *EMBO J* **24**(9): 1645–54.

Fairhurst, R. M. (2015). "Understanding artemisinin-resistant malaria: what a difference a year makes." *Curr Opin Infect Dis* **28**(5): 417–25.

Fang, L., Y. Y. Pu, X. C. Hu, L. J. Sun, H. M. Luo, S. K. Pan, J. Z. Gu, X. R. Cao and C. Q. Su (2010). "Antiangiogenesis gene armed tumor-targeting adenovirus yields multiple antitumor activities in human HCC xenografts in nude mice." *Hepatol Res* **40**(2): 216–28.

Fang, L., J. L. Stevens, A. J. Berk and K. R. Spindler (2004). "Requirement of Sur2 for efficient replication of mouse adenovirus type 1." *J Virol* **78**(23): 12888–900.

Farley, D. C., J. L. Brown and K. N. Leppard (2004). "Activation of the early-late switch in adenovirus type 5 major late transcription unit expression by L4 gene products." *J Virol* **78**(4): 1782–91.

Farmer, C., P. E. Morton, M. Snippe, G. Santis and M. Parsons (2009). "Coxsackie adenovirus receptor (CAR) regulates integrin function through activation of p44/42 MAPK." *Exp Cell Res* **315**(15): 2637–47.

Farrow, S. N., J. H. White, I. Martinou, T. Raven, K. T. Pun, C. J. Grinham, J. C. Martinou and R. Brown (1995). "Cloning of a bcl-2 homologue by interaction with adenovirus E1B 19K." *Nature* **374**(6524): 731–3.

Fasbender, A., J. Zabner, M. Chillon, T. O. Moninger, A. P. Puga, B. L. Davidson and M. J. Welsh (1997). "Complexes of adenovirus with polycationic polymers and cationic lipids increase the efficiency of gene transfer in vitro and in vivo." *J Biol Chem* **272**(10): 6479–89.

Fattaey, A. R., E. Harlow and K. Helin (1993). "Independent regions of adenovirus E1A are required for binding to and dissociation of E2F-protein complexes." *Mol Cell Biol* **13**(12): 7267–77.

Fejer, G., L. Drechsel, J. Liese, U. Schleicher, Z. Ruzsics, N. Imelli, U. F. Greber, S. Keck, B. Hildenbrand, A. Krug, C. Bogdan and M. A. Freudenberg (2008). "Key role of splenic myeloid DCs in the IFN-alphabeta response to adenoviruses in vivo." *PLoS Pathog* **4**(11): e1000208.

Fejer, G., K. Szalay, I. Gyory, M. Fejes, E. Kusz, S. Nedieanu, T. Pali, T. Schmidt, B. Siklodi, G. Lazar, Jr., G. Lazar, Sr. and E. Duda (2005). "Adenovirus infection dramatically augments lipopolysaccharide-induced TNF production and sensitizes to lethal shock." *J Immunol* **175**(3): 1498–506.

Felsani, A., A. M. Mileo and M. G. Paggi (2006). "Retinoblastoma family proteins as key targets of the small DNA virus oncoproteins." *Oncogene* **25**(38): 5277–85.

Fender, P., R. W. Ruigrok, E. Gout, S. Buffet and J. Chroboczek (1997). "Adenovirus dodecahedron, a new vector for human gene transfer." *Nat Biotechnol* **15**(1): 52–6.

Fender, P., G. Schoehn, J. Foucaud-Gamen, E. Gout, A. Garcel, E. Drouet and J. Chroboczek (2003). "Adenovirus dodecahedron allows large multimeric protein transduction in human cells." *J Virol* **77**(8): 4960–4.

Feng, Z., L. Hensley, K. L. McKnight, F. Hu, V. Madden, L. Ping, S. H. Jeong, C. Walker, R. E. Lanford and S. M. Lemon (2013). "A pathogenic picornavirus acquires an envelope by hijacking cellular membranes." *Nature* **496**(7445): 367–71.

Fernandes, P., A. I. Almeida, E. J. Kremer, P. M. Alves and A. S. Coroadinha (2015). "Canine helper-dependent vectors production: implications of Cre activity and co-infection on adenovirus propagation." *Sci Rep* **5**: 9135.

Fernandez-Lopez, S., H. S. Kim, E. C. Choi, M. Delgado, J. R. Granja, A. Khasanov, K. Kraehenbuehl, G. Long, D. A. Weinberger, K. M. Wilcoxen and M. R. Ghadiri (2001). "Antibacterial agents based on the cyclic D,L-alpha-peptide architecture." *Nature* **412**(6845): 452–5.

Ferrara, N. (2004). "Vascular endothelial growth factor: basic science and clinical progress." *Endocr Rev* **25**(4): 581–611.

Ferrara, N., H. P. Gerber and J. LeCouter (2003). "The biology of VEGF and its receptors." *Nat Med* **9**(6): 669–76.

Ferrari, R., D. Gou, G. Jawdekar, S. A. Johnson, M. Nava, T. Su, A. F. Yousef, N. R. Zemke, M. Pellegrini, S. K. Kurdistani and A. J. Berk (2014). "Adenovirus small E1A employs the lysine acetylases p300/CBP and tumor suppressor Rb to repress select host genes and promote productive virus infection." *Cell Host Microbe* **16**(5): 663–76.

Ferrari, R., M. Pellegrini, G. A. Horwitz, W. Xie, A. J. Berk and S. K. Kurdistani (2008). "Epigenetic reprogramming by adenovirus e1a." *Science* **321**(5892): 1086–8.

Ferrari, R., T. Su, B. Li, G. Bonora, A. Oberai, Y. Chan, R. Sasidharan, A. J. Berk, M. Pellegrini and S. K. Kurdistani (2012). "Reorganization of the host epigenome by a viral oncogene." *Genome Res* **22**(7): 1212–21.

Fessler, S. P. and C. S. Young (1999). "The role of the L4 33K gene in adenovirus infection." *Virology* **263**(2): 507–16.

Feucht, J., K. Opherk, P. Lang, S. Kayser, L. Hartl, W. Bethge, S. Matthes-Martin, P. Bader, M. H. Albert, B. Maecker-Kolhoff, J. Greil, H. Einsele, P. G. Schlegel, F. R. Schuster, B. Kremens, C. Rossig, B. Gruhn, R. Handgretinger and T. Feuchtinger (2015). "Adoptive T-cell therapy with hexon-specific Th1 cells as a treatment of refractory adenovirus infection after HSCT." *Blood* **125**(12): 1986–94.

Feuchtinger, T., J. Lucke, K. Hamprecht, C. Richard, R. Handgretinger, M. Schumm, J. Greil, T. Bock, D. Niethammer and P. Lang (2005). "Detection of adenovirus-specific T cells in children with adenovirus infection after allogeneic stem cell transplantation." *Br J Haematol* **128**(4): 503–9.

Feuchtinger, T., S. Matthes-Martin, C. Richard, T. Lion, M. Fuhrer, K. Hamprecht, R. Handgretinger, C. Peters, F. R. Schuster, R. Beck, M. Schumm, R. Lotfi, G. Jahn and P. Lang (2006). "Safe adoptive transfer of virus-specific T-cell immunity for the treatment of systemic adenovirus infection after allogeneic stem cell transplantation." *Br J Haematol* **134**(1): 64–76.

Feuchtinger, T., C. Richard, S. Joachim, M. H. Scheible, M. Schumm, K. Hamprecht, D. Martin, G. Jahn, R. Handgretinger and P. Lang (2008). "Clinical grade generation of hexon-specific T cells for adoptive T-cell transfer as a treatment of adenovirus infection after allogeneic stem cell transplantation." *J Immunother* **31**(2): 199–206.

Feuerbach, D., S. Etteldorf, C. Ebenau-Jehle, J. P. Abastado, D. Madden and H. G. Burgert (1994). "Identification of amino acids within the MHC molecule important for the interaction with the adenovirus protein E3/19K." *J Immunol* **153**(4): 1626–36.

Field, J., R. M. Gronostajski and J. Hurwitz (1984). "Properties of the adenovirus DNA polymerase." *J Biol Chem* **259**(15): 9487–95.

Finnen, R. L., J. F. Biddle and J. Flint (2001). "Truncation of the human adenovirus type 5 L4 33-kDa protein: evidence for an essential role of the carboxy-terminus in the viral infectious cycle." *Virology* **289**(2): 388–99.

Fitzgerald, D. W., H. Janes, M. Robertson, R. Coombs, I. Frank, P. Gilbert, M. Loufty, D. Mehrotra and A. Duerr; Step Study Protocol Team (2011). "An Ad5-vectored HIV-1 vaccine elicits cell-mediated immunity but does not affect disease progression in HIV-1-infected male subjects: results from a randomized placebo-controlled trial (the Step study)." *J Infect Dis* **203**(6): 765–72.

Flatt, J. W. and U. F. Greber (2015). "Misdelivery at the Nuclear Pore Complex-Stopping a Virus Dead in Its Tracks." *Cells* **4**(3): 277–96.

Flatt, J. W., R. Kim, J. G. Smith, G. R. Nemerow and P. L. Stewart (2013). "An intrinsically disordered region of the adenovirus capsid is implicated in neutralization by human alpha defensin 5." *PLoS One* **8**(4): e61571.

Fleischli, C., S. Verhaagh, M. Havenga, D. Sirena, W. Schaffner, R. Cattaneo, U. F. Greber and S. Hemmi (2005). "The distal short consensus repeats 1 and 2 of the membrane cofactor protein CD46 and their distance from the cell membrane determine productive entry of species B adenovirus serotype 35." *J Virol* **79**(15): 10013–22.

Flint, S. J., P. H. Gallimore and P. A. Sharp (1975). "Comparison of viral RNA sequences in adenovirus 2-transformed and lytically infected cells." *J. Mo. Biol.* **96**(1): 47–68.

Flint, S. J. and R. A. Gonzalez (2003). "Regulation of mRNA production by the adenoviral E1B 55kDa and E4 Orf6 proteins." *Curr. Top. Microbiol. Immunol.* **272**: 287–330.

Flint, S. J. and T. Shenk (1989). "Adenovirus E1A protein: paradigm viral transactivator." *Ann. Rev. Genet.* **23**: 141–161.

Flint, S. J. F., V. Racaniello, G. Rall and A. Skalka (2015). *Principles of Virology.* Washington, D.C., American Society for Microbiology Press.

Flomenberg, P., J. Szmulewicz, E. Gutierrez and H. Lupatkin (1992). "Role of the adenovirus E3-19k conserved region in binding major histocompatibility complex class I molecules." *J Virol* **66**(8): 4778–83.

Flomenberg, P. R., M. Chen and M. S. Horwitz (1987). "Characterization of a major histocompatibility complex class I antigen-binding glycoprotein from adenovirus type 35, a type associated with immunocompromised hosts." *J Virol* **61**(12): 3665–71.

Florescu, D. F. and J. A. Hoffman (2013). "Adenovirus in solid organ transplantation." *Am J Transplant* **13 Suppl 4**: 206–11.

Florescu, D. F., S. A. Pergam, M. N. Neely, F. Qiu, C. Johnston, S. Way, J. Sande, D. A. Lewinsohn, J. A. Guzman-Cottrill, M. L. Graham, G. Papanicolaou, J. Kurtzberg, J. Rigdon, W. Painter, H. Mommeja-Marin, R. Lanier, M. Anderson and C. van der Horst (2012). "Safety and efficacy of CMX001 as salvage therapy for severe adenovirus infections in immunocompromised patients." *Biol Blood Marrow Transplant* **18**(5): 731–8.

Fonseca, G. J., M. J. Cohen and J. S. Mymryk (2014). "Adenovirus E1A recruits the human Paf1 complex to enhance transcriptional elongation." *J Virol* **88**(10): 5630–7.

Fonseca, G. J., M. J. Cohen, A. C. Nichols, J. W. Barrett and J. S. Mymryk (2013). "Viral retasking of hBre1/RNF20 to recruit hPaf1 for transcriptional activation." *PLoS Pathog* **9**(6): e1003411.

Fonseca, G. J., G. Thillainadesan, A. F. Yousef, J. N. Ablack, K. L. Mossman, J. Torchia and J. S. Mymryk (2012). "Adenovirus evasion of interferon-mediated innate immunity by direct antagonism of a cellular histone posttranslational modification." *Cell Host Microbe* **11**(6): 597–606.

Forget, B. G. and S. M. Weissman (1967). "Low molecular weight RNA components from KB cells." *Nature* **213**(5079): 878–82.

Forrester, N. A., R. N. Patel, T. Speiseder, P. Groitl, G. G. Sedgwick, N. J. Shimwell, R. I. Seed, P. O. Catnaigh, C. J. McCabe, G. S. Stewart, T. Dobner, R. J. Grand, A. Martin and A. S. Turnell (2012). "Adenovirus E4orf3 targets transcriptional intermediary factor 1gamma for proteasome-dependent degradation during infection." *J Virol* **86**(6): 3167–79.

Fowlkes, D. M. and T. Shenk (1980). "Transcriptional control regions of the adenovirus VAI RNA gene." *Cell* **22**: 405–413.

Fox, J. P., C. E. Hall and M. K. Cooney (1977). "The Seattle Virus Watch. VII. Observations of adenovirus infections." *Am J Epidemiol* **105**(4): 362–86.

Fox J. P., C. D. Brandt, F. E. Wassermann, C. E. Hall, I. Spigland, A. Kogon and L. R. Elveback (1969). "The virus watch program: a continuing surveillance of viral infections in metropolitan New York families. VI. Observations of adenovirus infections: virus excretion patterns, antibody response, efficiency of surveillance, patterns of infections, and relation to illness." *Am J Epidemiol* **89**(1): 25–50.

Franklin, R. M., U. Pettersson, K. Akervall, B. Strandberg and L. Philipson (1971). "Structural proteins of adenovirus. V. Size and structure of the adenovirus type 2 hexon." *J Mol Biol* **57**(3): 383–95.

Fraser, N. W., J. R. Nevins, E. Ziff and J. E. J. Darnell (1979). "The major late adenovirus type-2 transcription unit: termination is downstream from the last poly(A) site." *J. Mol. Biol.* **129**: 643–656.

Fredman, J. N., S. C. Pettit, M. S. Horwitz and J. A. Engler (1991). "Linker insertion mutations in the adenovirus preterminal protein that affect DNA replication activity in vivo and in vitro." *J Virol* **65**(9): 4591–7.

Freimuth, P., K. Springer, C. Berard, J. Hainfeld, M. Bewley and J. Flanagan (1999). "Coxsackievirus and adenovirus receptor amino-terminal immunoglobulin V-related domain binds adenovirus type 2 and fiber knob from adenovirus type 12." *J Virol* **73**(2): 1392–8.

Freimuth, P. I. and H. S. Ginsberg (1986). "Codon insertion mutants of the adenovirus terminal protein." *Proc Natl Acad Sci U S A* **83**(20): 7816–20.

Freyer, G. A., Y. Katoh and R. J. Roberts (1984). "Characterization of the major mRNAs from adenovirus 2 early region 4 by cDNA cloning and sequencing." *Nucleic Acids Res* **12**(8): 3503–19.

Freytag, S. O., M. Khil, H. Stricker, J. Peabody, M. Menon, M. DePeralta-Venturina, D. Nafziger, J. Pegg, D. Paielli, S. Brown, K. Barton, M. Lu, E. Aguilar-Cordova and J. H. Kim (2002). "Phase I study of replication-competent adenovirus-mediated double suicide gene therapy for the treatment of locally recurrent prostate cancer." *Cancer Res* **62**(17): 4968–76.

Freytag, S. O., B. Movsas, I. Aref, H. Stricker, J. Peabody, J. Pegg, Y. Zhang, K. N. Barton, S. L. Brown, M. Lu, A. Savera and J. H. Kim (2007). "Phase I trial of replication-competent adenovirus-mediated suicide gene therapy combined with IMRT for prostate cancer." *Mol Ther* **15**(5): 1016–23.

Freytag, S. O., K. R. Rogulski, D. L. Paielli, J. D. Gilbert and J. H. Kim (1998). "A novel three-pronged approach to kill cancer cells selectively: concomitant viral, double suicide gene, and radiotherapy." *Hum Gene Ther* **9**(9): 1323–33.

Freytag, S. O., H. Stricker, M. Lu, M. Elshaikh, I. Aref, D. Pradhan, K. Levin, J. H. Kim, J. Peabody, F. Siddiqui, K. Barton, J. Pegg, Y. Zhang, J. Cheng, N. Oja-Tebbe, R. Bourgeois, N. Gupta, Z. Lane, R. Rodriguez, T. DeWeese and B. Movsas (2014). "Prospective randomized phase 2 trial of intensity modulated radiation therapy with or without oncolytic adenovirus-mediated cytotoxic gene therapy in intermediate-risk prostate cancer." *Int J Radiat Oncol Biol Phys* **89**(2): 268–76.

Freytag, S. O., H. Stricker, J. Peabody, J. Pegg, D. Paielli, B. Movsas, K. N. Barton, S. L. Brown, M. Lu and J. H. Kim (2007). "Five-year follow-up of trial of replication-competent adenovirus-mediated suicide gene therapy for treatment of prostate cancer." *Mol Ther* **15**(3): 636–42.

Freytag, S. O., H. Stricker, J. Pegg, D. Paielli, D. G. Pradhan, J. Peabody, M. DePeralta-Venturina, X. Xia, S. Brown, M. Lu and J. H. Kim (2003). "Phase I study of replication-competent adenovirus-mediated double-suicide gene therapy in combination with conventional-dose three-dimensional conformal radiation therapy for the treatment of newly diagnosed, intermediate- to high-risk prostate cancer." *Cancer Res* **63**(21): 7497–506.

Fridman, J. S. and S. W. Lowe (2003). "Control of apoptosis by p53." *Oncogene* **22**(56): 9030–40.

Friedman, J. M. and M. S. Horwitz (2002). "Inhibition of tumor necrosis factor alpha-induced NF-kappa B activation by the adenovirus E3–10.4/14.5K complex." *J Virol* **76**(11): 5515–21.

Friefeld, B. R., M. D. Krevolin and M. S. Horwitz (1983). "Effects of the adenovirus H5ts125 and H5ts107 DNA binding proteins on DNA replication in vitro." *Virology* **124**(2): 380–9.

Frolov, M. V. and N. J. Dyson (2004). "Molecular mechanisms of E2F-dependent activation and pRB-mediated repression." *J Cell Sci* **117**(Pt 11): 2173–81.

Fu, J. and M. Bouvier (2011). "Determinants of the endoplasmic reticulum (ER) lumenal-domain of the adenovirus serotype 2 E3-19K protein for association with and ER-retention of major histocompatibility complex class I molecules." *Mol Immunol* **48**(4): 532–8.

Fu, J., L. Li and M. Bouvier (2011). "Adenovirus E3-19K proteins of different serotypes and subgroups have similar, yet distinct, immunomodulatory functions toward major histocompatibility class I molecules." *J Biol Chem* **286**(20): 17631–9.

Fuchs, J. D., P. A. Bart, N. Frahm, C. Morgan, P. B. Gilbert, N. Kochar, S. C. DeRosa, G. D. Tomaras, T. M. Wagner, L. R. Baden, B. A. Koblin, N. G. Rouphael, S. A. Kalams, M. C. Keefer, P. A. Goepfert, M. E. Sobieszczyk, K. H. Mayer, E. Swann, H. X. Liao, B. F. Haynes, B. S. Graham, M. J. McElrath and N. H. V. T. Network (2015). "Safety and Immunogenicity of a Recombinant Adenovirus Serotype 35-Vectored HIV-1 Vaccine in Adenovirus Serotype 5 Seronegative and Seropositive Individuals." *J AIDS Clin Res* **6**(5).

Fuchsova, B., L. A. Serebryannyy and P. de Lanerolle (2015). "Nuclear actin and myosins in adenovirus infection." *Exp Cell Res* **338**(2): 170–82.

Fueyo, J., C. Gomez-Manzano, R. Alemany, P. S. Lee, T. J. McDonnell, P. Mitlianga, Y. X. Shi, V. A. Levin, W. K. Yung and A. P. Kyritsis (2000). "A mutant oncolytic adenovirus targeting the Rb pathway produces anti-glioma effect in vivo." *Oncogene* **19**(1): 2–12.

Fujimoto, L. M., R. Roth, J. E. Heuser and S. L. Schmid (2000). "Actin assembly plays a variable, but not obligatory role in receptor-mediated endocytosis in mammalian cells." *Traffic* **1**(2): 161–71.

Furuse, Y., D. A. Ornelles and B. R. Cullen (2013). "Persistently adenovirus-infected lymphoid cells express microRNAs derived from the viral VAI and especially VAII RNA." *Virology* **447**(1–2): 140–5.

Fuschiotti, P., G. Schoehn, P. Fender, C. M. Fabry, E. A. Hewat, J. Chroboczek, R. W. Ruigrok and J. F. Conway (2006). "Structure of the dodecahedral penton particle from human adenovirus type 3." *J Mol Biol* **356**(2): 510–20.

Gaggar, A., D. M. Shayakhmetov and A. Lieber (2003). "CD46 is a cellular receptor for group B adenoviruses." *Nat Med* **9**(11): 1408–12.

Gaggar, A., D. M. Shayakhmetov, M. K. Liszewski, J. P. Atkinson and A. Lieber (2005). "Localization of regions in CD46 that interact with adenovirus." *J Virol* **79**(12): 7503–13.

Gahery-Segard, H., F. Farace, D. Godfrin, J. Gaston, R. Lengagne, T. Tursz, P. Boulanger and J. G. Guillet (1998). "Immune response to recombinant capsid proteins of adenovirus in humans: antifiber and anti-penton base antibodies have a synergistic effect on neutralizing activity." *J Virol* **72**(3): 2388–97.

Gallagher, J. G. and N. Khoobyarian (1969). "Adenovirus susceptibility to interferon: sensitivity of types 2,7, and 12 to human interferon." *Proc Soc Exp Biol Med* **130**(1): 137–42.

Gallagher, J. G. and N. Khoobyarian (1971). "Sensitivity of adenovirus types 1, 3, 4, 5, 8, 11, and 18 to human interferon." *Proc Soc Exp Biol Med* **136**(3): 920–4.

Gallimore, P. H. (1972). "Tumour production in immunosuppressed rats with cells transformed in vitro by adenovirus type 2." *J Gen Virol* **16**(1): 99–102.

Gallimore, P. H. and C. Paraskeva (1980). "A study to determine the reasons for differences in the tumorigenicity of rat cell lines transformed by adenovirus 2 and adenovirus 12." *Cold Spring Harb Symp Quant Biol* **44 Pt 1**: 703–13.

Gallimore, P. H. and A. S. Turnell (2001). "Adenovirus E1A: remodelling the host cell, a life or death experience." *Oncogene* **20**(54): 7824–35.

Gambke, C. and W. Deppert (1981). "Late nonstructural 100,000- and 33,000-dalton proteins of adenovirus type 2 I. Subcellular localization during the course of infection." *J. Virol.* **40**: 585–593.

Gambke, C. and W. Deppert (1983). "Specific complex of the late nonstructural 100,000-dalton protein with newly synthesized hexon in adenovirus type 2-infected cells." *Virology* **124**(1): 1–12.

Ganem, D., J. R. Pollack and J. Tavis (1994). "Hepatitis B virus reverse transcriptase and its many roles in hepadnaviral genomic replication." *Infect Agents Dis* **3**(2–3): 85–93.

Ganly, I., D. Kirn, G. Eckhardt, G. I. Rodriguez, D. S. Soutar, R. Otto, A. G. Robertson, O. Park, M. L. Gulley, C. Heise, D. D. Von Hoff and S. B. Kaye (2000). "A phase I study of Onyx-015, an E1B attenuated adenovirus, administered intratumorally to patients with recurrent head and neck cancer." *Clin Cancer Res* **6**(3): 798–806.

Gaydos, C. A. and J. C. Gaydos (1995). "Adenovirus vaccines in the U.S. military." *Mil Med* **160**(6): 300–4.

Gazzola, M., C. J. Burckhardt, B. Bayati, M. Engelke, U. F. Greber and P. Koumoutsakos (2009). "A stochastic model for microtubule motors describes the in vivo cytoplasmic transport of human adenovirus." *PLoS Comput Biol* **5**(12): e1000623.

Geisbert, T. W., M. Bailey, L. Hensley, C. Asiedu, J. Geisbert, D. Stanley, A. Honko, J. Johnson, S. Mulangu, M. G. Pau, J. Custers, J. Vellinga, J. Hendriks, P. Jahrling, M. Roederer, J. Goudsmit, R. Koup and N. J. Sullivan (2011). "Recombinant adenovirus serotype 26 (Ad26) and Ad35 vaccine vectors bypass immunity to Ad5 and protect nonhuman primates against ebolavirus challenge." *J Virol* **85**(9): 4222–33.

Georgiades, J., T. Zielinski, A. Cicholska and E. Jordan (1959). "Research on the oncolytic effect of APC viruses in cancer of the cervix uteri; preliminary report." *Biul Inst Med Morsk Gdansk* **10**: 49–57.

Gerritsen, M. E., A. J. Williams, A. S. Neish, S. Moore, Y. Shi and T. Collins (1997). "CREB-binding protein/p300 are transcriptional coactivators of p65." *Proc Natl Acad Sci U S A* **94**(7): 2927–32.

Ghadge, G. D., P. Malhotra, M. R. Furtado, R. Dhar and B. Thimmapaya (1994). "In vitro analysis of virus-associated RNA I (VAI RNA): inhibition of the double-stranded RNA-activated protein kinase PKR by VAI RNA mutants correlates with the in vivo phenotype and the structural integrity of the central domain." *J Virol* **68**(7): 4137–51.

Ghadiri, M. R., J. R. Granja and L. K. Buehler (1994). "Artificial transmembrane ion channels from self-assembling peptide nanotubes." *Nature* **369**(6478): 301–4.

Ghosh, M. K. and M. L. Harter (2003). "A viral mechanism for remodeling chromatin structure in G0 cells." *Mol Cell* **12**(1): 255–69.

Ginn, S. L., I. E. Alexander, M. L. Edelstein, M. R. Abedi and J. Wixon (2013). "Gene therapy clinical trials worldwide to 2012 — an update." *J Gene Med* **15**(2): 65–77.

Ginsberg, H. S., U. Lundholm and T. Linne (1977). "Adenovirus DNA-binding protein in cells infected with wild-type 5 adenovirus and two DNA-minus, temperature-sensitive mutants, H5ts125 and H5ts149." *J Virol* **23**(1): 142–51.

Glozak, M. A., N. Sengupta, X. Zhang and E. Seto (2005). "Acetylation and deacetylation of non-histone proteins." *Gene* **363**: 15–23.

Gomez-Roman, V. R., G. J. Grimes, Jr., G. K. Potti, B. Peng, T. Demberg, L. Gravlin, J. Treece, R. Pal, E. M. Lee, W. G. Alvord, P. D. Markham and M. Robert-Guroff (2006). "Oral delivery of replication-competent adenovirus vectors is well tolerated by SIV- and SHIV-infected rhesus macaques." *Vaccine* **24**(23): 5064–72.

Gonzalez-Navajas, J. M., J. Lee, M. David and E. Raz (2012). "Immunomodulatory functions of type I interferons." *Nat Rev Immunol* **12**(2): 125–35.

Gonzalez, R., W. Huang, R. Finnen, C. Bragg and S. J. Flint (2006). "Adenovirus E1B 55-kilodalton protein is required for both regulation of mRNA export and efficient entry into the late phase of infection in normal human fibroblasts." *J Virol* **80**(2): 964–74.

Gonzalez, R. A. and S. J. Flint (2002). "Effects of mutations in the adenoviral E1B 55-kilodalton protein coding sequence on viral late mRNA metabolism." *J Virol* **76**(9): 4507–19.

Gooding, L. R., L. W. Elmore, A. E. Tollefson, H. A. Brady and W. S. Wold (1988). "A 14,700 MW protein from the E3 region of adenovirus inhibits cytolysis by tumor necrosis factor." *Cell* **53**(3): 341–6.

Goodman, A. L., A. M. Blagborough, S. Biswas, Y. Wu, A. V. Hill, R. E. Sinden and S. J. Draper (2011). "A viral vectored prime-boost immunization regime targeting the malaria Pfs25 antigen induces transmission-blocking activity." *PLoS One* **6**(12): e29428.

Goodrum, F. D. and D. A. Ornelles (1997). "The early region 1B 55-kilodalton oncoprotein of adenovirus relieves growth restrictions imposed on viral replication by the cell cycle." *J Virol* **71**(1): 548–61.

Goodrum, F. D. and D. A. Ornelles (1998). "p53 status does not determine outcome of E1B 55-kilodalton mutant adenovirus lytic infection." *J Virol* **72**(12): 9479–90.

Gordon, Y. J., T. P. Araullo-Cruz, Y. F. Johnson, E. G. Romanowski and P. R. Kinchington (1996). "Isolation of human adenovirus type 5 variants resistant to the antiviral cidofovir." *Invest Ophthalmol Vis Sci* **37**(13): 2774–8.

Grable, M. and P. Hearing (1990). "Adenovirus type 5 packaging domain is composed of a repeated element that is functionally redundant." *J Virol* **64**(5): 2047–56.

Granberg, F., C. Svensson, U. Pettersson and H. Zhao (2005). "Modulation of host cell gene expression during onset of the late phase of an adenovirus infection is focused on growth inhibition and cell architecture." *Virology* **343**(2): 236–45.

Granberg, F., C. Svensson, U. Pettersson and H. Zhao (2006). "Adenovirus-induced alterations in host cell gene expression prior to the onset of viral gene expression." *Virology* **353**(1): 1–5.

Grand, R. J., M. L. Grant and P. H. Gallimore (1994). "Enhanced expression of p53 in human cells infected with mutant adenoviruses." *Virology* **203**(2): 229–40.

Gray, G. C., P. R. Goswami, M. D. Malasig, A. W. Hawksworth, D. H. Trump, M. A. Ryan and D. P. Schnurr (2000). "Adult adenovirus infections: loss of orphaned vaccines precipitates military respiratory disease epidemics. For the Adenovirus Surveillance Group." *Clin Infect Dis* **31**(3): 663–70.

Gray, G. E., M. Allen, Z. Moodie, G. Churchyard, L. G. Bekker, M. Nchabeleng, K. Mlisana, B. Metch, G. de Bruyn, M. H. Latka, S. Roux, M. Mathebula, N. Naicker, C. Ducar, D. K. Carter, A. Puren, N. Eaton, M. J. McElrath, M. Robertson, L. Corey, J. G. Kublin and H. P. s. team (2011). "Safety and efficacy of the HVTN 503/Phambili study of a clade-B-based HIV-1 vaccine in South Africa: a double-blind, randomised, placebo-controlled test-of-concept phase 2b study." *Lancet Infect Dis* **11**(7): 507–15.

Gray, G. E., Z. Moodie, B. Metch, P. B. Gilbert, L. G. Bekker, G. Churchyard, M. Nchabeleng, K. Mlisana, F. Laher, S. Roux, K. Mngadi, C. Innes, M. Mathebula, M. Allen, M. J. McElrath, M. Robertson, J. Kublin, L. Corey and H. P. s. team (2014). "Recombinant adenovirus type 5 HIV gag/pol/nef vaccine in South Africa: unblinded, long-term follow-up of the phase 2b HVTN 503/Phambili study." *Lancet Infect Dis* **14**(5): 388–96.

Graziano, V., G. Luo, P. C. Blainey, A. J. Perez-Berna, W. J. McGrath, S. J. Flint, C. San Martin, X. S. Xie and W. F. Mangel (2013). "Regulation of a viral proteinase by a peptide and DNA in one-dimensional space: II. adenovirus proteinase is activated in an unusual one-dimensional biochemical reaction." *J Biol Chem* **288**(3): 2068–80.

Greber, U. F., M. Suomalainen, R. P. Stidwill, K. Boucke, M. W. Ebersold and A. Helenius (1997). "The role of the nuclear pore complex in adenovirus DNA entry." *EMBO J* **16**(19): 5998–6007.

Greber, U. F., P. Webster, J. Weber and A. Helenius (1996). "The role of the adenovirus protease on virus entry into cells." *EMBO J* **15**(8): 1766–77.

Greber, U. F., M. Willetts, P. Webster and A. Helenius (1993). "Stepwise dismantling of adenovirus 2 during entry into cells." *Cell* **75**(3): 477–86.

Green, D. R. and G. Kroemer (2009). "Cytoplasmic functions of the tumour suppressor p53." *Nature* **458**(7242): 1127–30.

Green, N. K., C. W. Herbert, S. J. Hale, A. B. Hale, V. Mautner, R. Harkins, T. Hermiston, K. Ulbrich, K. D. Fisher and L. W. Seymour (2004). "Extended plasma circulation time and decreased toxicity of polymer-coated adenovirus." *Gene Ther* **11**(16): 1256–63.

Green, N. M., N. G. Wrigley, W. C. Russell, S. R. Martin and A. D. McLachlan (1983). "Evidence for a repeating cross-beta sheet structure in the adenovirus fibre." *EMBO J* **2**(8): 1357–65.

Gros, A., J. Martinez-Quintanilla, C. Puig, S. Guedan, D. G. Mollevi, R. Alemany and M. Cascallo (2008). "Bioselection of a gain of function mutation that enhances adenovirus 5 release and improves its antitumoral potency." *Cancer Res* **68**(21): 8928–37.

Guan, H., J. Jiao and R. P. Ricciardi (2008). "Tumorigenic adenovirus type 12 E1A inhibits phosphorylation of NF-kappaB by PKAc, causing loss of DNA binding and transactivation." *J Virol* **82**(1): 40–8.

Guimet, D. and P. Hearing (2013). "The adenovirus L4-22K protein has distinct functions in the posttranscriptional regulation of gene expression and encapsidation of the viral genome." *J Virol* **87**(13): 7688–99.

Gunther, P. S., J. K. Peper, B. Faist, S. Kayser, L. Hartl, T. Feuchtinger, G. Jahn, M. Neuenhahn, D. H. Busch, S. Stevanovic and K. M. Dennehy (2015). "Identification of a Novel Immunodominant HLA-B*07: 02-restricted Adenoviral Peptide Epitope and Its Potential in Adoptive Transfer Immunotherapy." *J Immunother* **38**(7): 267–75.

Gupta, V., W. Wang, B. A. Sosnowski, F. M. Hofman and T. C. Chen (2006). "Fibroblast growth factor-2-retargeted adenoviral vector for selective transduction of primary glioblastoma multiforme endothelial cells." *Neurosurg Focus* **20**(4): E26.

Gurcel, L., L. Abrami, S. Girardin, J. Tschopp and F. G. van der Goot (2006). "Caspase-1 activation of lipid metabolic pathways in response to bacterial pore-forming toxins promotes cell survival." *Cell* **126**(6): 1135–45.

Gurwith, M., M. Lock, E. M. Taylor, G. Ishioka, J. Alexander, T. Mayall, J. E. Ervin, R. N. Greenberg, C. Strout, J. J. Treanor, R. Webby and P. F. Wright (2013).

"Safety and immunogenicity of an oral, replicating adenovirus serotype 4 vector vaccine for H5N1 influenza: a randomised, double-blind, placebo-controlled, phase 1 study." *Lancet Infect Dis* **13**(3): 238–50.

Guse, K., I. Diaconu, M. Rajecki, M. Sloniecka, T. Hakkarainen, A. Ristimaki, A. Kanerva, S. Pesonen and A. Hemminki (2009). "Ad5/3-9HIF-Delta24-VEGFR-1-Ig, an infectivity enhanced, dual-targeted and antiangiogenic oncolytic adenovirus for kidney cancer treatment." *Gene Ther* **16**(8): 1009–20.

Gustin, K. E. and M. J. Imperiale (1998). "Encapsidation of viral DNA requires the adenovirus L1 52/55-kilodalton protein." *J Virol* **72**(10): 7860–70.

Gutch, M. J. and N. C. Reich (1991). "Repression of the interferon signal transduction pathway by the adenovirus E1A oncogene." *Proc Natl Acad Sci U S A* **88**(18): 7913–7.

Gwizdek, C., B. Ossareh-Nazari, A. M. Brownawell, A. Doglio, E. Bertrand, I. G. Macara and C. Dargemont (2003). "Exportin-5 mediates nuclear export of minihelix-containing RNAs." *J Biol Chem* **278**(8): 5505–8.

Hahn, W. C. and M. Meyerson (2001). "Telomerase activation, cellular immortalization and cancer." *Ann Med* **33**(2): 123–9.

Haisma, H. J., J. A. Kamps, G. K. Kamps, J. A. Plantinga, M. G. Rots and A. R. Bellu (2008). "Polyinosinic acid enhances delivery of adenovirus vectors in vivo by preventing sequestration in liver macrophages." *J Gen Virol* **89**(Pt 5): 1097–105.

Halestrap, A. P., E. Doran, J. P. Gillespie and A. O'Toole (2000). "Mitochondria and cell death." *Biochem Soc Trans* **28**(2): 170–7.

Hall, A. R., B. R. Dix, S. J. O'Carroll and A. W. Braithwaite (1998). "p53-dependent cell death/apoptosis is required for a productive adenovirus infection." *Nat Med* **4**(9): 1068–72.

Hallden, G. and G. Portella (2012). "Oncolytic virotherapy with modified adenoviruses and novel therapeutic targets." *Expert Opin Ther Targets* **16**(10): 945–58.

Hammer, K., A. Kazcorowski, L. Liu, M. Behr, P. Schemmer, I. Herr and D. M. Nettelbeck (2015). "Engineered adenoviruses combine enhanced oncolysis with improved virus production by mesenchymal stromal carrier cells." *Int J Cancer* **137**(4): 978–90.

Hammer, S. M., M. E. Sobieszczyk, H. Janes, S. T. Karuna, M. J. Mulligan, D. Grove, B. A. Koblin, S. P. Buchbinder, M. C. Keefer, G. D. Tomaras, N. Frahm, J. Hural, C. Anude, B. S. Graham, M. E. Enama, E. Adams, E. DeJesus, R. M. Novak, I. Frank, C. Bentley, S. Ramirez, R. Fu, R. A. Koup, J. R. Mascola, G. J. Nabel, D. C. Montefiori, J. Kublin, M. J. McElrath, L. Corey, P. B. Gilbert and H. S. Team (2013). "Efficacy trial of a DNA/rAd5 HIV-1 preventive vaccine." *N Engl J Med* **369**(22): 2083–92.

Hammond, J. M., E. S. Jansen, C. J. Morrissy, B. van der Heide, W. V. Goff, M. M. Williamson, P. T. Hooper, L. A. Babiuk, S. K. Tikoo and M. A. Johnson (2001). "Vaccination of pigs with a recombinant porcine adenovirus expressing the gD gene from pseudorabies virus." *Vaccine* **19**(27): 3752–8.

Han, J., D. Modha and E. White (1998). "Interaction of E1B 19K with Bax is required to block Bax-induced loss of mitochondrial membrane potential and apoptosis." *Oncogene* **17**(23): 2993–3005.

Han, J., P. Sabbatini, D. Perez, L. Rao, D. Modha and E. White (1996). "The E1B 19K protein blocks apoptosis by interacting with and inhibiting the p53-inducible and death-promoting Bax protein." *Genes Dev* **10**(4): 461–77.

Han, M., V. A. Kickhoefer, G. R. Nemerow and L. H. Rome (2011). "Targeted vault nanoparticles engineered with an endosomolytic peptide deliver biomolecules to the cytoplasm." *ACS Nano* **5**(8): 6128–37.

Hara, H., A. Kobayashi, K. Yoshida, M. Ohashi, S. Ohnami, E. Uchida, E. Higashihara, T. Yoshida, K. Aoki (2007). "Local interferon-alpha gene therapy elicits systemic immunity in a syngeneic pancreatic cancer model in hamster." *Cancer Sci* **98**(3): 455–63.

Harada, J. N. and A. J. Berk (1999). "p53-Independent and -dependent requirements for E1B-55K in adenovirus type 5 replication." *J Virol* **73**(7): 5333–44.

Harada, J. N., A. Shevchenko, D. C. Pallas and A. J. Berk (2002). "Analysis of the adenovirus E1B-55K-anchored proteome reveals its link to ubiquitination machinery." *J. Virol.* **76**(18): 9194–9206.

Harfst, E. and K. N. Leppard (1999). "A comparative analysis of the phosphorylation and biochemical properties of wild type and host range variant DNA binding proteins of human adenovirus 5." *Virus Genes* **18**(2): 97–106.

Harlow, E., B. Franza, Jr. and C. Schley (1985). "Monoclonal antibodies specific for adenovirus early region 1A proteins: extensive heterogeneity in early region 1A products." *J. Virol.* **55**: 533–546.

Harlow, E., P. Whyte, B. R. Franza, Jr. and C. Schley (1986). "Association of adenovirus early-region 1A proteins with cellular polypeptides." *Mol Cell Biol* **6**(5): 1579–89.

Harrach, B., M. Benko, G. W. Both, M. Brown, A. J. Davison, M. Echavarria, M. Hess, M. S. Jones, A. E. Kajon, H. D. Lehmkuhl, V. Mautner, S. K. Mittal and G. Wadell (2012). *Adenoviridae*. Virus Taxonomy. In: Ninth Report of the International Committee on Taxonomy of Viruses. A. M. Q. King, M. J. Adams, E. B. Larstens and E. J. Lefkowitz (eds), Academic Press, Elsevier.

Hartley, J. W. and W. P. Rowe (1960). "A new mouse virus apparently related to the adenovirus group." *Virology* **11**: 645–647.

Hartman, Z. C., D. M. Appledorn and A. Amalfitano (2008). "Adenovirus vector induced innate immune responses: impact upon efficacy and toxicity in gene therapy and vaccine applications." *Virus Res* **132**(1–2): 1–14.

Haruki, H., B. Gyurcsik, M. Okuwaki and K. Nagata (2003). "Ternary complex formation between DNA-adenovirus core protein VII and TAF-Ibeta/SET, an acidic molecular chaperone." *FEBS Lett* **555**(3): 521–7.

Harvey, B. G., P. L. Leopold, N. R. Hackett, T. M. Grasso, P. M. Williams, A. L. Tucker, R. J. Kaner, B. Ferris, I. Gonda, T. D. Sweeney, R. Ramalingam, I. Kovesdi, S. Shak and R. G. Crystal (1999). "Airway epithelial CFTR mRNA expression in cystic fibrosis patients after repetitive administration of a recombinant adenovirus." *J Clin Invest* **104**(9): 1245–55.

Hasson, T. B., D. A. Ornelles and T. Shenk (1992). "Adenovirus L1 52- and 55-kilodalton proteins are present within assembling virions and colocalize with nuclear structures distinct from replication centers." *J Virol* **66**(10): 6133–42.

Hatakeyama, S. (2011). "TRIM proteins and cancer." *Nat Rev Cancer* **11**(11): 792–804.

Hay, J., D. Carter, A. Lieber and A. L. Astier (2014). "Recombinant Ad35 adenoviral proteins as potent modulators of human T cell activation." *Immunology*.

Hay, R. T. (1985). "The origin of adenovirus DNA replication: minimal DNA sequence requirement in vivo." *EMBO J* **4**(2): 421–6.

Hay, R. T. (1996). Adenovirus DNA replication. In: *DNA Replication in Eukaryotic Cells.* M. L. DePamphilis (ed). Cold Spring Harbor, NY, Cold Spring Harbor Laboratory Press: 699–719.

Hayes, B. W., G. C. Telling, M. M. Myat, J. F. Williams and S. J. Flint (1990). "The adenovirus L4 100-kilodalton protein is necessary for efficient translation of viral late mRNA species." *J. Virol.* **64**: 2732–2742.

He, G., W. Lei, S. Wang, R. Xiao, K. Guo, Y. Xia, X. Zhou, K. Zhang, X. Liu and Y. Wang (2012). "Overexpression of tumor suppressor TSLC1 by a survivin-regulated oncolytic adenovirus significantly inhibits hepatocellular carcinoma growth." *J Cancer Res Clin Oncol* **138**(4): 657–70.

He, T. C., S. Zhou, L. T. da Costa, J. Yu, K. W. Kinzler and B. Vogelstein (1998). "A simplified system for generating recombinant adenoviruses." *Proc Natl Acad Sci U S A* **95**(5): 2509–14.

He, Z., A. P. Wlazlo, D. W. Kowalczyk, J. Cheng, Z. Q. Xiang, W. Giles-Davis and H. C. Ertl (2000). "Viral recombinant vaccines to the E6 and E7 antigens of HPV-16." *Virology* **270**(1): 146–61.

Hearing, P., R. J. Samulski, W. L. Wishart and T. Shenk (1987). "Identification of a repeated sequence element required for efficient encapsidation of the adenovirus type 5 chromosome." *J Virol* **61**(8): 2555–8.

Hearing, P. and T. Shenk (1983a). "Functional analysis of the nucleotide sequence surrounding the cap site for adenovirus type 5 region E1A messenger RNAs." *J Mol Biol* **167**: 809–822.

Hearing, P. and T. Shenk (1983b). "The adenovirus type 5 E1A transcriptional control region contains a duplicated enhancer element." *Cell* **33**: 695–703.

Hearing, P. and T. Shenk (1986). "The adenovirus type 5 enhancer contains two functionally distinct domains: one is a duplicated enhancer element." *Cell* **45**: 229–236.

Hecht, J. R., R. Bedford, J. L. Abbruzzese, S. Lahoti, T. R. Reid, R. M. Soetikno, D. H. Kirn and S. M. Freeman (2003). "A phase I/II trial of intratumoral endoscopic ultrasound injection of ONYX-015 with intravenous gemcitabine in unresectable pancreatic carcinoma." *Clin Cancer Res* **9**(2): 555–61.

Heemskerk, B., A. C. Lankester, T. van Vreeswijk, M. F. Beersma, E. C. Claas, L. A. Veltrop-Duits, A. C. Kroes, J. M. Vossen, M. W. Schilham and M. J. van Tol (2005). "Immune reconstitution and clearance of human adenovirus viremia in pediatric stem-cell recipients." *J Infect Dis* **191**(4): 520–30.

Heemskerk, B., T. van Vreeswijk, L. A. Veltrop-Duits, C. C. Sombroek, K. Franken, R. M. Verhoosel, P. S. Hiemstra, D. van Leeuwen, M. E. Ressing, R. E. Toes, M. J. van Tol and M. W. Schilham (2006). "Adenovirus-specific CD4+ T cell clones recognizing endogenous antigen inhibit viral replication in vitro through cognate interaction." *J Immunol* **177**(12): 8851–9.

Heise, C., T. Hermiston, L. Johnson, G. Brooks, A. Sampson-Johannes, A. Williams, L. Hawkins and D. Kirn (2000). "An adenovirus E1A mutant that demonstrates potent and selective systemic anti-tumoral efficacy." *Nat Med* **6**(10): 1134–9.

Heise, C., M. Lemmon and D. Kirn (2000). "Efficacy with a replication-selective adenovirus plus cisplatin-based chemotherapy: dependence on sequencing but not p53 functional status or route of administration." *Clin Cancer Res* **6**(12): 4908–14.

Helt, A. M. and D. A. Galloway (2003). "Mechanisms by which DNA tumor virus oncoproteins target the Rb family of pocket proteins." *Carcinogenesis* **24**(2): 159–69.

Hendrickx, R., N. Stichling, J. Koelen, L. Kuryk, A. Lipiec and U. F. Greber (2014). "Innate immunity to adenovirus." *Hum Gene Ther* **25**(4): 265–84.

Henry, L. J., D. Xia, M. E. Wilke, J. Deisenhofer and R. D. Gerard (1994). "Characterization of the knob domain of the adenovirus type 5 fiber protein expressed in Escherichia coli." *J Virol* **68**(8): 5239–46.

Hentze, M. W. (1997). "eIF4G: a multipurpose ribosome adapter?" *Science* **275**(5299): 500–1.

Herbst, R. S., M. Pelletier, E. M. Boczko and L. E. Babiss (1990). "The state of cellular differentiation determines the activity of the adenovirus E1A enhancer element: evidence for negative regulation of enhancer function." *J Virol* **64**(1): 161–72.

Hermiston, T. W., R. Hellwig, J. C. Hierholzer and W. S. Wold (1993). "Sequence and functional analysis of the human adenovirus type 7 E3-gp19K protein from 17 clinical isolates." *Virology* **197**(2): 593–600.

Hernandez-Alcoceba, R., M. Pihalja, D. Qian and M. F. Clarke (2002). "New oncolytic adenoviruses with hypoxia- and estrogen receptor-regulated replication." *Hum Gene Ther* **13**(14): 1737–50.

Heyward, C. Y., R. Patel, E. M. Mace, J. T. Grier, H. Guan, A. P. Makrigiannis, J. S. Orange and R. P. Ricciardi (2012). "Tumorigenic adenovirus 12 cells evade NK cell lysis by reducing the expression of NKG2D ligands." *Immunol Lett* **144**(1–2): 16–23.

Hidalgo, P. and R. A. Gonzalez (2015). "Isolation of Viral Replication Compartment-enriched Sub-nuclear Fractions from Adenovirus-infected Normal Human Cells." *J Vis Exp*(105) doi: 10.3791/53296.

Higashino, F., K. Yoshida, Y. Fujinaga and e. al. (1993). "Isolation of a cDNA encoding the adenovirus E1A enhancer binding protein: a new human member of the ets oncogene family." *Nucleic Acids Res.* **21**: 547–553.

Higginbotham, J. M. and C. C. O'Shea (2015). "Adenovirus E4-ORF3 Targets PIAS3 and Together with E1B-55K Remodels SUMO Interactions in the Nucleus and at Virus Genome Replication Domains." *J Virol* **89**(20): 10260–72.

Hilgendorf, A., J. Lindberg, Z. Ruzsics, S. Honing, A. Elsing, M. Lofqvist, H. Engelmann and H. G. Burgert (2003). "Two distinct transport motifs in the adenovirus E3/10.4–14.5 proteins act in concert to down-modulate apoptosis receptors and the epidermal growth factor receptor." *J Biol Chem* **278**(51): 51872–84.

Hilleman, M. R. and J. H. Werner (1954). "Recovery of a new agent from patients with acute respiratory illness." *Proc Soc Exp Biol and Med* **85**: 183–0.

Hippenmeyer, P. J., P. G. Ruminski, J. G. Rico, H. S. Lu and D. W. Griggs (2002). "Adenovirus inhibition by peptidomimetic integrin antagonists." *Antiviral Res* **55**(1): 169–78.

Hodgson, S. H., K. J. Ewer, C. M. Bliss, N. J. Edwards, T. Rampling, N. A. Anagnostou, E. de Barra, T. Havelock, G. Bowyer, I. D. Poulton, S. de Cassan, R. Longley, J. J. Illingworth, A. D. Douglas, P. B. Mange, K. A. Collins, R. Roberts, S. Gerry, E. Berrie, S. Moyle, S. Colloca, R. Cortese, R. E. Sinden, S. C. Gilbert, P. Bejon, A. M. Lawrie, A. Nicosia, S. N. Faust and A. V. Hill (2015). "Evaluation of the efficacy of ChAd63-MVA vectored vaccines expressing circumsporozoite protein and ME-TRAP against controlled human malaria infection in malaria-naive individuals." *J Infect Dis* **211**(7): 1076–86.

Hoeben, R. C. and T. G. Uil (2013). "Adenovirus DNA replication." *Cold Spring Harb Perspect Biol* **5**(3): a013003.

Hoffmann, D., B. Meyer and O. Wildner (2007). "Improved glioblastoma treatment with Ad5/35 fiber chimeric conditionally replicating adenoviruses." *J Gene Med* **9**(9): 764–78.

Hoffmann, D. and O. Wildner (2006). "Restriction of adenoviral replication to the transcriptional intersection of two different promoters for colorectal and pancreatic cancer treatment." *Mol Cancer Ther* **5**(2): 374–81.

Hofherr, S. E., H. Mok, F. C. Gushiken, J. A. Lopez and M. A. Barry (2007). "Polyethylene glycol modification of adenovirus reduces platelet activation, endothelial cell activation, and thrombocytopenia." *Hum Gene Ther* **18**(9): 837–48.

Hofherr, S. E., E. V. Shashkova, E. A. Weaver, R. Khare and M. A. Barry (2008). "Modification of adenoviral vectors with polyethylene glycol modulates in vivo tissue tropism and gene expression." *Mol Ther* **16**(7): 1276–82.

Hofmann, C., P. Loser, G. Cichon, W. Arnold, G. W. Both and M. Strauss (1999). "Ovine adenovirus vectors overcome preexisting humoral immunity against human adenoviruses in vivo." *J Virol* **73**(8): 6930–6.

Hoft, D. F., A. Blazevic, J. Stanley, B. Landry, D. Sizemore, E. Kpamegan, J. Gearhart, A. Scott, S. Kik, M. G. Pau, J. Goudsmit, J. B. McClain and J. Sadoff (2012). "A recombinant adenovirus expressing immunodominant TB antigens can significantly enhance BCG-induced human immunity." *Vaccine* **30**(12): 2098–108.

Holterman, L., R. Vogels, R. van der Vlugt, M. Sieuwerts, J. Grimbergen, J. Kaspers, E. Geelen, E. van der Helm, A. Lemckert, G. Gillissen, S. Verhaagh, J. Custers, D. Zuijdgeest, B. Berkhout, M. Bakker, P. Quax, J. Goudsmit and M. Havenga (2004). "Novel replication-incompetent vector derived from adenovirus type 11

(Ad11) for vaccination and gene therapy: low seroprevalence and non-cross-reactivity with Ad5." *J Virol* **78**(23): 13207–15.

Hong, S. S., N. A. Habib, L. Franqueville, S. Jensen and P. A. Boulanger (2003). "Identification of adenovirus (ad) penton base neutralizing epitopes by use of sera from patients who had received conditionally replicative ad (addl1520) for treatment of liver tumors." *J Virol* **77**(19): 10366–75.

Hoppe, A., S. J. Beech, J. Dimmock and K. N. Leppard (2006). "Interaction of the adenovirus type 5 E4 Orf3 protein with promyelocytic leukemia protein isoform II is required for ND10 disruption." *J Virol* **80**(6): 3042–9.

Horne, R. W., S. Bonner, A. P. Waterson and P. Wildy (1959). "The icosahedral form of an adenovirus." *J. Mol. Biol.* **1**: 84–86.

Horne, W. S., C. M. Wiethoff, C. Cui, K. M. Wilcoxen, M. Amorin, M. R. Ghadiri and G. R. Nemerow (2005). "Antiviral cyclic D,L-alpha-peptides: targeting a general biochemical pathway in virus infections." *Bioorg Med Chem* **13**(17): 5145–53.

Horton, T. M., T. S. Ranheim, L. Aquino, D. I. Kusher, S. K. Saha, C. F. Ware, W. S. Wold and L. R. Gooding (1991). "Adenovirus E3 14.7K protein functions in the absence of other adenovirus proteins to protect transfected cells from tumor necrosis factor cytolysis." *J Virol* **65**(5): 2629–39.

Horwitz, M. S. (1978). "Temperature-sensitive replication of H5ts125 adenovirus DNA in vitro." *Proc Natl Acad Sci U S A* **75**(9): 4291–5.

Horwitz, M. S. (2004). "Function of adenovirus E3 proteins and their interactions with immunoregulatory cell proteins." *J Gene Med* **6 Suppl 1**: S172–83.

Hourcade, D., V. M. Holers and J. P. Atkinson (1989). "The regulators of complement activation (RCA) gene cluster." *Adv Immunol* **45**: 381–416.

Howe, J. A., J. S. Mymryk, C. Egan, P. E. Branton and S. T. Bayley (1990). "Retinoblastoma growth suppressor and a 300-kDa protein appear to regulate cellular DNA synthesis." *Proc Natl Acad Sci U S A* **87**(15): 5883–7.

Huang, G., W. Yao, W. Yu, L. Mao, H. Sun, W. Yao, J. Tian, L. Wang, Z. Bo, Z. Zhu, Y. Zhang, Z. Zhao and W. Xu (2014). "Outbreak of epidemic keratoconjunctivitis caused by human adenovirus type 56, China, 2012." *PLoS One* **9**(10): e110781.

Huang J. and R. J. Schneider (1991). "Adenovirus inhibition of cellular protein synthesis involves inactivation of cap-binding protein." *Cell* **65**: 271–280.

Huang, M. M. and P. Hearing (1989). "Adenovirus early region 4 encodes two gene products with redundant effects in lytic infection." *J Virol* **63**(6): 2605–15.

Huang, S., D. Stupack, P. Mathias, Y. Wang and G. Nemerow (1997). "Growth arrest of Epstein-Barr virus immortalized B lymphocytes by adenovirus-delivered ribozymes." *Proc Natl Acad Sci U S A* **94**(15): 8156–61.

Huang W, Kiefer J, Whalen D and S. Flint (2003). "DNA synthesis-dependent relief of transcription from the adenoviral type 2 IVa$_2$ promoter by a cellular protein." *Virology* **314**: 394–402.

Huang, W. and S. J. Flint (1998). "The tripartite leader sequence of subgroup C adenovirus major late mRNAs can increase the efficiency of mRNA export." *J. Virol.* **72**(1): 225–235.

Huang, W. and S. J. Flint (2003). "Unusual Properties of Adenoviral E2E Transcription by RNA Polymerase III." *J. Virol.* **77**: 4015–4024.

Huang, W., R. Pruzan and S. J. Flint (1994). "*In vivo* transcription from the adenovirus E2E promoter by RNA polymerase III." *Proc. Natl. Acad. Sci. USA* **91**: 1265–1269.

Huang, X., M. Griffiths, J. Wu, R. V. Farese, Jr. and D. Sheppard (2000). "Normal development, wound healing, and adenovirus susceptibility in beta5-deficient mice." *Mol Cell Biol* **20**(3): 755–9.

Huebner, R. J., W. P. Rowe, W. E. Schatten, R. R. Smith and L. B. Thomas (1956). "Studies on the use of viruses in the treatment of carcinoma of the cervix." *Cancer* **9**(6): 1211–8.

Hull, R. N., J. R. Minner and C. C. Mascoli (1958). "New viral agents recovered from tissue cultures of monkey cells. III. Additional agents both from cultures of monkey tissues and directly from tissues and excreta." *Amer. J. Hyg.* **68**: 31–44.

Hutnick, N. A., D. G. Carnathan, S. A. Dubey, K. S. Cox, L. Kierstead, G. Makadonas, S. J. Ratcliffe, M. O. Lasaro, M. N. Robertson, D. R. Casimiro, H. C. Ertl and M. R. Betts (2010). "Vaccination with Ad5 vectors expands Ad5-specific CD8 T cells without altering memory phenotype or functionality." *PLoS One* **5**(12): e14385.

Hutnick, N. A., D. G. Carnathan, S. A. Dubey, G. Makedonas, K. S. Cox, L. Kierstead, S. J. Ratcliffe, M. N. Robertson, D. R. Casimiro, H. C. Ertl and M. R. Betts (2009). "Baseline Ad5 serostatus does not predict Ad5 HIV vaccine-induced expansion of adenovirus-specific CD4+ T cells." *Nat Med* **15**(8): 876–8.

Hynes, R. O. (1992). "Integrins: versatility, modulation, and signaling in cell adhesion." *Cell* **69**(1): 11–25.

Iacobelli-Martinez, M. and G. R. Nemerow (2007). "Preferential activation of Toll-like receptor nine by CD46-utilizing adenoviruses." *J Virol* **81**(3): 1305–12.

Iacobelli-Martinez, M., R. R. Nepomuceno, J. Connolly and G. R. Nemerow (2005). "CD46-utilizing adenoviruses inhibit C/EBPbeta-dependent expression of proinflammatory cytokines." *J Virol* **79**(17): 11259–68.

Ibrisimovic, M., D. Kneidinger, T. Lion and R. Klein (2013). "An adenoviral vector-based expression and delivery system for the inhibition of wild-type adenovirus replication by artificial microRNAs." *Antiviral Res* **97**(1): 10–23.

Iftode, C. and S. J. Flint (2004). "Viral synthesis-dependent titration of a cellular repressor activates transcription of the human adenovirus type 2 IVa2 Gene." *Pro. Natl. Acad. Sci. USA* **101**(51): 17831–17836.

Ikeda, J. E., T. Enomoto and J. Hurwitz (1982). "Adenoviral protein-primed initiation of DNA chains in vitro." *Proc Natl Acad Sci Usa* **79**: 2442–0.

Ikeda, M. A. and J. R. Nevins (1993). "Identification of distinct roles for separate E1A domains in disruption of E2F complexes." *Mol Cell Biol* **13**(11): 7029–35.

Imelli, N., Z. Ruzsics, D. Puntener, M. Gastaldelli and U. F. Greber (2009). "Genetic reconstitution of the human adenovirus type 2 temperature-sensitive 1 mutant defective in endosomal escape." *Virol J* **6**: 174.

Irving, J., Z. Wang, S. Powell, C. O'Sullivan, M. Mok, B. Murphy, L. Cardoza, J. S. Lebkowski and A. S. Majumdar (2004). "Conditionally replicative adenovirus driven by the human telomerase promoter provides broad-spectrum antitumor activity without liver toxicity." *Cancer Gene Ther* **11**(3): 174–85.

Ishibashi, M. and J. V. Maizel, Jr. (1974). "The polypeptides of adenovirus. V. Young virions, structural intermediate between top components and aged virions." *Virology* **57**(2): 409–24.

Ishov, A. M. and G. G. Maul (1996). "The periphery of nuclear domain 10 (ND10) as site of DNA virus deposition." *J Cell Biol* **134**(4): 815–26.

Ivashkiv, L. B. and L. T. Donlin (2014). "Regulation of type I interferon responses." *Nat Rev Immunol* **14**(1): 36–49.

Iwakura, A., M. Fujita, M. Ikemoto, K. Hasegawa, R. Nohara, S. Sasayama, S. Miyamoto, A. Yamazato, K. Tambara and M. Komeda (2000). "Myocardial ischemia enhances the expression of acidic fibroblast growth factor in human pericardial fluid." *Heart Vessels* **15**(3): 112–6.

Iwamoto, S., F. Eggerding, E. Falck-Pederson and J. E. Darnell, Jr. (1986). "Transcription unit mapping in adenovirus: regions of termination." *J. Virol.* **59**: 112–119.

Izzedine, H., V. Launay-Vacher and G. Deray (2005). "Antiviral drug-induced nephrotoxicity." *Am J Kidney Dis* **45**(5): 804–17.

Jaiswal, S., N. Khanna and S. Swaminathan (2003). "Replication-defective adenoviral vaccine vector for the induction of immune responses to dengue virus type 2." *J Virol* **77**(23): 12907–13.

Jakubczak, J. L., P. Ryan, M. Gorziglia, L. Clarke, L. K. Hawkins, C. Hay, Y. Huang, M. Kaloss, A. Marinov, S. Phipps, A. Pinkstaff, P. Shirley, Y. Skripchenko, D. Stewart, S. Forry-Schaudies and P. L. Hallenbeck (2003). "An oncolytic adenovirus selective for retinoblastoma tumor suppressor protein pathway-defective tumors: dependence on E1A, the E2F-1 promoter, and viral replication for selectivity and efficacy." *Cancer Res* **63**(7): 1490–9.

James, L. C., A. H. Keeble, Z. Khan, D. A. Rhodes and J. Trowsdale (2007). "Structural basis for PRYSPRY-mediated tripartite motif (TRIM) protein function." *Proc Natl Acad Sci U S A* **104**(15): 6200–5.

Jansen-Durr, P., G. Mondésert and C. Kedinger (1989). "Replication-dependent activation of the adenovirus major late promoter is mediated by the increased binding of a transcription factor to sequences in the first intron." *J. Virol.* **63**: 5124–5132.

Jia, H., C. Li, Y. Zhang, L. Yu, D. Xiang, J. Liu, F. Chen and X. Han (2015). "Immunostimulatory activities of dendritic cells loaded with adenovirus vector carrying HBcAg/HBsAg." *Int J Clin Exp Med* **8**(3): 3456–64.

Jiang, H., C. Gomez-Manzano, Y. Rivera-Molina, F. F. Lang, C. A. Conrad and J. Fueyo (2015). "Oncolytic adenovirus research evolution: from cell-cycle checkpoints to immune checkpoints." *Curr Opin Virol* **13**: 33–9.

Jiang, H., Z. Wang, D. Serra, M. M. Frank and A. Amalfitano (2004). "Recombinant adenovirus vectors activate the alternative complement pathway, leading to the binding of human complement protein C3 independent of anti-ad antibodies." *Mol Ther* **10**(6): 1140–2.

Jiang, H., E. J. White, C. I. Rios-Vicil, J. Xu, C. Gomez-Manzano and J. Fueyo (2011). "Human adenovirus type 5 induces cell lysis through autophagy and autophagy-triggered caspase activity." *J Virol* **85**(10): 4720–9.

Jiang, W., J. Chang, J. Jakana, P. Weigele, J. King and W. Chiu (2006). "Structure of epsilon15 bacteriophage reveals genome organization and DNA packaging/injection apparatus." *Nature* **439**(7076): 612–6.

Jiang, Z., G. Schiedner, N. van Rooijen, C. C. Liu, S. Kochanek and P. R. Clemens (2004). "Sustained muscle expression of dystrophin from a high-capacity adenoviral vector with systemic gene transfer of T cell costimulatory blockade." *Mol Ther* **10**(4): 688–96.

Jiang, Z. K., S. B. Koh, M. Sato, I. C. Atanasov, M. Johnson, Z. H. Zhou, T. J. Deming and L. Wu (2013). "Engineering polypeptide coatings to augment gene transduction and in vivo stability of adenoviruses." *J Control Release* **166**(1): 75–85.

Jiao, J., H. Guan, A. M. Lippa and R. P. Ricciardi (2010). "The N terminus of adenovirus type 12 E1A inhibits major histocompatibility complex class I expression by preventing phosphorylation of NF-kappaB p65 Ser276 through direct binding." *J Virol* **84**(15): 7668–74.

Jin, H., S. Lv, J. Yang, X. Wang, H. Hu, C. Su, C. Zhou, J. Li, Y. Huang, L. Li, X. Liu, M. Wu and Q. Qian (2011). "Use of microRNA Let-7 to control the replication specificity of oncolytic adenovirus in hepatocellular carcinoma cells." *PLoS One* **6**(7): e21307.

Johansson, C., M. Jonsson, M. Marttila, D. Persson, X. L. Fan, J. Skog, L. Frangsmyr, G. Wadell and N. Arnberg (2007). "Adenoviruses use lactoferrin as a bridge for CAR-independent binding to and infection of epithelial cells." *J Virol* **81**(2): 954–63.

Johansson, S. M., N. Arnberg, M. Elofsson, G. Wadell and J. Kihlberg (2005). "Multivalent HSA conjugates of 3'-sialyllactose are potent inhibitors of adenoviral cell attachment and infection." *Chembiochem* **6**(2): 358–64.

Johnson, D. G. and R. Schneider-Broussard (1998). "Role of E2F in cell cycle control and cancer." *Front Biosci* **3**: d447–8.

Johnson, J. E. (2010). "Virus particle maturation: insights into elegantly programmed nanomachines." *Curr Opin Struct Biol* **20**(2): 210–6.

Johnson, J. S., Y. N. Osheim, Y. Xue, M. R. Emanuel, P. W. Lewis, A. Bankovich, A. L. Beyer and D. A. Engel (2004). "Adenovirus protein VII condenses DNA, represses transcription, and associates with transcriptional activator E1A." *J Virol* **78**(12): 6459–68.

Johnson, M. J., N. K. Bjorkstrom, C. Petrovas, F. Liang, J. G. Gall, K. Lore and R. A. Koup (2014). "Type I interferon-dependent activation of NK cells by rAd28 or rAd35, but not rAd5, leads to loss of vector-insert expression." *Vaccine* **32**(6): 717–24.

Jones, M. S., 2nd, B. Harrach, R. D. Ganac, M. M. Gozum, W. P. Dela Cruz, B. Riedel, C. Pan, E. L. Delwart and D. P. Schnurr (2007). "New adenovirus species found in a patient presenting with gastroenteritis." *J Virol* **81**(11): 5978–84.

Jones, M. S., N. R. Hudson, C. Gibbins and S. L. Fischer (2011). "Evaluation of type-specific real-time PCR assays using the LightCycler and J.B.A.I.D.S. for detection of adenoviruses in species HAdV-C." *PLoS One* **6**(10): e26862.

Jones, N. and T. Shenk (1979). "An adenovirus type 5 early gene function regulates expression of other early viral genes." *Proc. Natl. Acad. Sci. USA* **76**: 3665–3669.

Jörnvall, H., G. Akusjarvi, P. Alestrom, H. von Bahr-Lindstrom, U. Pettersson, E. Appella, A. V. Fowler and L. Philipson (1981). "The adenovirus hexon protein. The primary structure of the polypeptide and its correlation with the hexon gene." *J Biol Chem* **256**(12): 6181–6.

Joshi, A., J. Tang, M. Kuzma, J. Wagner, B. Mookerjee, J. Filicko, M. Carabasi, N. Flomenberg and P. Flomenberg (2009). "Adenovirus DNA polymerase is recognized by human CD8+ T cells." *J Gen Virol* **90**(Pt 1): 84–94.

Joshi, A., B. Zhao, C. Romanowski, D. Rosen and P. Flomenberg (2011). "Comparison of human memory CD8 T cell responses to adenoviral early and late proteins in peripheral blood and lymphoid tissue." *PLoS One* **6**(5): e20068.

Joung, I. and J. A. Engler (1992). "Mutations in two cysteine-histidine-rich clusters in adenovirus type 2 DNA polymerase affect DNA binding." *J Virol* **66**(10): 5788–96.

Joung, I., M. S. Horwitz and J. A. Engler (1991). "Mutagenesis of conserved region I in the DNA polymerase from human adenovirus serotype 2." *Virology* **184**(1): 235–41.

Joyce, C. M. and T. A. Steitz (1994). "Function and structure relationships in DNA polymerases." *Annu Rev Biochem* **63**: 777–822.

Jung, Y., H. J. Park, P. H. Kim, J. Lee, W. Hyung, J. Yang, H. Ko, J. H. Sohn, J. H. Kim, Y. M. Huh, C. O. Yun and S. Haam (2007). "Retargeting of adenoviral gene delivery via Herceptin-PEG-adenovirus conjugates to breast cancer cells." *J Control Release* **123**(2): 164–71.

Kahl, C. A., J. Bonnell, S. Hiriyanna, M. Fultz, C. Nyberg-Hoffman, P. Chen, C. R. King and J. G. Gall (2010). "Potent immune responses and in vitro pro-inflammatory cytokine suppression by a novel adenovirus vaccine vector based on rare human serotype 28." *Vaccine* **28**(35): 5691–702.

Kajon, A. E., M. Echavarria and J. C. de Jong (2013). "Designation of human adenovirus types based on sequence data: an unfinished debate." *J Clin Virol* **58**(4): 743–4.

Kajon, A. E., D. Lamson, M. Shudt, Z. Oikonomopoulou, B. Fisher, S. Klieger, K. St George and R. L. Hodinka (2014). "Identification of a novel intertypic recombinant species D human adenovirus in a pediatric stem cell transplant recipient." *J Clin Virol* **61**(4): 496–502.

Kajon, A. E., J. M. Moseley, D. Metzgar, H. S. Huong, A. Wadleigh, M. A. Ryan and K. L. Russell (2007). "Molecular epidemiology of adenovirus type 4 infections in US military recruits in the postvaccination era (1997–2003)." *J Infect Dis* **196**(1): 67–75.

Kalin, S., B. Amstutz, M. Gastaldelli, N. Wolfrum, K. Boucke, M. Havenga, F. DiGennaro, N. Liska, S. Hemmi and U. F. Greber (2010). "Macropinocytotic uptake and infection of human epithelial cells with species B2 adenovirus type 35." *J Virol* **84**(10): 5336–50.

Kalvakolanu, D. V., S. K. Bandyopadhyay, M. L. Harter and G. C. Sen (1991). "Inhibition of interferon-inducible gene expression by adenovirus E1A proteins: block in transcriptional complex formation." *Proc Natl Acad Sci U S A* **88**(17): 7459–63.

Kalyuzhniy, O., N. C. Di Paolo, M. Silvestry, S. E. Hofherr, M. A. Barry, P. L. Stewart and D. M. Shayakhmetov (2008). "Adenovirus serotype 5 hexon is critical for virus infection of hepatocytes in vivo." *Proc Natl Acad Sci U S A* **105**(14): 5483–8.

Kamel, W., B. Segerman, D. Oberg, T. Punga and G. Akusjarvi (2013). "The adenovirus VA RNA-derived miRNAs are not essential for lytic virus growth in tissue culture cells." *Nucleic Acids Res* **41**(9): 4802–12.

Kaminsky, S. M., T. K. Rosengart, J. Rosenberg, M. J. Chiuchiolo, B. Van de Graaf, D. Sondhi and R. G. Crystal (2013). "Gene therapy to stimulate angiogenesis to treat diffuse coronary artery disease." *Hum Gene Ther* **24**(11): 948–63.

Kampe, O., D. Bellgrau, U. Hammerling, P. Lind, S. Paabo, L. Severinsson and P. A. Peterson (1983). "Complex formation of class I transplantation antigens and a viral glycoprotein." *J Biol Chem* **258**(17): 10594–8.

Kamtekar, S., A. J. Berman, J. Wang, J. M. Lazaro, M. de Vega, L. Blanco, M. Salas and T. A. Steitz (2004). "Insights into strand displacement and processivity from the crystal structure of the protein-primed DNA polymerase of bacteriophage phi29." *Mol Cell* **16**(4): 609–18.

Kaneko, H., K. Aoki, S. Ohno, H. Ishiko, T. Fujimoto, M. Kikuchi, S. Harada, G. Gonzalez, K. O. Koyanagi, H. Watanabe and T. Suzutani (2011). "Complete genome analysis of a novel intertypic recombinant human adenovirus causing epidemic keratoconjunctivitis in Japan." *J Clin Microbiol* **49**(2): 484–90.

Kaneko, H., S. Mori, O. Suzuki, T. Iida, S. Shigeta, M. Abe, S. Ohno, K. Aoki and T. Suzutani (2004). "The cotton rat model for adenovirus ocular infection: antiviral activity of cidofovir." *Antiviral Res* **61**(1): 63–6.

Kaneko, H., T. Suzutani, K. Aoki, N. Kitaichi, S. Ishida, H. Ishiko, T. Ohashi, S. Okamoto, H. Nakagawa, R. Hinokuma, Y. Asato, S. Oniki, T. Hashimoto, T. Iida and S. Ohno (2011). "Epidemiological and virological features of epidemic keratoconjunctivitis due to new human adenovirus type 54 in Japan." *Br J Ophthalmol* **95**(1): 32–6.

Kanerva, A., A. Koski, I. Liikanen, M. Oksanen, T. Joensuu, O. Hemminki, J. Palmgren, K. Hemminki and A. Hemminki (2015). "Case-control estimation of the impact of oncolytic adenovirus on the survival of patients with refractory solid tumors." *Mol Ther* **23**(2): 321–9.

Kanerva, A., M. Wang, G. J. Bauerschmitz, J. T. Lam, R. A. Desmond, S. M. Bhoola, M. N. Barnes, R. D. Alvarez, G. P. Siegal, D. T. Curiel and A. Hemminki (2002). "Gene transfer to ovarian cancer versus normal tissues with fiber-modified adenoviruses." *Mol Ther* **5**(6): 695–704.

Kannan, K., N. Amariglio, G. Rechavi, J. Jakob-Hirsch, I. Kela, N. Kaminski, G. Getz, E. Domany and D. Givol (2001). "DNA microarrays identification of primary and secondary target genes regulated by p53." *Oncogene* **20**(18): 2225–34.

Kanopka, A., O. Muhlemann, S. Petersen-Mahrt, C. Estmer, C. Ohrmalm and G. Akusjarvi (1998). "Regulation of adenovirus alternative RNA splicing by dephosphorylation of SR proteins." *Nature* **393**(6681): 185–7.

Kapsenberg, J. G. (1959). "Relationship of infectious canine hepatitis virus to human adenovirus." *Proc. Soc. Exp. Biol. Med.* **101**: 611–614.

Karen, K. A. and P. Hearing (2011). "Adenovirus core protein VII protects the viral genome from a DNA damage response at early times after infection." *J Virol* **85**(9): 4135–42.

Karen, K. A., P. J. Hoey, C. S. Young and P. Hearing (2009). "Temporal regulation of the Mre11-Rad50-Nbs1 complex during adenovirus infection." *J Virol* **83**(9): 4565–73.

Karp, C. L., M. Wysocka, L. M. Wahl, J. M. Ahearn, P. J. Cuomo, B. Sherry, G. Trinchieri and D. E. Griffin (1996). "Mechanism of suppression of cell-mediated immunity by measles virus." *Science* **273**(5272): 228–31.

Kasai, Y., H. Chen and S. J. Flint (1992). "Anatomy of an unusual RNA polymerase II promoter containing a downstream TATA element." *Mol Cell Biol* **12**(6): 2884–97.

Katahira, J. (2015). "Nuclear export of messenger RNA." *Genes (Basel)* **6**(2): 163–84.

Kato, S. E., J. S. Chahal and S. J. Flint (2012). "Reduced infectivity of adenovirus type 5 particles and degradation of entering viral genomes associated with incomplete processing of the preterminal protein." *J Virol* **86**(24): 13554–65.

Katze, M. G., B. M. Detjen, B. Safer and R. M. Krug (1986). "Translational control by influenza virus: suppression of the kinase that phosphorylates the alpha subunit of initiation factor eIF-2 and selective translation of influenza viral mRNAs." *Mol Cell Biol* **6**(5): 1741–50.

Kauffman, R. S. and H. S. Ginsberg (1976). "Characterization of a temperature-sensitive, hexon transport mutant of type 5 adenovirus." *J Virol* **19**(2): 643–58.

Kawashima, T., S. Kagawa, N. Kobayashi, Y. Shirakiya, T. Umeoka, F. Teraishi, M. Taki, S. Kyo, N. Tanaka and T. Fujiwara (2004). "Telomerase-specific replication-selective virotherapy for human cancer." *Clin Cancer Res* **10**(1 Pt 1): 285–92.

Kemper, C., A. C. Chan, J. M. Green, K. A. Brett, K. M. Murphy and J. P. Atkinson (2003). "Activation of human CD4+ cells with CD3 and CD46 induces a T-regulatory cell 1 phenotype." *Nature* **421**(6921): 388–92.

Kennedy, M. A. and R. J. Parks (2009). "Adenovirus virion stability and the viral genome: size matters." *Mol Ther* **17**(10): 1664–6.

Kerem, E. and B. Kerem (1995). "The relationship between genotype and phenotype in cystic fibrosis." *Curr Opin Pulm Med* **1**(6): 450–6.

Khanam, S., R. Pilankatta, N. Khanna and S. Swaminathan (2009). "An adenovirus type 5 (AdV5) vector encoding an envelope domain III-based tetravalent antigen elicits immune responses against all four dengue viruses in the presence of prior AdV5 immunity." *Vaccine* **27**(43): 6011–21.

Khare, R., C. Y. Chen, E. A. Weaver and M. A. Barry (2011). "Advances and future challenges in adenoviral vector pharmacology and targeting." *Curr Gene Ther* **11**(4): 241–58.

Khare, R., V. S. Reddy, G. R. Nemerow and M. A. Barry (2012). "Identification of adenovirus serotype 5 hexon regions that interact with scavenger receptors." *J Virol* **86**(4): 2293–301.

Khuri, F. R., J. Nemunaitis, I. Ganly, J. Arseneau, I. F. Tannock, L. Romel, M. Gore, J. Ironside, R. H. MacDougall, C. Heise, B. Randlev, A. M. Gillenwater, P. Bruso, S. B. Kaye, W. K. Hong and D. H. Kirn (2000). "a controlled trial of intratumoral ONYX-015, a selectively-replicating adenovirus, in combination with cisplatin and 5-fluorouracil in patients with recurrent head and neck cancer." *Nat Med* **6**(8): 879–85.

Kiang, A., Z. C. Hartman, R. S. Everett, D. Serra, H. Jiang, M. M. Frank and A. Amalfitano (2006). "Multiple innate inflammatory responses induced after systemic adenovirus vector delivery depend on a functional complement system." *Mol Ther* **14**(4): 588–98.

Kickhoefer, V. A., Y. Garcia, Y. Mikyas, E. Johansson, J. C. Zhou, S. Raval-Fernandes, P. Minoofar, J. I. Zink, B. Dunn, P. L. Stewart and L. H. Rome (2005). "Engineering of vault nanocapsules with enzymatic and fluorescent properties." *Proc Natl Acad Sci U S A* **102**(12): 4348–52.

Kickhoefer, V. A., M. Han, S. Raval-Fernandes, M. J. Poderycki, R. J. Moniz, D. Vaccari, M. Silvestry, P. L. Stewart, K. A. Kelly and L. H. Rome (2009). "Targeting vault nanoparticles to specific cell surface receptors." *ACS Nano* **3**(1): 27–36.

Kim, E., J. H. Kim, H. Y. Shin, H. Lee, J. M. Yang, J. Kim, J. H. Sohn, H. Kim and C. O. Yun (2003). "Ad-mTERT-delta19, a conditional replication-competent adenovirus driven by the human telomerase promoter, selectively replicates in and elicits cytopathic effect in a cancer cell-specific manner." *Hum Gene Ther* **14**(15): 1415–28.

Kim, J., J. H. Kim, K. J. Choi, P. H. Kim and C. O. Yun (2007). "E1A- and E1B-Double mutant replicating adenovirus elicits enhanced oncolytic and antitumor effects." *Hum Gene Ther* **18**(9): 773–86.

Kim, J., P. H. Kim, J. Y. Yoo, A. R. Yoon, H. J. Choi, J. Seong, I. W. Kim, J. H. Kim and C. O. Yun (2009). "Double E1B 19 kDa- and E1B 55 kDa-deleted oncolytic adenovirus in combination with radiotherapy elicits an enhanced anti-tumor effect." *Gene Ther* **16**(9): 1111–21.

Kim, J. and R. G. Roeder (2009). "Direct Bre1-Paf1 complex interactions and RING finger-independent Bre1-Rad6 interactions mediate histone H2B ubiquitylation in yeast." *J Biol Chem* **284**(31): 20582–92.

Kim, K. H., I. P. Dmitriev, S. Saddekni, E. A. Kashentseva, R. D. Harris, R. Aurigemma, S. Bae, K. P. Singh, G. P. Siegal, D. T. Curiel and R. D. Alvarez (2013). "A phase I clinical trial of Ad5/3-Delta24, a novel serotype-chimeric, infectivity-enhanced, conditionally-replicative adenovirus (CRAd), in patients with recurrent ovarian cancer." *Gynecol Oncol* **130**(3): 518–24.

Kim, N. W. (1997). "Clinical implications of telomerase in cancer." *Eur J Cancer* **33**(5): 781–6.

Kimani, D., Y. J. Jagne, M. Cox, E. Kimani, C. M. Bliss, E. Gitau, C. Ogwang, M. O. Afolabi, G. Bowyer, K. A. Collins, N. Edwards, S. H. Hodgson, C. J. Duncan,

A. J. Spencer, M. G. Knight, A. Drammeh, N. A. Anagnostou, E. Berrie, S. Moyle, S. C. Gilbert, P. Soipei, J. Okebe, S. Colloca, R. Cortese, N. K. Viebig, R. Roberts, A. M. Lawrie, A. Nicosia, E. B. Imoukhuede, P. Bejon, R. Chilengi, K. Bojang, K. L. Flanagan, A. V. Hill, B. C. Urban and K. J. Ewer (2014). "Translating the immunogenicity of prime-boost immunization with ChAd63 and MVA ME-TRAP from malaria naive to malaria-endemic populations." *Mol Ther* **22**(11): 1992–2003.

Kimball, K. J., M. A. Preuss, M. N. Barnes, M. Wang, G. P. Siegal, W. Wan, H. Kuo, S. Saddekni, C. R. Stockard, W. E. Grizzle, R. D. Harris, R. Aurigemma, D. T. Curiel and R. D. Alvarez (2010). "A phase I study of a tropism-modified conditionally replicative adenovirus for recurrent malignant gynecologic diseases." *Clin Cancer Res* **16**(21): 5277–87.

Kimelman, D., J. S. Miller, D. Porter and B. E. Roberts (1985). *J Virol* **53**: 399–409.

Kindsmüller, K., P. Groitl, B. Hartl, P. Blanchette, J. Hauber and T. Dobner (2007). "Intranuclear targeting and nuclear export of the adenovirus E1B-55K protein are regulated by SUMO1 conjugation." *Proc Natl Acad Sci U S A* **104**(16): 6684–9.

King, A. J. and P. C. van der Vliet (1994). "A precursor terminal protein-trinucleotide intermediate during initiation of adenovirus DNA replication: regeneration of molecular ends in vitro by a jumping back mechanism." *EMBO J* **13**(23): 5786–92.

King, J. and W. Chiu (1997). The Procapsid-to-Capsid Transition in Double-Stranded DNA Bacteriophages. *Structural Biology of Viruses*. W. Chiu, R. M. Burnett and R. L. Garcea. New York, Oxfore University Press: 288–311.

King, W. J. and P. H. Krebsbach (2013). "Cyclic-RGD peptides increase the adenoviral transduction of human mesenchymal stem cells." *Stem Cells Dev* **22**(4): 679–86.

Kitajewski, J., R. J. Schneider, B. Safer, S. M. Munemitsu, C. E. Samuel, B. Thimmappaya and T. Shenk (1986). "Adenovirus VAI RNA antagonizes the antiviral action of interferon by preventing activation of the interferon-induced eIF-2 alpha kinase." *Cell* **45**(2): 195–200.

Kitchingman, G. R. (1985). "Sequence of the DNA-binding protein of a human subgroup E adenovirus (type 4): comparisons with subgroup A (type 12), subgroup B (type 7), and subgroup C (type 5)." *Virology* **146**(1): 90–101.

Kitchingman, G. R. (1995). "Mutations in the adenovirus-encoded single-stranded DNA binding protein that result in altered accumulation of early and late viral RNAs." *Virology* **212**(1): 91–101.

Kjellen, L. and H. G. Pereira (1968). "Role of adenovirus antigens in the induction of virus neutralizing antibody." *J Gen Virol* **2**(1): 177–85.

Kleinberger, T. and T. Shenk (1993). "Adenovirus E4orf4 protein binds to protein phosphatase 2A, and the complex down regulates E1A-enhanced junB transcription." *J Virol* **67**(12): 7556–60.

Klose, T. and M. G. Rossmann (2014). "Structure of large dsDNA viruses." *Biol Chem* **395**(7–8): 711–9.

Kneidinger, D., M. Ibrisimovic, T. Lion and R. Klein (2012). "Inhibition of adenovirus multiplication by short interfering RNAs directly or indirectly targeting the viral DNA replication machinery." *Antiviral Res* **94**(3): 195–207.

Knopf, C. W. (1998). "Evolution of viral DNA-dependent DNA polymerases." *Virus Genes* **16**(1): 47–58.

Kobinger, G. P., H. Feldmann, Y. Zhi, G. Schumer, G. Gao, F. Feldmann, S. Jones and J. M. Wilson (2006). "Chimpanzee adenovirus vaccine protects against Zaire Ebola virus." *Virology* **346**(2): 394–401.

Koblin, B. A., K. H. Mayer, E. Noonan, C. Y. Wang, M. Marmor, J. Sanchez, S. J. Brown, M. N. Robertson and S. P. Buchbinder (2012). "Sexual risk behaviors, circumcision status, and preexisting immunity to adenovirus type 5 among men who have sex with men participating in a randomized HIV-1 vaccine efficacy trial: step study." *J Acquir Immune Defic Syndr* **60**(4): 405–13.

Kochanek, S. (1999). "High-capacity adenoviral vectors for gene transfer and somatic gene therapy." *Hum Gene Ther* **10**(15): 2451–9.

Kohler, A. and E. Hurt (2007). "Exporting RNA from the nucleus to the cytoplasm." *Nat Rev Mol Cell Biol* **8**(10): 761–73.

Komatsu, T., H. Haruki and K. Nagata (2011). "Cellular and viral chromatin proteins are positive factors in the regulation of adenovirus gene expression." *Nucleic Acids Res* **39**(3): 889–901.

Komatsu, T. and K. Nagata (2012). "Replication-uncoupled histone deposition during adenovirus DNA replication." *J Virol* **86**(12): 6701–11.

Komatsu, T., K. Nagata and H. Wodrich (2015). "An Adenovirus DNA Replication Factor, but Not Incoming Genome Complexes, Targets PML Nuclear Bodies." *J Virol* **90**(3): 1657–67.

Koonin, E. V., T. G. Senkevich and V. I. Chernos (1993). "Gene A32 product of vaccinia virus may be an ATPase involved in viral DNA packaging as indicated by sequence comparisons with other putative viral ATPases." *Virus Genes* **7**(1): 89–94.

Kopycinski, J., P. Hayes, A. Ashraf, H. Cheeseman, F. Lala, J. Czyzewska-Khan, A. Spentzou, D. K. Gill, M. C. Keefer, J. L. Excler, P. Fast, J. Cox and J. Gilmour (2014). "Broad HIV epitope specificity and viral inhibition induced by multigenic HIV-1 adenovirus subtype 35 vector vaccine in healthy uninfected adults." *PLoS One* **9**(3): e90378.

Korokhov, N., G. Mikheeva, A. Krendelshchikov, N. Belousova, V. Simonenko, V. Krendelshchikova, A. Pereboev, A. Kotov, O. Kotova, P. L. Triozzi, W. A. Aldrich, J. T. Douglas, K. M. Lo, P. T. Banerjee, S. D. Gillies, D. T. Curiel and V. Krasnykh (2003). "Targeting of adenovirus via genetic modification of the viral capsid combined with a protein bridge." *J Virol* **77**(24): 12931–40.

**Koski, A., L. Kangasniemi, S. Escutenaire, S. Pesonen, V. Cerullo, I. Diaconu, P. Nokisalmi, M. Raki, M. Rajecki, K. Guse, T. Ranki, M. Oksanen, S. L. Holm, E. Haavisto, A. Karioja-Kallio, L. Laasonen, K. Partanen, M. Ugolini,

A. Helminen, E. Karli, P. Hannuksela, T. Joensuu, A. Kanerva and A. Hemminki (2010). "Treatment of cancer patients with a serotype 5/3 chimeric oncolytic adenovirus expressing GMCSF." *Mol Ther* **18**(10): 1874–84.

Kostense, S., W. Koudstaal, M. Sprangers, G. J. Weverling, G. Penders, N. Helmus, R. Vogels, M. Bakker, B. Berkhout, M. Havenga and J. Goudsmit (2004). "Adenovirus types 5 and 35 seroprevalence in AIDS risk groups supports type 35 as a vaccine vector." *AIDS* **18**(8): 1213–6.

Kotha, P. L., P. Sharma, A. O. Kolawole, R. Yan, M. S. Alghamri, T. L. Brockman, J. Gomez-Cambronero and K. J. Excoffon (2015). "Adenovirus entry from the apical surface of polarized epithelia is facilitated by the host innate immune response." *PLoS Pathog* **11**(3): e1004696.

Koup, R. A., M. Roederer, L. Lamoreaux, J. Fischer, L. Novik, M. C. Nason, B. D. Larkin, M. E. Enama, J. E. Ledgerwood, R. T. Bailer, J. R. Mascola, G. J. Nabel, B. S. Graham, V. R. C. S. Team and V. R. C. S. Team (2010). "Priming immunization with DNA augments immunogenicity of recombinant adenoviral vectors for both HIV-1 specific antibody and T-cell responses." *PLoS One* **5**(2): e9015.

Kovacs, G. M., S. E. LaPatra, J. C. D'Halluin and M. Benko (2003). "Phylogenetic analysis of the hexon and protease genes of a fish adenovirus isolated from white sturgeon (Acipenser transmontanus) supports the proposal for a new adenovirus genus." *Virus Res* **98**(1): 27–34.

Kovesdi, I., R. Reichel and J. R. Nevins (1986). "Identification of a cellular transcription factor involved in E1A trans-activation." *Cell* **45**: 219–228.

Kralli, A., R. Ge, U. Graeven, R. P. Ricciardi and R. Weinmann (1992). "Negative regulation of the major histocompatibility complex class I enhancer in adenovirus type 12-transformed cells via a retinoic acid response element." *J Virol* **66**(12): 6979–88.

Kremer, E. J. (2004). "CAR chasing: canine adenovirus vectors-all bite and no bark?" *J Gene Med* **6 Suppl 1**: S139–51.

Kremer, E. J. and G. R. Nemerow (2015). "Adenovirus tales: from the cell surface to the nuclear pore complex." *PLoS Pathog* **11**(6): e1004821.

Kreppel, F. (2015). Production of High-Capacity Adenovirus Vectors. *Adenovirus Methods and Protocols*. M. Chillón and A. Bosch, Humana Press: 211.

Kreppel, F. and S. Kochanek (2004). "Long-term transgene expression in proliferating cells mediated by episomally maintained high-capacity adenovirus vectors." *J Virol* **78**(1): 9–22.

Kreppel, F. and S. Kochanek (2008). "Modification of adenovirus gene transfer vectors with synthetic polymers: a scientific review and technical guide." *Mol Ther* **16**(1): 16–29.

Kristiansen, H., H. H. Gad, S. Eskildsen-Larsen, P. Despres and R. Hartmann (2011). "The oligoadenylate synthetase family: an ancient protein family with multiple anti-viral activities." *J Interferon Cytokine Res* **31**(1): 41–7.

Kuersten, S., M. Ohno and I. W. Mattaj (2001). "Nucleocytoplasmic transport: Ran, beta and beyond." *Trends Cell Biol* **11**(12): 497–503.

Kurachi, S., N. Koizumi, F. Sakurai, K. Kawabata, H. Sakurai, S. Nakagawa, T. Hayakawa and H. Mizuguchi (2007). "Characterization of capsid-modified adenovirus vectors containing heterologous peptides in the fiber knob, protein IX, or hexon." *Gene Ther* **14**(3): 266–74.

Kushner, D. B., D. S. Pereira, X. Liu, F. L. Graham and R. P. Ricciardi (1996). "The first exon of Ad12 E1A excluding the transactivation domain mediates differential binding of COUP-TF and NF-kappa B to the MHC class I enhancer in transformed cells." *Oncogene* **12**(1): 143–51.

Kushner, D. B. and R. P. Ricciardi (1999). "Reduced phosphorylation of p50 is responsible for diminished NF-kappaB binding to the major histocompatibility complex class I enhancer in adenovirus type 12-transformed cells." *Mol Cell Biol* **19**(3): 2169–79.

Kwon, O. J., P. H. Kim, S. Huyn, L. Wu, M. Kim and C. O. Yun (2010). "A hypoxia- and {alpha}-fetoprotein-dependent oncolytic adenovirus exhibits specific killing of hepatocellular carcinomas." *Clin Cancer Res* **16**(24): 6071–82.

La Thangue, N. B. (1996). "E2F and the molecular mechanisms of early cell-cycle control." *Biochem Soc Trans* **24**(1): 54–9.

La Thangue, N. B. (2003). "The yin and yang of E2F-1: balancing life and death." *Nat Cell Biol* **5**(7): 587–9.

Laborda, E., C. Puig-Saus, A. Rodriguez-Garcia, R. Moreno, M. Cascallo, J. Pastor and R. Alemany (2014). "A pRb-responsive, RGD-modified, and hyaluronidase-armed canine oncolytic adenovirus for application in veterinary oncology." *Mol Ther* **22**(5): 986–98.

Lai, C. Y., C. M. Wiethoff, V. A. Kickhoefer, L. H. Rome and G. R. Nemerow (2009). "Vault nanoparticles containing an adenovirus-derived membrane lytic protein facilitate toxin and gene transfer." *ACS Nano* **3**(3): 691–9.

Lakdawala, S. S., R. A. Schwartz, K. Ferenchak, C. T. Carson, B. P. McSharry, G. W. Wilkinson and M. D. Weitzman (2008). "Differential requirements of the C terminus of Nbs1 in suppressing adenovirus DNA replication and promoting concatemer formation." *J Virol* **82**(17): 8362–72.

Lam, E. and E. Falck-Pedersen (2014). "Unabated adenovirus replication following activation of the cGAS/STING-dependent antiviral response in human cells." *J Virol* **88**(24): 14426–39.

Lam, E., S. Stein and E. Falck-Pedersen (2014). "Adenovirus detection by the cGAS/STING/TBK1 DNA sensing cascade." *J Virol* **88**(2): 974–81.

Lamfers, M. L., D. Gianni, C. H. Tung, S. Idema, F. H. Schagen, J. E. Carette, P. H. Quax, V. W. Van Beusechem, W. P. Vandertop, C. M. Dirven, E. A. Chiocca and W. R. Gerritsen (2005). "Tissue inhibitor of metalloproteinase-3 expression from an oncolytic adenovirus inhibits matrix metalloproteinase activity in vivo without affecting antitumor efficacy in malignant glioma." *Cancer Res* **65**(20): 9398–405.

Lanciotti, J., A. Song, J. Doukas, B. Sosnowski, G. Pierce, R. Gregory, S. Wadsworth and C. O'Riordan (2003). "Targeting adenoviral vectors using heterofunctional polyethylene glycol FGF2 conjugates." *Mol Ther* **8**(1): 99–107.

Lander, G. C., L. Tang, S. R. Casjens, E. B. Gilcrease, P. Prevelige, A. Poliakov, C. S. Potter, B. Carragher and J. E. Johnson (2006). "The Structure of an Infectious P22 Virion Shows the Signal for Headful DNA Packaging." *Science* **312**(5781): 1791–1795.

Lanier, L. L. (2008). "Evolutionary struggles between NK cells and viruses." *Nat Rev Immunol* **8**(4): 259–68.

Lanman, J., J. Crum, T. J. Deerinck, G. M. Gaietta, A. Schneemann, G. E. Sosinsky, M. H. Ellisman and J. E. Johnson (2008). "Visualizing flock house virus infection in Drosophila cells with correlated fluorescence and electron microscopy." *J Struct Biol* **161**(3): 439–46.

Larson, C., B. Oronsky, J. Scicinski, G. R. Fanger, M. Stirn, A. Oronsky and T. R. Reid (2015). "Going viral: a review of replication-selective oncolytic adenoviruses." *Oncotarget* **6**(24): 19976–89.

Larsson, S., A. Bellett and G. Akusjarvi (1986). "VA RNAs from avian and human adenoviruses: dramatic differences in length, sequence, and gene location." *J Virol* **58**(2): 600–609.

Larsson, S., C. Svensson and G. Akusjarvi (1992). "Control of adenovirus major late gene expression at multiple levels." *J Mol Biol* **225**(2): 287–98.

Lassar, A. B., P. L. Martin and R. G. Roeder (1983). "Transcription of class III genes: formation of preinitiation complexes." *Science* **222**: 740–748.

Lau, L., E. E. Gray, R. L. Brunette and D. B. Stetson (2015). "DNA tumor virus oncogenes antagonize the cGAS-STING DNA-sensing pathway." *Science* **350**(6260): 568–71.

Launer-Felty, K., C. J. Wong and J. L. Cole (2015). "Structural analysis of adenovirus VAI RNA defines the mechanism of inhibition of PKR." *Biophys J* **108**(3): 748–57.

Launer-Felty, K., C. J. Wong, A. M. Wahid, G. L. Conn and J. L. Cole (2010). "Magnesium-dependent interaction of PKR with adenovirus VAI." *J Mol Biol* **402**(4): 638–44.

Lavilla-Alonso, S., G. Bauerschmitz, U. Abo-Ramadan, J. Halavaara, S. Escutenaire, I. Diaconu, T. Tatlisumak, A. Kanerva, A. Hemminki and S. Pesonen (2010). "Adenoviruses with an alphavbeta integrin targeting moiety in the fiber shaft or the HI-loop increase tumor specificity without compromising antitumor efficacy in magnetic resonance imaging of colorectal cancer metastases." *J Transl Med* **8**: 80.

Lavin, M. F. (2007). "ATM and the Mre11 complex combine to recognize and signal DNA double-strand breaks." *Oncogene* **26**(56): 7749–58.

Lavoie, J. N., M. Nguyen, R. C. Marcellus, P. E. Branton and G. C. Shore (1998). "E4orf4, a novel adenovirus death factor that induces p53-independent apoptosis by a pathway that is not inhibited by zVAD-fmk." *J Cell Biol* **140**(3): 637–45.

Lawrence, W. C. and H. S. Ginsberg (1967). "Intracellular uncoating of type 5 adenovirus deoxyribonucleic acid." *J Virol* **1**(5): 851–67.

Le Friec, G., D. Sheppard, P. Whiteman, C. M. Karsten, S. A. Shamoun, A. Laing, L. Bugeon, M. J. Dallman, T. Melchionna, C. Chillakuri, R. A. Smith, C. Drouet,

L. Couzi, V. Fremeaux-Bacchi, J. Kohl, S. N. Waddington, J. M. McDonnell, A. Baker, P. A. Handford, S. M. Lea and C. Kemper (2012). "The CD46-Jagged1 interaction is critical for human TH1 immunity." *Nat Immunol* **13**(12): 1213–21.

Lechner, R. L. and T. J. Kelly, Jr. (1977). "The structure of replicating adenovirus 2 DNA molecules." *Cell* **12**(4): 1007–1020.

Ledgerwood, J. E., P. Costner, N. Desai, L. Holman, M. E. Enama, G. Yamshchikov, S. Mulangu, Z. Hu, C. A. Andrews, R. A. Sheets, R. A. Koup, M. Roederer, R. Bailer, J. R. Mascola, M. G. Pau, N. J. Sullivan, J. Goudsmit, G. J. Nabel, B. S. Graham and V. R. C. S. Team (2010). "A replication defective recombinant Ad5 vaccine expressing Ebola virus GP is safe and immunogenic in healthy adults." *Vaccine* **29**(2): 304–13.

Ledgerwood, J. E., A. D. DeZure, D. A. Stanley, L. Novik, M. E. Enama, N. M. Berkowitz, Z. Hu, G. Joshi, A. Ploquin, S. Sitar, I. J. Gordon, S. A. Plummer, L. A. Holman, C. S. Hendel, G. Yamshchikov, F. Roman, A. Nicosia, S. Colloca, R. Cortese, R. T. Bailer, R. M. Schwartz, M. Roederer, J. R. Mascola, R. A. Koup, N. J. Sullivan, B. S. Graham and the V. R. C. S. T. (2014). "Chimpanzee Adenovirus Vector Ebola Vaccine — Preliminary Report." *N Engl J Med*.

Ledinko, N. and C. K. Fong (1969). "Kinetics of nucleic acid synthesis in human embryonic kidney cultures infected with adenovirus 2 or 12: inhibition of cellular deoxyribonucleic acid synthesis." *J Virol* **4**(2): 123–32.

Lee, J. I., J. Y. Chung, T. H. Han, M. O. Song and E. S. Hwang (2007). "Detection of human bocavirus in children hospitalized because of acute gastroenteritis." *J. Infect Dis.* **196**(7): 994–7.

Lee, K. A., T. Y. Hai, L. SivaRaman, B. Thimmappaya, H. C. Hurst, N. C. Jones and M. R. Green (1987). "A cellular protein, activating transcription factor, activates transcription of multiple E1A-inducible adenovirus early promoters." *Proc Natl Acad Sci U S A* **84**(23): 8355–9.

Lee, M. G., M. A. Abina, H. Haddada and M. Perricaudet (1995). "The constitutive expression of the immunomodulatory gp19k protein in E1-, E3- adenoviral vectors strongly reduces the host cytotoxic T cell response against the vector." *Gene Ther* **2**(4): 256–62.

Leen, A. M., C. M. Bollard, A. M. Mendizabal, E. J. Shpall, P. Szabolcs, J. H. Antin, N. Kapoor, S. Y. Pai, S. D. Rowley, P. Kebriaei, B. R. Dey, B. J. Grilley, A. P. Gee, M. K. Brenner, C. M. Rooney and H. E. Heslop (2013). "Multicenter study of banked third-party virus-specific T cells to treat severe viral infections after hematopoietic stem cell transplantation." *Blood* **121**(26): 5113–23.

Leen, A. M., A. Christin, M. Khalil, H. Weiss, A. P. Gee, M. K. Brenner, H. E. Heslop, C. M. Rooney and C. M. Bollard (2008). "Identification of hexon-specific CD4 and CD8 T-cell epitopes for vaccine and immunotherapy." *J Virol* **82**(1): 546–54.

Leen, A. M., A. Christin, G. D. Myers, H. Liu, C. R. Cruz, P. J. Hanley, A. A. Kennedy-Nasser, K. S. Leung, A. P. Gee, R. A. Krance, M. K. Brenner, H. E. Heslop, C. M. Rooney and C. M. Bollard (2009). "Cytotoxic T lymphocyte therapy with donor T cells prevents and treats adenovirus and Epstein-Barr virus infections after haploidentical and matched unrelated stem cell transplantation." *Blood* **114**(19): 4283–92.

Leen, A. M., U. Sili, E. F. Vanin, A. M. Jewell, W. Xie, D. Vignali, P. A. Piedra, M. K. Brenner and C. M. Rooney (2004). "Conserved CTL epitopes on the adenovirus hexon protein expand subgroup cross-reactive and subgroup-specific CD8+ T cells." *Blood* **104**(8): 2432–40.

Lehrer, R. I., A. Barton, K. A. Daher, S. S. Harwig, T. Ganz and M. E. Selsted (1989). "Interaction of human defensins with Escherichia coli. Mechanism of bactericidal activity." *J Clin Invest* **84**(2): 553–61.

Lehrmann, H. and M. Cotten (1999). "Characterization of CELO virus proteins that modulate the pRb/E2F pathway." *J Virol* **73**(8): 6517–25.

Lei, J., Q. H. Li, J. L. Yang, F. Liu, L. Wang, W. M. Xu and W. X. Zhao (2015). "The antitumor effects of oncolytic adenovirus H101 against lung cancer." *Int J Oncol* **47**(2): 555–62.

Lei, M., Y. Liu and C. E. Samuel (1998). "Adenovirus VAI RNA antagonizes the RNA-editing activity of the ADAR adenosine deaminase." *Virology* **245**(2): 188–96.

Lei, N., F. B. Shen, J. H. Chang, L. Wang, H. Li, C. Yang, J. Li and D. C. Yu (2009). "An oncolytic adenovirus expressing granulocyte macrophage colony-stimulating factor shows improved specificity and efficacy for treating human solid tumors." *Cancer Gene Ther* **16**(1): 33–43.

Lemarchand, P., H. A. Jaffe, C. Danel, M. C. Cid, H. K. Kleinman, L. D. Stratford-Perricaudet, M. Perricaudet, A. Pavirani, J. P. Lecocq and R. G. Crystal (1992). "Adenovirus-mediated transfer of a recombinant human alpha 1-antitrypsin cDNA to human endothelial cells." *Proc Natl Acad Sci U S A* **89**(14): 6482–6.

Lemckert, A. A., J. Grimbergen, S. Smits, E. Hartkoorn, L. Holterman, B. Berkhout, D. H. Barouch, R. Vogels, P. Quax, J. Goudsmit and M. J. Havenga (2006). "Generation of a novel replication-incompetent adenoviral vector derived from human adenovirus type 49: manufacture on PER.C6 cells, tropism and immunogenicity." *J Gen Virol* **87**(Pt 10): 2891–9.

Lemckert, A. A., S. M. Sumida, L. Holterman, R. Vogels, D. M. Truitt, D. M. Lynch, A. Nanda, B. A. Ewald, D. A. Gorgone, M. A. Lifton, J. Goudsmit, M. J. Havenga and D. H. Barouch (2005). "Immunogenicity of heterologous prime-boost regimens involving recombinant adenovirus serotype 11 (Ad11) and Ad35 vaccine vectors in the presence of anti-ad5 immunity." *J Virol* **79**(15): 9694–701.

Lenman, A., A. M. Liaci, Y. Liu, C. Ardahl, A. Rajan, E. Nilsson, W. Bradford, L. Kaeshammer, M. S. Jones, L. Frangsmyr, T. Feizi, T. Stehle and N. Arnberg (2015). "Human adenovirus 52 uses sialic acid-containing glycoproteins and the coxsackie and adenovirus receptor for binding to target cells." *PLoS Pathog* **11**(2): e1004657.

Leonard, G. T. and G. C. Sen (1996). "Effects of adenovirus E1A protein on interferon-signaling." *Virology* **224**(1): 25–33.

Leong, K., W. Lee and A. J. Berk (1990). "High-level transcription from the adenovirus major late promoter requires downstream binding sites for late-phase-specific factors." *J. Virol.* **64**: 51–60.

Leopold, P. L., G. Kreitzer, N. Miyazawa, S. Rempel, K. K. Pfister, E. Rodriguez-Boulan and R. G. Crystal (2000). "Dynein- and microtubule-mediated translocation of adenovirus serotype 5 occurs after endosomal lysis." *Hum Gene Ther* **11**(1): 151–65.

Leopold, P. L., R. L. Wendland, T. Vincent and R. G. Crystal (2006). "Neutralized adenovirus-immune complexes can mediate effective gene transfer via an Fc receptor-dependent infection pathway." *J Virol* **80**(20): 10237–47.

Leppard, K. N., E. Emmott, M. S. Cortese and T. Rich (2009). "Adenovirus type 5 E4 Orf3 protein targets promyelocytic leukaemia (PML) protein nuclear domains for disruption via a sequence in PML isoform II that is predicted as a protein interaction site by bioinformatic analysis." *J Gen Virol* **90**(Pt 1): 95–104.

Levine, A. J. (2009). "The common mechanisms of transformation by the small DNA tumor viruses: The inactivation of tumor suppressor gene products: p53." *Virology* **384**(2): 285–93.

Levine, A. J., J. Momand and C. A. Finlay (1991). "The p53 tumour suppressor gene." *Nature* **351**(6326): 453–6.

Levine, B. and G. Kroemer (2009). "Autophagy in aging, disease and death: the true identity of a cell death impostor." *Cell Death Differ* **16**(1): 1–2.

Li, E., S. L. Brown, D. G. Stupack, X. S. Puente, D. A. Cheresh and G. R. Nemerow (2001). "Integrin alpha(v)beta1 is an adenovirus coreceptor." *J Virol* **75**(11): 5405–9.

Li, E., S. L. Brown, D. J. Von Seggern, G. B. Brown and G. R. Nemerow (2000). "Signaling antibodies complexed with adenovirus circumvent CAR and integrin interactions and improve gene delivery." *Gene Ther* **7**(18): 1593–9.

Li, E., D. Stupack, G. M. Bokoch and G. R. Nemerow (1998). "Adenovirus endocytosis requires actin cytoskeleton reorganization mediated by Rho family GTPases." *J Virol* **72**(11): 8806–12.

Li, E., D. Stupack, R. Klemke, D. A. Cheresh and G. R. Nemerow (1998). "Adenovirus endocytosis via alpha(v) integrins requires phosphoinositide-3-OH kinase." *J Virol* **72**(3): 2055–61.

Li, E., D. G. Stupack, S. L. Brown, R. Klemke, D. D. Schlaepfer and G. R. Nemerow (2000). "Association of p130CAS with phosphatidylinositol-3-OH kinase mediates adenovirus cell entry." *J Biol Chem* **275**(19): 14729–35.

Li, G., H. Kawashima, A. Ogose, T. Ariizumi, Y. Xu, T. Hotta, Y. Urata, T. Fujiwara and N. Endo (2011). "Efficient virotherapy for osteosarcoma by telomerase-specific oncolytic adenovirus." *J Cancer Res Clin Oncol* **137**(6): 1037–51.

Li, G., J. Sham, J. Yang, C. Su, H. Xue, D. Chua, L. Sun, Q. Zhang, Z. Cui, M. Wu and Q. Qian (2005). "Potent antitumor efficacy of an E1B 55kDa-deficient adenovirus carrying murine endostatin in hepatocellular carcinoma." *Int J Cancer* **113**(4): 640–8.

Li, H., E. G. Rhee, K. Masek-Hammerman, J. E. Teigler, P. Abbink and D. H. Barouch (2012). "Adenovirus serotype 26 utilizes CD46 as a primary cellular receptor and only transiently activates T lymphocytes following vaccination of rhesus monkeys." *J Virol* **86**(19): 10862–5.

Li, J. L., H. L. Liu, X. R. Zhang, J. P. Xu, W. K. Hu, M. Liang, S. Y. Chen, F. Hu and D. T. Chu (2009). "A phase I trial of intratumoral administration of recombinant oncolytic adenovirus overexpressing HSP70 in advanced solid tumor patients." *Gene Ther* **16**(3): 376–82.

Li, L., Y. Muzahim and M. Bouvier (2012). "Crystal structure of adenovirus E3-19K bound to HLA-A2 reveals mechanism for immunomodulation." *Nat Struct Mol Biol* **19**(11): 1176–81.

Li, X., Y. Liu, Z. Wen, C. Li, H. Lu, M. Tian, K. Jin, L. Sun, P. Gao, E. Yang, X. Xu, S. Kan, Z. Wang, Y. Wang and N. Jin (2010). "Potent anti-tumor effects of a dual specific oncolytic adenovirus expressing apoptin in vitro and in vivo." *Mol Cancer* **9**: 10.

Li, X., Q. Mao, D. Wang, W. Zhang and H. Xia (2012). "A fiber chimeric CRAd vector Ad5/11-D24 double-armed with TRAIL and arresten for enhanced glioblastoma therapy." *Hum Gene Ther* **23**(6): 589–96.

Li, X. L., H. J. Ezelle, T. Y. Hsi and B. A. Hassel (2011). "A central role for RNA in the induction and biological activities of type 1 interferons." *Wiley Interdiscip Rev RNA* **2**(1): 58–78.

Li, Y., J. Kang, J. Friedman, L. Tarassishin, J. Ye, A. Kovalenko, D. Wallach and M. S. Horwitz (1999). "Identification of a cell protein (FIP-3) as a modulator of NF-kappaB activity and as a target of an adenovirus inhibitor of tumor necrosis factor alpha-induced apoptosis." *Proc Natl Acad Sci U S A* **96**(3): 1042–7.

Li, Y., J. Kang and M. S. Horwitz (1997). "Interaction of an adenovirus 14.7-kilodalton protein inhibitor of tumor necrosis factor alpha cytolysis with a new member of the GTPase superfamily of signal transducers." *J Virol* **71**(2): 1576–82.

Li, Y., J. Kang and M. S. Horwitz (1998). "Interaction of an adenovirus E3 14.7-kilodalton protein with a novel tumor necrosis factor alpha-inducible cellular protein containing leucine zipper domains." *Mol Cell Biol* **18**(3): 1601–10.

Li, Y., B. Zhang, H. Zhang, X. Zhu, D. Feng, D. Zhang, B. Zhuo, L. Li and J. Zheng (2013). "Oncolytic adenovirus armed with shRNA targeting MYCN gene inhibits neuroblastoma cell proliferation and in vivo xenograft tumor growth." *J Cancer Res Clin Oncol* **139**(6): 933–41.

Liberali, P., E. Kakkonen, G. Turacchio, C. Valente, A. Spaar, G. Perinetti, R. A. Bockmann, D. Corda, A. Colanzi, V. Marjomaki and A. Luini (2008). "The closure of Pak1-dependent macropinosomes requires the phosphorylation of CtBP1/BARS." *EMBO J* **27**(7): 970–81.

Lichtenstein, D. L., K. Doronin, K. Toth, M. Kuppuswamy, W. S. Wold and A. E. Tollefson (2004). "Adenovirus E3–6.7K protein is required in conjunction with the E3-RID protein complex for the internalization and degradation of TRAIL receptor 2." *J Virol* **78**(22): 12297–307.

Lichtenstein, D. L., P. Krajcsi, D. J. Esteban, A. E. Tollefson and W. S. Wold (2002). "Adenovirus RIDbeta subunit contains a tyrosine residue that is critical for RID-mediated receptor internalization and inhibition of Fas- and TRAIL-induced apoptosis." *J Virol* **76**(22): 11329–42.

Lichtenstein, D. L., K. Toth, K. Doronin, A. E. Tollefson and W. S. Wold (2004). "Functions and mechanisms of action of the adenovirus E3 proteins." *Int Rev Immunol* **23**(1–2): 75–111.

Lichy, J. F., J. Field, M. S. Horowitz and J. Hurwitz (1982). "Separation of the adenoviral terminal protein precursor from its associated DNA polymerase: Role of both proteins in the initiation of adenovirus DNA replication." *Proc Natl Acad Sci Usa* **79**: 5225–5229.

Liebowitz, D., J. D. Lindbloom, J. R. Brandl, S. J. Garg and S. N. Tucker (2015). "High titre neutralising antibodies to influenza after oral tablet immunisation: a phase 1, randomised, placebo-controlled trial." *Lancet Infect Dis* **15**(9): 1041–8.

Lillie, J. W. and M. R. Green (1989). "Transcription activation by the adenovirus E1a protein." *Nature* **338**(6210): 39–44.

Lim, R. Y., N. P. Huang, J. Koser, J. Deng, K. H. Lau, K. Schwarz-Herion, B. Fahrenkrog and U. Aebi (2006). "Flexible phenylalanine-glycine nucleoporins as entropic barriers to nucleocytoplasmic transport." *Proc Natl Acad Sci U S A* **103**(25): 9512–7.

Lin, H. J. and S. J. Flint (2000). "Identification of a cellular repressor of transcription of the adenoviral late IVa$_2$ gene that is unaltered in activity in infected cells." *Virology* **277**(2): 397–410.

Lindenbaum, J. O., J. Field and J. Hurwitz (1986). "The adenovirus DNA binding protein and adenovirus DNA polymerase interact to catalyze elongation of primed DNA templates." *J Biol Chem* **261**(22): 10218–27.

Lindert, S., M. Silvestry, T. M. Mullen, G. R. Nemerow and P. L. Stewart (2009). "Cryo-electron microscopy structure of an adenovirus-integrin complex indicates conformational changes in both penton base and integrin." *J Virol* **83**(22): 11491–501.

Lion, T. (2014). "Adenovirus infections in immunocompetent and immunocompromised patients." *Clin Microbiol Rev* **27**(3): 441–62.

Lisewski, U., Y. Shi, U. Wrackmeyer, R. Fischer, C. Chen, A. Schirdewan, R. Juttner, F. Rathjen, W. Poller, M. H. Radke and M. Gotthardt (2008). "The tight junction protein CAR regulates cardiac conduction and cell-cell communication." *J Exp Med* **205**(10): 2369–79.

Liszewski, M. K., T. W. Post and J. P. Atkinson (1991). "Membrane cofactor protein (MCP or CD46): newest member of the regulators of complement activation gene cluster." *Annu Rev Immunol* **9**: 431–55.

Liu, E. B., D. A. Wadford, J. Seto, M. Vu, N. R. Hudson, L. Thrasher, S. Torres, D. W. Dyer, J. Chodosh, D. Seto and M. S. Jones (2012). "Computational and serologic analysis of novel and known viruses in species human adenovirus D in which serology and genomics do not correlate." *PLoS One* **7**(3): e33212.

Liu, F. and M. R. Green (1994). "Promoter targeting by adenovirus E1a through interaction with different cellular DNA-binding domains." *Nature* **368**(6471): 520–5.

Liu, G. Q., L. E. Babiss, F. C. Volkert, C. S. Young and H. S. Ginsberg (1985). "A thermolabile mutant of adenovirus 5 resulting from a substitution mutation in the protein VIII gene." *J Virol* **53**(3): 920–5.

Liu, H., J. Fu and M. Bouvier (2007). "Allele- and locus-specific recognition of class I MHC molecules by the immunomodulatory E3-19K protein from adenovirus." *J Immunol* **178**(7): 4567–75.

Liu, H., L. Jin, S. B. Koh, I. Atanasov, S. Schein, L. Wu and Z. H. Zhou (2010). "Atomic structure of human adenovirus by cryo-EM reveals interactions among protein networks." *Science* **329**(5995): 1038–43.

Liu, H., J. H. Naismith and R. T. Hay (2000). "Identification of conserved residues contributing to the activities of adenovirus DNA polymerase." *J Virol* **74**(24): 11681–9.

Liu, H., J. H. Naismith and R. T. Hay (2003). "Adenovirus DNA replication." *Curr Top Microbiol Immunol* **272**: 131–64.

Liu, H., W. F. Stafford and M. Bouvier (2005). "The endoplasmic reticulum lumenal domain of the adenovirus type 2 E3-19K protein binds to peptide-filled and peptide-deficient HLA-A*1101 molecules." *J Virol* **79**(21): 13317–25.

Liu, H., L. Wu and Z. H. Zhou (2011). "Model of the trimeric fiber and its interactions with the pentameric penton base of human adenovirus by cryo-electron microscopy." *J Mol Biol* **406**(5): 764–74.

Liu, J., K. L. O'Brien, D. M. Lynch, N. L. Simmons, A. La Porte, A. M. Riggs, P. Abbink, R. T. Coffey, L. E. Grandpre, M. S. Seaman, G. Landucci, D. N. Forthal, D. C. Montefiori, A. Carville, K. G. Mansfield, M. J. Havenga, M. G. Pau, J. Goudsmit and D. H. Barouch (2009). "Immune control of an SIV challenge by a T-cell-based vaccine in rhesus monkeys." *Nature* **457**(7225): 87–91.

Liu, X., R. Ge, S. Westmoreland, A. J. Cooney, S. Y. Tsai, M. J. Tsai and R. P. Ricciardi (1994). "Negative regulation by the R2 element of the MHC class I enhancer in adenovirus-12 transformed cells correlates with high levels of COUP-TF binding." *Oncogene* **9**(8): 2183–90.

Liu, X. and R. Marmorstein (2007). "Structure of the retinoblastoma protein bound to adenovirus E1A reveals the molecular basis for viral oncoprotein inactivation of a tumor suppressor." *Genes Dev* **21**(21): 2711–6.

Liu, Y., A. L. Colosimo, X. J. Yang and D. Liao (2000). "Adenovirus E1B 55-kilodalton oncoprotein inhibits p53 acetylation by PCAF." *Mol Cell Biol* **20**(15): 5540–53.

Liu, Y. and A. Deisseroth (2006). "Oncolytic adenoviral vector carrying the cytosine deaminase gene for melanoma gene therapy." *Cancer Gene Ther* **13**(9): 845–55.

Liu, Y., A. Shevchenko, A. Shevchenko and A. J. Berk (2005). "Adenovirus exploits the cellular aggresome response to accelerate inactivation of the MRN complex." *J Virol* **79**(22): 14004–16.

Liu, Z. X., S. Govindarajan, S. Okamoto and G. Dennert (2000). "NK cells cause liver injury and facilitate the induction of T cell-mediated immunity to a viral liver infection." *J Immunol* **164**(12): 6480–6.

Logan, J. and T. Shenk (1984). "Adenovirus tripartite leader sequence enhances translation of mRNAs late after infection." *Proc. Natl. Acad. Sci. USA* **81**: 3655–3659.

Lomonosova, E., T. Subramanian and G. Chinnadurai (2002). "Requirement of BAX for efficient adenovirus-induced apoptosis." *J Virol* **76**(22): 11283–90.

Lomonosova, E., T. Subramanian and G. Chinnadurai (2005). "Mitochondrial localization of p53 during adenovirus infection and regulation of its activity by E1B-19K." *Oncogene* **24**(45): 6796–808.

Lonberg-Holm, K. and L. Philipson (1969). "Early events of virus-cell interaction in an adenovirus system." *J Virol* **4**(4): 323–38.

Look, D. C., W. T. Roswit, A. G. Frick, Y. Gris-Alevy, D. M. Dickhaus, M. J. Walter and M. J. Holtzman (1998). "Direct suppression of Stat1 function during adenoviral infection." *Immunity* **9**(6): 871–80.

Lortat-Jacob, H., E. Chouin, S. Cusack and M. J. van Raaij (2001). "Kinetic analysis of adenovirus fiber binding to its receptor reveals an avidity mechanism for trimeric receptor-ligand interactions." *J Biol Chem* **276**(12): 9009–15.

Loskog, A. (2015). "Immunostimulatory Gene Therapy Using Oncolytic Viruses as Vehicles." *Viruses* **7**(11): 5780–91.

Loustalot, F., E. J. Kremer and S. Salinas (2015). "The Intracellular Domain of the Coxsackievirus and Adenovirus Receptor Differentially Influences Adenovirus Entry." *J Virol* **89**(18): 9417–26.

Lowe, S. W. and H. E. Ruley (1993). "Stabilization of the p53 tumor suppressor is induced by adenovirus 5 E1A and accompanies apoptosis." *Genes Dev* **7**(4): 535–45.

Lu, S. and B. R. Cullen (2004). "Adenovirus VA1 noncoding RNA can inhibit small interfering RNA and MicroRNA biogenesis." *J Virol* **78**(23): 12868–76.

Lu, W., S. Zheng, X. F. Li, J. J. Huang, X. Zheng and Z. Li (2004). "Intra-tumor injection of H101, a recombinant adenovirus, in combination with chemotherapy in patients with advanced cancers: a pilot phase II clinical trial." *World J Gastroenterol* **10**(24): 3634–8.

Lu, Z. Z., H. Wang, Y. Zhang, H. Cao, Z. Li, P. Fender and A. Lieber (2013). "Penton-dodecahedral particles trigger opening of intercellular junctions and facilitate viral spread during adenovirus serotype 3 infection of epithelial cells." *PLoS Pathog* **9**(10): e1003718.

Lucas, T., K. Benihoud, F. Vigant, C. Q. Schmidt, A. Wortmann, M. G. Bachem, T. Simmet and S. Kochanek (2015). "Hexon modification to improve the activity of oncolytic adenovirus vectors against neoplastic and stromal cells in pancreatic cancer." *PLoS One* **10**(2): e0117254.

Luisoni, S., M. Suomalainen, K. Boucke, L. B. Tanner, M. R. Wenk, X. L. Guan, M. Grzybek, U. Coskun and U. F. Greber (2015). "Co-option of Membrane Wounding Enables Virus Penetration into Cells." *Cell Host Microbe* **18**(1): 75–85.

Lukashev, A. N., O. E. Ivanova, T. P. Eremeeva and R. D. Iggo (2008). "Evidence of frequent recombination among human adenoviruses." *J Gen Virol* **89**(Pt 2): 380–8.

Lutz, P. and C. Kedinger (1996). "Properties of the adenovirus IVa2 gene product, an effector of late-phase-dependent activation of the major late promoter." *J Virol* **70**(3): 1396–405.

Lyons, M., D. Onion, N. K. Green, K. Aslan, R. Rajaratnam, M. Bazan-Peregrino, S. Phipps, S. Hale, V. Mautner, L. W. Seymour and K. D. Fisher (2006).

"Adenovirus type 5 interactions with human blood cells may compromise systemic delivery." *Mol Ther* **14**(1): 118–28.

Lyons, R. H., B. Q. Ferguson and M. Rosenberg (1987). "Pentapeptide nuclear localization signal in adenovirus E1a." *Mol Cell Biol* **7**(7): 2451–6.

Ma, Y. and M. B. Mathews (1993). "Comparative analysis of the structure and function of adenovirus virus-associated RNAs." *J Virol* **67**(11): 6605–17.

Mabit, H., M. Y. Nakano, U. Prank, B. Saam, K. Dohner, B. Sodeik and U. F. Greber (2002). "Intact microtubules support adenovirus and herpes simplex virus infections." *J Virol* **76**(19): 9962–71.

Mac Sweeney, A., P. Grosche, D. Ellis, K. Combrink, P. Erbel, N. Hughes, F. Sirockin, S. Melkko, A. Bernardi, P. Ramage, N. Jarousse and E. Altmann (2014). "Discovery and structure-based optimization of adenain inhibitors." *ACS Med Chem Lett* **5**(8): 937–41.

Macara, I. G. (2001). "Transport into and out of the nucleus." *Microbiol Mol Biol Rev* **65**(4): 570–94, table of contents.

Macintyre, W. M., H. G. Pereira and W. C. Russell (1969). "Crystallographic data for the hexon of adenovirus type 5." *Nature* **222**(5199): 1165–6.

Magovern, C. J., C. A. Mack and K. T. Budenbender (1996). "Gene transfer utilizing a replication-deficient adenovirus vector expressing vascular endothelial growth factor protects against acute arterial occlusion in the setting of chronic ischemia." *Circulation* **94**: I-636-I-637.

Maheshwari, G., G. Brown, D. A. Lauffenburger, A. Wells and L. G. Griffith (2000). "Cell adhesion and motility depend on nanoscale RGD clustering." *J Cell Sci* **113** **(Pt 10)**: 1677–86.

Maier, O., D. L. Galan, H. Wodrich and C. M. Wiethoff (2010). "An N-terminal domain of adenovirus protein VI fragments membranes by inducing positive membrane curvature." *Virology* **402**(1): 11–9.

Maier, O., S. A. Marvin, H. Wodrich, E. M. Campbell and C. M. Wiethoff (2012). "Spatiotemporal dynamics of adenovirus membrane rupture and endosomal escape." *J Virol* **86**(19): 10821–8.

Maier, O. and C. M. Wiethoff (2010). "N-terminal alpha-helix-independent membrane interactions facilitate adenovirus protein VI induction of membrane tubule formation." *Virology* **408**(1): 31–8.

Majhen, D., H. Calderon, N. Chandra, C. A. Fajardo, A. Rajan, R. Alemany and J. Custers (2014). "Adenovirus-based vaccines for fighting infectious diseases and cancer: progress in the field." *Hum Gene Ther* **25**(4): 301–17.

Majhen, D., J. Richardson, B. Vukelic, I. Dodig, M. Cindric, K. Benihoud and A. Ambriovic-Ristov (2012). "The disulfide bond of an RGD4C motif inserted within the Hi loop of the adenovirus type 5 fiber protein is critical for retargeting to alphav -integrins." *J Gene Med* **14**(12): 788–97.

Makedonas, G., N. Hutnick, D. Haney, A. C. Amick, J. Gardner, G. Cosma, A. R. Hersperger, D. Dolfi, E. J. Wherry, G. Ferrari and M. R. Betts (2010). "Perforin and IL-2 upregulation define qualitative differences among highly functional virus-specific human CD8 T cells." *PLoS Pathog* **6**(3): e1000798.

Makimura, M., S. Miyake, N. Akino, K. Takamori, Y. Matsuura, T. Miyamura and I. Saito (1996). "Induction of antibodies against structural proteins of hepatitis C virus in mice using recombinant adenovirus." *Vaccine* **14**(1): 28–36.

Mallery, D. L., W. A. McEwan, S. R. Bidgood, G. J. Towers, C. M. Johnson and L. C. James (2010). "Antibodies mediate intracellular immunity through tripartite motif-containing 21 (TRIM21)." *Proc Natl Acad Sci U S A* **107**(46): 19985–90.

Mangel, W. F., M. L. Baniecki and W. J. McGrath (2003). "Specific interactions of the adenovirus proteinase with the viral DNA, an 11-amino-acid viral peptide, and the cellular protein actin." *Cell Mol Life Sci* **60**(11): 2347–55.

Mangel, W. F., W. J. McGrath, D. L. Toledo and C. W. Anderson (1993). "Viral DNA and a viral peptide can act as cofactors of adenovirus virion proteinase activity." *Nature* **361**(6409): 274–5.

Mangel, W. F., W. J. McGrath, K. Xiong, V. Graziano and P. C. Blainey (2016). "Molecular sled is an eleven-amino acid vehicle facilitating biochemical interactions via sliding components along DNA." *Nat Commun* **7**.

Mangel, W. F. and C. San Martin (2014). "Structure, function and dynamics in adenovirus maturation." *Viruses* **6**(11): 4536–70.

Mangel, W. F., D. L. Toledo, M. T. Brown, J. H. Martin and W. J. McGrath (1996). "Characterization of three components of human adenovirus proteinase activity in vitro." *J Biol Chem* **271**(1): 536–43.

Mann, K. P., E. A. Weiss and J. R. Nevins (1993). "Alternative poly(A) site utilization during adenovirus infection coincides with a decrease in the activity of a poly(A) site processing factor." *Mol Cell Biol* **13**(4): 2411–9.

Marcellus, R. C., H. Chan, D. Paquette, S. Thirlwell, D. Boivin and P. E. Branton (2000). "Induction of p53-independent apoptosis by the adenovirus E4orf4 protein requires binding to the Balpha subunit of protein phosphatase 2A." *J Virol* **74**(17): 7869–77.

Marcellus, R. C., J. N. Lavoie, D. Boivin, G. C. Shore, G. Ketner and P. E. Branton (1998). "The early region 4 orf4 protein of human adenovirus type 5 induces p53-independent cell death by apoptosis." *J Virol* **72**(9): 7144–53.

Marcellus, R. C., J. G. Teodoro, T. Wu, D. E. Brough, G. Ketner, G. C. Shore and P. E. Branton (1996). "Adenovirus type 5 early region 4 is responsible for E1A-induced p53-independent apoptosis." *J Virol* **70**(9): 6207–15.

Martin-Fernandez, M., S. V. Longshaw, I. Kirby, G. Santis, M. J. Tobin, D. T. Clarke and G. R. Jones (2004). "Adenovirus type-5 entry and disassembly followed in living cells by FRET, fluorescence anisotropy, and FLIM." *Biophys J* **87**(2): 1316–27.

Martin, K. J., J. W. Lillie and M. R. Green (1990). "Evidence for interaction of different eukaryotic transcriptional activators with distinct cellular targets." *Nature* **346**(6280): 147–52.

Martinez, R., P. Schellenberger, D. Vasishtan, C. Aknin, S. Austin, D. Dacheux, F. Rayne, A. Siebert, Z. Ruzsics, K. Gruenewald and H. Wodrich (2015). "The amphipathic helix of adenovirus capsid protein VI contributes to penton release and postentry sorting." *J Virol* **89**(4): 2121–35.

Marton, M. J., S. Baim, D. A. Ornelles and T. Shenk (1990). "The adenovirus E4 17-kilodalton protein complexes with the cellular transcription factor E2F, altering its DNA binding properties and stimulating E1A-independent accumulation of E2 mRNA." *J Virol* **64**: 2345–2359.

Marttila, M., D. Persson, D. Gustafsson, M. K. Liszewski, J. P. Atkinson, G. Wadell and N. Arnberg (2005). "CD46 is a cellular receptor for all species B adenoviruses except types 3 and 7." *J Virol* **79**(22): 14429–36.

Mascola, J. R., A. Sambor, K. Beaudry, S. Santra, B. Welcher, M. K. Louder, T. C. Vancott, Y. Huang, B. K. Chakrabarti, W. P. Kong, Z. Y. Yang, L. Xu, D. C. Montefiori, G. J. Nabel and N. L. Letvin (2005). "Neutralizing antibodies elicited by immunization of monkeys with DNA plasmids and recombinant adenoviral vectors expressing human immunodeficiency virus type 1 proteins." *J Virol* **79**(2): 771–9.

Masek-Hammerman, K., H. Li, J. Liu, P. Abbink, A. La Porte, K. L. O'Brien, J. B. Whitney, A. Carville, K. G. Mansfield and D. H. Barouch (2010). "Mucosal trafficking of vector-specific CD4+ T lymphocytes following vaccination of rhesus monkeys with adenovirus serotype 5." *J Virol* **84**(19): 9810–6.

Mathew, S. S. and E. Bridge (2007). "The cellular Mre11 protein interferes with adenovirus E4 mutant DNA replication." *Virology* **365**(2): 346–55.

Mathews, M. B. (1975). "Genes for VA RNA in adenovirus 2." *Cell* **6**: 223–229.

Mathews, M. B. and U. Pettersson (1978). "The low molecular weight RNAs of adenovirus 2 infected cells." *J. Mol. Biol.* **119**: 293–328.

Mathias, P., T. Wickham, M. Moore and G. Nemerow (1994). "Multiple adenovirus serotypes use alpha v integrins for infection." *J Virol* **68**(10): 6811–4.

Matsui, M., O. Moriya and T. Akatsuka (2003). "Enhanced induction of hepatitis C virus-specific cytotoxic T lymphocytes and protective efficacy in mice by DNA vaccination followed by adenovirus boosting in combination with the interleukin-12 expression plasmid." *Vaccine* **21**(15): 1629–39.

Matsushima, Y., H. Shimizu, A. Kano, E. Nakajima, Y. Ishimaru, S. K. Dey, Y. Watanabe, F. Adachi, K. Mitani, T. Fujimoto, T. G. Phan and H. Ushijima (2013). "Genome sequence of a novel virus of the species human adenovirus d associated with acute gastroenteritis." *Genome Announc* **1**(1).

Matthes-Martin, S., T. Feuchtinger, P. J. Shaw, D. Engelhard, H. H. Hirsch, C. Cordonnier and P. Ljungman (2012). "European guidelines for diagnosis and treatment of adenovirus infection in leukemia and stem cell transplantation: summary of ECIL-4 (2011)." *Transpl Infect Dis* **14**(6): 555–63.

Mautner, V. and H. N. Willcox (1974). "Adenovirus antigens: a model system in mice for subunit vaccination." *J Gen Virol* **25**(3): 325–36.

McCarthy, T., M. G. Lebeck, A. W. Capuano, D. P. Schnurr and G. C. Gray (2009). "Molecular typing of clinical adenovirus specimens by an algorithm which permits detection of adenovirus coinfections and intermediate adenovirus strains." *J Clin Virol* **46**(1): 80–4.

McCormick, F. (2003). "Cancer-specific viruses and the development of ONYX-015." *Cancer Biol Ther* **2**(4 Suppl 1): S157–60.

McCoy, K., N. Tatsis, B. Korioth-Schmitz, M. O. Lasaro, S. E. Hensley, S. W. Lin, Y. Li, W. Giles-Davis, A. Cun, D. Zhou, Z. Xiang, N. L. Letvin and H. C. Ertl (2007). "Effect of preexisting immunity to adenovirus human serotype 5 antigens on the immune responses of nonhuman primates to vaccine regimens based on human- or chimpanzee-derived adenovirus vectors." *J Virol* **81**(12): 6594–604.

McEwan, W. A., F. Hauler, C. R. Williams, S. R. Bidgood, D. L. Mallery, R. A. Crowther and L. C. James (2012). "Regulation of virus neutralization and the persistent fraction by TRIM21." *J Virol* **86**(16): 8482–91.

McEwan, W. A. and L. C. James (2015). "TRIM21-dependent intracellular antibody neutralization of virus infection." *Prog Mol Biol Transl Sci* **129**: 167–87.

McEwan, W. A., J. C. Tam, R. E. Watkinson, S. R. Bidgood, D. L. Mallery and L. C. James (2013). "Intracellular antibody-bound pathogens stimulate immune signaling via the Fc receptor TRIM21." *Nat Immunol* **14**(4): 327–36.

McGrath, W. J., J. Ding, A. Didwania, R. M. Sweet and W. F. Mangel (2003). "Crystallographic structure at 1.6-A resolution of the human adenovirus proteinase in a covalent complex with its 11-amino-acid peptide cofactor: insights on a new fold." *Biochim Biophys Acta* **1648**(1–2): 1–11.

McGrath, W. J., V. Graziano, K. Zabrocka and W. F. Mangel (2013). "First generation inhibitors of the adenovirus proteinase." *FEBS Lett* **587**(15): 2332–9.

McGuire, K. A., A. U. Barlan, T. M. Griffin and C. M. Wiethoff (2011). "Adenovirus type 5 rupture of lysosomes leads to cathepsin B-dependent mitochondrial stress and production of reactive oxygen species." *J Virol* **85**(20): 10806–13.

McKenna, S. A., D. A. Lindhout, T. Shimoike, C. E. Aitken and J. D. Puglisi (2007). "Viral dsRNA inhibitors prevent self-association and autophosphorylation of PKR." *J Mol Biol* **372**(1): 103–13.

McNally, L. R., E. L. Rosenthal, W. Zhang and D. J. Buchsbaum (2009). "Therapy of head and neck squamous cell carcinoma with replicative adenovirus expressing tissue inhibitor of metalloproteinase-2 and chemoradiation." *Cancer Gene Ther* **16**(3): 246–55.

McSharry, B. P., H. G. Burgert, D. P. Owen, R. J. Stanton, V. Prod'homme, M. Sester, K. Koebernick, V. Groh, T. Spies, S. Cox, A. M. Little, E. C. Wang, P. Tomasec and G. W. Wilkinson (2008). "Adenovirus E3/19K promotes evasion of NK cell recognition by intracellular sequestration of the NKG2D ligands major histocompatibility complex class I chain-related proteins A and B." *J Virol* **82**(9): 4585–94.

Meckes, D. G., Jr. (2015). "Exosomal communication goes viral." *J Virol* **89**(10): 5200–3.

Meier, O., K. Boucke, S. V. Hammer, S. Keller, R. P. Stidwill, S. Hemmi and U. F. Greber (2002). "Adenovirus triggers macropinocytosis and endosomal leakage together with its clathrin-mediated uptake." *J Cell Biol* **158**(6): 1119–31.

Menendez D., A. Inga and M. A. Resnick (2009). "The expanding universe of p53 targets." *Nat Rev Cancer* **9**(10): 724–37.

Mennechet, F. J., T. T. Tran, K. Eichholz, P. van de Perre and E. J. Kremer (2015). "Ebola virus vaccine: benefit and risks of adenovirus-based vectors." *Expert Rev Vaccines* **14**(11): 1471–8.

Mercer, J. and U. F. Greber (2013). "Virus interactions with endocytic pathways in macrophages and dendritic cells." *Trends Microbiol* **21**(8): 380–8.

Mercier, G. T., J. A. Campbell, J. D. Chappell, T. Stehle, T. S. Dermody and M. A. Barry (2004). "A chimeric adenovirus vector encoding reovirus attachment protein sigma1 targets cells expressing junctional adhesion molecule 1." *Proc Natl Acad Sci U S A* **101**(16): 6188–93.

Merle, N. S., R. Noe, L. Halbwachs-Mecarelli, V. Fremeaux-Bacchi and L. T. Roumenina (2015). "Complement System Part II: Role in Immunity." *Front Immunol* **6**: 257.

Meunier-Durmort, C., R. Picart, T. Ragot, M. Perricaudet, B. Hainque and C. Forest (1997). "Mechanism of adenovirus improvement of cationic liposome-mediated gene transfer." *Biochim Biophys Acta* **1330**(1): 8–16.

Michou, A. I., H. Lehrmann, M. Saltik and M. Cotten (1999). "Mutational analysis of the avian adenovirus CELO, which provides a basis for gene delivery vectors." *J Virol* **73**(2): 1399–410.

Mikyas, Y., M. Makabi, S. Raval-Fernandes, L. Harrington, V. A. Kickhoefer, L. H. Rome and P. L. Stewart (2004). "Cryoelectron microscopy imaging of recombinant and tissue derived vaults: localization of the MVP N termini and VPARP." *J Mol Biol* **344**(1): 91–105.

Miles, B. D., R. B. Luftig, J. A. Weatherbee, R. R. Weihing and J. Weber (1980). "Quantitation of the interaction between adenovirus types 2 and 5 and microtubules inside infected cells." *Virology* **105**(1): 265–9.

Miller, D. L. (2006). Control of cellular gene expression by adenovirus and myc. *Molecular Biology*. Princeton, Princeton University. **Ph.D.:** 155.

Miller, D. L., C. L. Myers, B. Rickards, H. A. Coller and S. J. Flint (2007). "Adenovirus type 5 exerts genome-wide control over cellular programs governing proliferation, quiescence, and survival." *Genome Biol* **8**(4): R58.

Miller, D. L., B. Rickards, M. Mashiba, W. Huang and S. J. Flint (2009). "The adenoviral E1B 55-kilodalton protein controls expression of immune response genes but not p53-dependent transcription." *J Virol* **83**(8): 3591–603.

Miller, L. H., H. C. Ackerman, X. Z. Su and T. E. Wellems (2013). "Malaria biology and disease pathogenesis: insights for new treatments." *Nat Med* **19**(2): 156–67.

Miller, M. S., P. Pelka, G. J. Fonseca, M. J. Cohen, J. N. Kelly, S. D. Barr, R. J. Grand, A. S. Turnell, P. Whyte and J. S. Mymryk (2012). "Characterization of the 55-residue protein encoded by the 9S E1A mRNA of species C adenovirus." *J Virol* **86**(8): 4222–33.

Minamitani, T., D. Iwakiri and K. Takada (2011). "Adenovirus virus-associated RNAs induce type I interferon expression through a RIG-I-mediated pathway." *J Virol* **85**(8): 4035–40.

Mistchenko, A. S., E. R. Koch, A. E. Kajon, F. Tibaldi, A. F. Maffey and R. A. Diez (1998). "Lymphocyte subsets and cytokines in adenoviral infection in children." *Acta Paediatr* **87**(9): 933–9.

Mitraki, A., A. Barge, J. Chroboczek, J. P. Andrieu, J. Gagnon and R. W. Ruigrok (1999). "Unfolding studies of human adenovirus type 2 fibre trimers. Evidence for a stable domain." *Eur J Biochem* **264**(2): 599–606.

Miyazawa, N., P. L. Leopold, N. R. Hackett, B. Ferris, S. Worgall, E. Falck-Pedersen and R. G. Crystal (1999). "Fiber swap between adenovirus subgroups B and C alters intracellular trafficking of adenovirus gene transfer vectors." *J Virol* **73**(7): 6056–65.

Mok, H., D. J. Palmer, P. Ng and M. A. Barry (2005). "Evaluation of polyethylene glycol modification of first-generation and helper-dependent adenoviral vectors to reduce innate immune responses." *Mol Ther* **11**(1): 66–79.

Molin, M. and G. Akusjarvi (2000). "Overexpression of essential splicing factor ASF/SF2 blocks the temporal shift in adenovirus pre-mRNA splicing and reduces virus progeny formation." *J Virol* **74**(19): 9002–9.

Moll, U. M., S. Wolff, D. Speidel and W. Deppert (2005). "Transcription-independent pro-apoptotic functions of p53." *Curr Opin Cell Biol* **17**(6): 631–6.

Monaghan, A., A. Webster and R. T. Hay (1994). "Adenovirus DNA binding protein: helix destabilising properties." *Nucleic Acids Res* **22**(5): 742–8.

Moodie, Z., B. Metch, L. G. Bekker, G. Churchyard, M. Nchabeleng, K. Mlisana, F. Laher, S. Roux, K. Mngadi, C. Innes, M. Mathebula, M. Allen, C. Bentley, P. B. Gilbert, M. Robertson, J. Kublin, L. Corey and G. E. Gray (2015). "Continued Follow-Up of Phambili Phase 2b Randomized HIV-1 Vaccine Trial Participants Supports Increased HIV-1 Acquisition among Vaccinated Men." *PLoS One* **10**(9): e0137666.

Moore, M., J. Schaack, S. B. Baim, R. I. Morimoto and T Shenk (1987). "Induced heat shock mRNAs escape the nucleocytoplasmic transport block in adenovirus-infected HeLa cells." *Mol Cell Biol* **7**(12): 4505–12.

Moran, E. (1993). "DNA tumor virus transforming proteins and the cell cycle." *Curr. Opin. Genet. Dev.* **3**(1): 63–70.

Morfin, F., S. Dupuis-Girod, E. Frobert, S. Mundweiler, D. Carrington, P. Sedlacek, M. Bierings, P. Cetkovsky, A. C. Kroes, M. J. van Tol and D. Thouvenot (2009). "Differential susceptibility of adenovirus clinical isolates to cidofovir and ribavirin is not related to species alone." *Antivir Ther* **14**(1): 55–61.

Morin, N. and P. Boulanger (1986). "Hexon trimerization occurring in an assembly-defective, 100K temperature-sensitive mutant of adenovirus 2." *Virology* **152**(1): 11–31.

Morris, S. J. and K. N. Leppard (2009). "Adenovirus serotype 5 L4-22K and L4-33K proteins have distinct functions in regulating late gene expression." *J Virol* **83**(7): 3049–58.

Morris, S. J., G. E. Scott and K. N. Leppard (2010). "Adenovirus late-phase infection is controlled by a novel L4 promoter." *J Virol* **84**(14): 7096–104.

Morrow, C. D. and A. Dasgupta (1983). "Antibody to a synthetic nonapeptide corresponding to the NH2 terminus of poliovirus genome-linked protein VPg reacts with native VPg and inhibits in vitro replication of poliovirus RNA." *J Virol* **48**(2): 429–39.

Morsy, M. A., E. L. Alford, A. Bett, F. L. Graham and C. T. Caskey (1993). "Efficient adenoviral-mediated ornithine transcarbamylase expression in deficient mouse and human hepatocytes." *J Clin Invest* **92**(3): 1580–6.

Moyer, C. L., E. S. Besser and G. R. Nemerow (2015). "A Single Maturation Cleavage Site in Adenovirus Impacts Cell Entry and Capsid Assembly." *J Virol* **90**(1): 521–32.

Moyer, C. L., C. M. Wiethoff, O. Maier, J. G. Smith and G. R. Nemerow (2011). "Functional genetic and biophysical analyses of membrane disruption by human adenovirus." *J Virol* **85**(6): 2631–41.

Moyne, G., E. Pichard and W. Bernhard (1978). " Localization of simian adenovirus 7 (SA7) transcription and replication in lytic infection: an ultracytochemical and autoradiographical study." *J. Gen. Virol.* **40**: 77–92.

Muhlemann, O., B. G. Yue, S. Petersen-Mahrt and G. Akusjarvi (2000). "A novel type of splicing enhancer regulating adenovirus pre-mRNA splicing." *Mol Cell Biol* **20**(7): 2317–25.

Mul, Y. M. and P. C. van der Vliet (1993). "The adenovirus DNA binding protein effects the kinetics of DNA replication by a mechanism distinct from NFI or Oct-1." *Nucleic Acids Res* **21**(3): 641–7.

Mul, Y. M., C. P. Verrijzer and P. C. van der Vliet (1990). "Transcription factors NFI and NFIII/oct-1 function independently, employing different mechanisms to enhance adenovirus DNA replication." *J Virol* **64**(11): 5510–8.

Muller-McNicoll, M. and K. M. Neugebauer (2013). "How cells get the message: dynamic assembly and function of mRNA-protein complexes." *Nat Rev Genet* **14**(4): 275–87.

Murali, V. K., D. A. Ornelles, L. R. Gooding, H. T. Wilms, W. Huang, A. E. Tollefson, W. S. Wold and C. Garnett-Benson (2014). "Adenovirus death protein (ADP) is required for lytic infection of human lymphocytes." *J Virol* **88**(2): 903–12.

Muruve, D. A., M. J. Barnes, I. E. Stillman and T. A. Libermann (1999). "Adenoviral gene therapy leads to rapid induction of multiple chemokines and acute neutrophil-dependent hepatic injury in vivo." *Hum Gene Ther* **10**(6): 965–76.

Muruve, D. A., V. Petrilli, A. K. Zaiss, L. R. White, S. A. Clark, P. J. Ross, R. J. Parks and J. Tschopp (2008). "The inflammasome recognizes cytosolic microbial and host DNA and triggers an innate immune response." *Nature* **452**(7183): 103–7.

Muthana, M., A. Giannoudis, S. D. Scott, H. Y. Fang, S. B. Coffelt, F. J. Morrow, C. Murdoch, J. Burton, N. Cross, B. Burke, R. Mistry, F. Hamdy, N. J. Brown, L. Georgopoulos, P. Hoskin, M. Essand, C. E. Lewis and N. J. Maitland (2011). "Use of macrophages to target therapeutic adenovirus to human prostate tumors." *Cancer Res* **71**(5): 1805–15.

Myers, G. D., C. M. Bollard, M. F. Wu, H. Weiss, C. M. Rooney, H. E. Heslop and A. M. Leen (2007). "Reconstitution of adenovirus-specific cell-mediated immunity in pediatric patients after hematopoietic stem cell transplantation." *Bone Marrow Transplant* **39**(11): 677–86.

Myers, G. D., R. A. Krance, H. Weiss, I. Kuehnle, G. Demmler, H. E. Heslop and C. M. Bollard (2005). "Adenovirus infection rates in pediatric recipients of alternate donor allogeneic bone marrow transplants receiving either antithymocyte globulin (ATG) or alemtuzumab (Campath)." *Bone Marrow Transplant* **36**(11): 1001–8.

Mysiak, M. E., M. H. Bleijenberg, C. Wyman, P. E. Holthuizen and P. C. van der Vliet (2004). "Bending of adenovirus origin DNA by nuclear factor I as shown by scanning force microscopy is required for optimal DNA replication." *J Virol* **78**(4): 1928–35.

Mysiak, M. E., C. Wyman, P. E. Holthuizen and P. C. van der Vliet (2004). "NFI and Oct-1 bend the Ad5 origin in the same direction leading to optimal DNA replication." *Nucleic Acids Res* **32**(21): 6218–25.

Nagarajan, S. R., B. Devadas, J. W. Malecha, H. F. Lu, P. G. Ruminski, J. G. Rico, T. E. Rogers, L. D. Marrufo, J. T. Collins, H. P. Kleine, M. K. Lantz, J. Zhu, N. F. Green, M. A. Russell, B. H. Landis, L. M. Miller, D. M. Meyer, T. D. Duffin, V. W. Engleman, M. B. Finn, S. K. Freeman, D. W. Griggs, M. L. Williams, M. A. Nickols, J. A. Pegg, K. E. Shannon, C. Steininger, M. M. Westlin, G. A. Nickols and J. L. Keene (2007). "R-isomers of Arg-Gly-Asp (RGD) mimics as potent alphavbeta3 inhibitors." *Bioorg Med Chem* **15**(11): 3783–800.

Nagata, K., R. A. Guggenheimer and J. Hurwitz (1983). "Adenovirus DNA replication in vitro: synthesis of full length DNA with purified proteins." *Proc Natl Acad Sci Usa* **80**: 4266–0.

Nakano, M. Y., K. Boucke, M. Suomalainen, R. P. Stidwill and U. F. Greber (2000). "The first step of adenovirus type 2 disassembly occurs at the cell surface, independently of endocytosis and escape to the cytosol." *J Virol* **74**(15): 7085–95.

Nanda, A., D. M. Lynch, J. Goudsmit, A. A. Lemckert, B. A. Ewald, S. M. Sumida, D. M. Truitt, P. Abbink, M. G. Kishko, D. A. Gorgone, M. A. Lifton, L. Shen, A. Carville, K. G. Mansfield, M. J. Havenga and D. H. Barouch (2005). "Immunogenicity of recombinant fiber-chimeric adenovirus serotype 35 vector-based vaccines in mice and rhesus monkeys." *J Virol* **79**(22): 14161–8.

Nandi, S., I. V. Ulasov, C. E. Rolle, Y. Han and M. S. Lesniak (2009). "A chimeric adenovirus with an Ad3 fiber knob modification augments glioma virotherapy." *J Gene Med* **11**(11): 1005–11.

Neale, G. A. and G. R. Kitchingman (1990). "Conserved region 3 of the adenovirus type 5 DNA-binding protein is important for interaction with single-stranded DNA." *J Virol* **64**(2): 630–8.

Nemerow, G. R. and D. A. Cheresh (2002). "Herpesvirus hijacks an integrin." *Nat Cell Biol* **4**(4): E69–71.

Nemunaitis, J., C. Cunningham, A. Tong, L. Post, G. Netto, A. S. Paulson, D. Rich, A. Blackburn, B. Sands, B. Gibson, B. Randlev and S. Freeman (2003). "Pilot trial of intravenous infusion of a replication-selective adenovirus (ONYX-015) in combination with chemotherapy or IL-2 treatment in refractory cancer patients." *Cancer Gene Ther* **10**(5): 341–52.

Nemunaitis, J., A. W. Tong, M. Nemunaitis, N. Senzer, A. P. Phadke, C. Bedell, N. Adams, Y. A. Zhang, P. B. Maples, S. Chen, B. Pappen, J. Burke, D. Ichimaru, Y. Urata and T. Fujiwara (2010). "A phase I study of telomerase-specific replication competent oncolytic adenovirus (telomelysin) for various solid tumors." *Mol Ther* **18**(2): 429–34.

Neumann, R., J. Chroboczek and B. Jacrot (1988). "Determination of the nucleotide sequence for the penton-base gene of human adenovirus type 5." *Gene* **69**(1): 153–7.

Nevins, J. R. (1981). "Mechanism of activation of early viral transcription by the adenovirus E1A gene product." *Cell* **26**: 213–220.

Nevins, J. R. (1995). "Adenovirus E1A: transcription regulation and alteration of cell growth control." *Curr Top Microbiol Immunol* **199** (Pt 3): 25–32.

Nevins, J. R. and J. E. Darnell (1978). "Groups of adenovirus type 2 mRNA's derived from a large primary transcript: probable nuclear origin and possible common 3′ ends." *J Virol* **25**(3): 811–23.

Nevins, J. R. and J. E. Darnell, Jr. (1978). "Steps in the processing of Ad2 mRNA: poly(A)+ nuclear sequences are conserved and poly(A) addition precedes splicing." *Cell* **15**(4): 1477–93.

Nevins, J. R., H. S. Ginsberg, J. M. Blanchard, M. C. Wilson and J. Darnell, Jr. (1979). "Regulation of the primary expression of the early adenovirus transcription units." *J Virol* **32**: 727–733.

Nevins, J. R. and M. C. Wilson (1981). "Regulation of adenovirus-2 gene expression at the level of transcription termination and RNA processing." *Nature* **280**: 113–118.

Newcomb, W. W., J. W. Boring and J. C. Brown (1984). "Ion etching of human adenovirus 2: structure of the core." *J Virol* **51**(1): 52–6.

Newcomb, W. W., R. M. Juhas, D. R. Thomsen, F. L. Homa, A. D. Burch, S. K. Weller and J. C. Brown (2001). "The UL6 gene product forms the portal for entry of DNA into the herpes simplex virus capsid." *J Virol* **75**(22): 10923–32.

Nguyen, E. K., G. R. Nemerow and J. G. Smith (2010). "Direct evidence from single-cell analysis that human {alpha}-defensins block adenovirus uncoating to neutralize infection." *J Virol* **84**(8): 4041–9.

Nichols, G. J., J. Schaack and D. A. Ornelles (2009). "Widespread phosphorylation of histone H2AX by species C adenovirus infection requires viral DNA replication." *J Virol* **83**(12): 5987–98.

Nicklin, S. A., E. Wu, G. R. Nemerow and A. H. Baker (2005). "The influence of adenovirus fiber structure and function on vector development for gene therapy." *Mol Ther* **12**(3): 384–93.

Nielsch, U., S. G. Zimmer and L. E. Babiss (1991). "Changes in NF-kappa B and ISGF3 DNA binding activities are responsible for differences in MHC and beta-IFN gene expression in Ad5- versus Ad12-transformed cells." *EMBO J* **10**(13): 4169–75.

Nilsson, E. C., R. J. Storm, J. Bauer, S. M. Johansson, A. Lookene, J. Angstrom, M. Hedenstrom, T. L. Eriksson, L. Frangsmyr, S. Rinaldi, H. J. Willison, F. Pedrosa Domellof, T. Stehle and N. Arnberg (2011). "The GD1a glycan is a cellular receptor for adenoviruses causing epidemic keratoconjunctivitis." *Nat Med* **17**(1): 105–9.

Nisole, S., J. P. Stoye and A. Saib (2005). "TRIM family proteins: retroviral restriction and antiviral defence." *Nat Rev Microbiol* **3**(10): 799–808.

Nociari, M., O. Ocheretina, M. Murphy and E. Falck-Pedersen (2009). "Adenovirus induction of IRF3 occurs through a binary trigger targeting Jun N-terminal kinase and TBK1 kinase cascades and type I interferon autocrine signaling." *J Virol* **83**(9): 4081–91.

Nociari, M., O. Ocheretina, J. W. Schoggins and E. Falck-Pedersen (2007). "Sensing infection by adenovirus: Toll-like receptor-independent viral DNA recognition signals activation of the interferon regulatory factor 3 master regulator." *J Virol* **81**(8): 4145–57.

Nokisalmi, P., S. Pesonen, S. Escutenaire, M. Sarkioja, M. Raki, V. Cerullo, L. Laasonen, R. Alemany, J. Rojas, M. Cascallo, K. Guse, M. Rajecki, L. Kangasniemi, E. Haavisto, A. Karioja-Kallio, P. Hannuksela, M. Oksanen, A. Kanerva, T. Joensuu, L. Ahtiainen and A. Hemminki (2010). "Oncolytic adenovirus ICOVIR-7 in patients with advanced and refractory solid tumors." *Clin Cancer Res* **16**(11): 3035–43.

Noureddini, S. C. and D. T. Curiel (2005). "Genetic targeting strategies for adenovirus." *Mol Pharm* **2**(5): 341–7.

Nunes, F. A., E. E. Furth, J. M. Wilson and S. E. Raper (1999). "Gene transfer into the liver of nonhuman primates with E1-deleted recombinant adenoviral vectors: safety of readministration." *Hum Gene Ther* **10**(15): 2515–26.

Nwanegbo, E., E. Vardas, W. Gao, H. Whittle, H. Sun, D. Rowe, P. D. Robbins and A. Gambotto (2004). "Prevalence of neutralizing antibodies to adenoviral serotypes 5 and 35 in the adult populations of The Gambia, South Africa, and the United States." *Clin Diagn Lab Immunol* **11**(2): 351–7.

O'Brien, K. L., J. Liu, S. L. King, Y. H. Sun, J. E. Schmitz, M. A. Lifton, N. A. Hutnick, M. R. Betts, S. A. Dubey, J. Goudsmit, J. W. Shiver, M. N. Robertson, D. R. Casimiro and D. H. Barouch (2009). "Adenovirus-specific immunity after immunization with an Ad5 HIV-1 vaccine candidate in humans." *Nat Med* **15**(8): 873–5.

O'Hara, G. A., C. J. Duncan, K. J. Ewer, K. A. Collins, S. C. Elias, F. D. Halstead, A. L. Goodman, N. J. Edwards, A. Reyes-Sandoval, P. Bird, R. Rowland, S. H. Sheehy, I. D. Poulton, C. Hutchings, S. Todryk, L. Andrews, A. Folgori, E. Berrie, S. Moyle, A. Nicosia, S. Colloca, R. Cortese, L. Siani, A. M. Lawrie, S. C. Gilbert and A. V. Hill (2012). "Clinical assessment of a recombinant simian adenovirus ChAd63: a potent new vaccine vector." *J Infect Dis* **205**(5): 772–81.

O'Malley, R. P., T. M. Mariano, J. Siekierka and M. B. Mathews (1986). "A mechanism for the control of protein synthesis by adenovirus VA RNAI." *Cell* **44**(3): 391–400.

O'Riordan, C. R., A. Lachapelle, C. Delgado, V. Parkes, S. C. Wadsworth, A. E. Smith and G. E. Francis (1999). "PEGylation of adenovirus with retention of infectivity and protection from neutralizing antibody in vitro and in vivo." *Hum Gene Ther* **10**(8): 1349–58.

O'Shea, C., L. Johnson, B. Bagus, S. Choi, C. Nicholas, A. Shen, L. Boyle, K. Pandey, C. Soria, J. Kunich, Y. Shen, G. Habets, D. Ginzinger and F. McCormick (2004). "Late viral RNA export, rather than p53 inactivation, determines ONYX-015 tumor selectivity." *Cancer Cell* **6**(6): 611–623.

O'Shea, C. C., S. Choi, F. McCormick and D. Stokoe (2005). "Adenovirus overrides cellular checkpoints for protein translation." *Cell Cycle* **4**(7): 883–8.

Oberg, D., E. Yanover, V. Adam, K. Sweeney, C. Costas, N. R. Lemoine and G. Hallden (2010). "Improved potency and selectivity of an oncolytic E1ACR2 and E1B19K deleted adenoviral mutant in prostate and pancreatic cancers." *Clin Cancer Res* **16**(2): 541–53.

Obert, S., R. J. O'Connor, S. Schmid and P. Hearing (1994). "The adenovirus E4–6/7 protein transactivates the E2 promoter by inducing dimerization of a heteromeric E2F complex." *Mol. Cell. Biol.* **14**(2): 1333–1346.

Ogwang, C., D. Kimani, N. J. Edwards, R. Roberts, J. Mwacharo, G. Bowyer, C. Bliss, S. H. Hodgson, P. Njuguna, N. K. Viebig, A. Nicosia, E. Gitau, S. Douglas, J. Illingworth, K. Marsh, A. Lawrie, E. B. Imoukhuede, K. Ewer, B. C. Urban, S. H. AV, P. Bejon and M. Group (2015). "Prime-boost vaccination with chimpanzee adenovirus and modified vaccinia Ankara encoding TRAP provides partial protection against Plasmodium falciparum infection in Kenyan adults." *Sci Transl Med* **7**(286): 286re5.

Oh, I. K., H. Mok and T. G. Park (2006). "Folate immobilized and PEGylated adenovirus for retargeting to tumor cells." *Bioconjug Chem* **17**(3): 721–7.

Ohe, K. and S. M. Weissman (1970). "Nucleotide sequence of an RNA from cells infected with adenovirus 2." *Science* **167**(3919): 879–81.

Olive, M., L. Eisenlohr, N. Flomenberg, S. Hsu and P. Flomenberg (2002). "The adenovirus capsid protein hexon contains a highly conserved human CD4+ T-cell epitope." *Hum Gene Ther* **13**(10): 1167–78.

Onion, D., L. J. Crompton, D. W. Milligan, P. A. Moss, S. P. Lee and V. Mautner (2007). "The CD4+ T-cell response to adenovirus is focused against conserved residues within the hexon protein." *J Gen Virol* **88**(Pt 9): 2417–25.

Oosterom-Dragon, E. A. and H. S. Ginsberg (1981). "Characterization of two temperature-sensitive mutants of type 5 adenovirus with mutations in the 100,000-dalton protein gene." *J Virol* **40**(2): 491–500.

Orazio, N. I., C. M. Naeger, J. Karlseder and M. D. Weitzman (2011). "The adenovirus E1b55K/E4orf6 complex induces degradation of the Bloom helicase during infection." *J Virol* **85**(4): 1887–92.

Ornelles, D. and T. Shenk (1991). "Location of the adenovirus early region 1B 55 kilodalton protein during lytic infection: association with nuclear viral inclusions requires the early region 4 34 kilodalton protein." *J Virol* **65**: 424–439.

Orvedahl, A. and B. Levine (2009). "Autophagy in Mammalian antiviral immunity." *Curr Top Microbiol Immunol* **335**: 267–85.

Osborne, T. F., R. B. Gaynor and A. J. Berk (1982). "The TATA homology and the mRNA 5′ untranslated sequence are not required for expression of essential adenovirus E1A functions." *Cell* **29**: 139–148.

Osley, M. A. (2004). "H2B ubiquitylation: the end is in sight." *Biochim Biophys Acta* **1677**(1–3): 74–8.

Ostapchuk, P., M. Almond and P. Hearing (2011). "Characterization of empty adenovirus particles assembled in the absence of a functional adenovirus IVa2 protein." *J Virol* **85**(11): 5524–31.

Ostapchuk, P., M. E. Anderson, S. Chandrasekhar and P. Hearing (2006). "The L4 22-kilodalton protein plays a role in packaging of the adenovirus genome." *J Virol* **80**(14): 6973–81.

Ostapchuk, P. and P. Hearing (2005). "Control of adenovirus packaging." *J Cell Biochem* **96**(1): 25–35.

Ostapchuk, P. and P. Hearing (2008). "Adenovirus IVa2 protein binds ATP." *J Virol* **82**(20): 10290–4.

Ostapchuk, P., J. Yang, E. Auffarth and P. Hearing (2005). "Functional interaction of the adenovirus IVa2 protein with adenovirus type 5 packaging sequences." *J Virol* **79**(5): 2831–8.

Ostberg, S., H. Tormanen Persson and G. Akusjarvi (2012). "Serine 192 in the tiny RS repeat of the adenoviral L4-33K splicing enhancer protein is essential for function and reorganization of the protein to the periphery of viral replication centers." *Virology* **433**(2): 273–81.

Ou, H. D., W. Kwiatkowski, T. J. Deerinck, A. Noske, K. Y. Blain, H. S. Land, C. Soria, C. J. Powers, A. P. May, X. Shu, R. Y. Tsien, J. A. Fitzpatrick, J. A. Long, M. H. Ellisman, S. Choe and C. C. O'Shea (2012). "A structural basis for the assembly and functions of a viral polymer that inactivates multiple tumor suppressors." *Cell* **151**(2): 304–19.

Paabo, S., F. Weber, T. Nilsson, W. Schaffner and P. A. Peterson (1986). "Structural and functional dissection of an MHC class I antigen-binding adenovirus glycoprotein." *EMBO J* **5**(8): 1921–7.

Pache, L., S. Venkataraman, G. R. Nemerow and V. S. Reddy (2008). "Conservation of fiber structure and CD46 usage by subgroup B2 adenoviruses." *Virology* **375**(2): 573–9.

Pache, L., S. Venkataraman, V. S. Reddy and G. R. Nemerow (2008). "Structural variations in species B adenovirus fibers impact CD46 association." *J Virol* **82**(16): 7923–31.

Pahl, J. H., D. H. Verhoeven, K. M. Kwappenberg, J. Vellinga, A. C. Lankester, M. J. van Tol and M. W. Schilham (2012). "Adenovirus type 35, but not type 5, stimulates NK cell activation via plasmacytoid dendritic cells and TLR9 signaling." *Mol Immunol* **51**(1): 91–100.

Painter, W., A. Robertson, L. C. Trost, S. Godkin, B. Lampert and G. Painter (2012). "First pharmacokinetic and safety study in humans of the novel lipid antiviral conjugate

CMX001, a broad-spectrum oral drug active against double-stranded DNA viruses." *Antimicrob Agents Chemother* **56**(5): 2726–34.

Palazzo, A. F. and A. Akef (2012). "Nuclear export as a key arbiter of "mRNA identity" in eukaryotes." *Biochim Biophys Acta* **1819**(6): 566–77.

Palmer, D. and P. Ng (2003). "Improved system for helper-dependent adenoviral vector production." *Mol Ther* **8**(5): 846–52.

Pang, Y. P., K. Xu, T. M. Kollmeyer, E. Perola, W. J. McGrath, D. T. Green and W. F. Mangel (2001). "Discovery of a new inhibitor lead of adenovirus proteinase: steps toward selective, irreversible inhibitors of cysteine proteinases." *FEBS Lett* **502**(3): 93–7.

Pante, N. and M. Kann (2002). "Nuclear pore complex is able to transport macromolecules with diameters of about 39 nm." *Mol Biol Cell* **13**(2): 425–34.

Panto, L., Podgorski, II, M. Janoska, O. Marko and B. Harrach (2015). "Taxonomy proposal for Old World monkey adenoviruses: characterisation of several non-human, non-ape primate adenovirus lineages." *Arch Virol* **160**(12): 3165–77.

Papadopoulou, A., U. Gerdemann, U. L. Katari, I. Tzannou, H. Liu, C. Martinez, K. Leung, G. Carrum, A. P. Gee, J. F. Vera, R. A. Krance, M. K. Brenner, C. M. Rooney, H. E. Heslop and A. M. Leen (2014). "Activity of broad-spectrum T cells as treatment for AdV, EBV, CMV, BKV, and HHV6 infections after HSCT." *Sci Transl Med* **6**(242): 242ra83.

Pardo-Mateos, A. and C. S. Young (2004). "Adenovirus IVa2 protein plays an important role in transcription from the major late promoter in vivo." *Virology* **327**(1): 50–59.

Parker, A. L., S. N. Waddington, C. G. Nicol, D. M. Shayakhmetov, S. M. Buckley, L. Denby, G. Kemball-Cook, S. Ni, A. Lieber, J. H. McVey, S. A. Nicklin and A. H. Baker (2006). "Multiple vitamin K-dependent coagulation zymogens promote adenovirus-mediated gene delivery to hepatocytes." *Blood* **108**(8): 2554–61.

Parker, E. J., C. H. Botting, A. Webster and R. T. Hay (1998). "Adenovirus DNA polymerase: domain organisation and interaction with preterminal protein." *Nucleic Acids Res* **26**(5): 1240–7.

Parks, R. J., J. L. Bramson, Y. Wan, C. L. Addison and F. L. Graham (1999). "Effects of stuffer DNA on transgene expression from helper-dependent adenovirus vectors." *J Virol* **73**(10): 8027–34.

Parks, R. J., L. Chen, M. Anton, U. Sankar, M. A. Rudnicki and F. L. Graham (1996). "A helper-dependent adenovirus vector system: removal of helper virus by Cre-mediated excision of the viral packaging signal." *Proc Natl Acad Sci U S A* **93**(24): 13565–70.

Patel, S. S., B. J. Belmont, J. M. Sante and M. F. Rexach (2007). "Natively unfolded nucleoporins gate protein diffusion across the nuclear pore complex." *Cell* **129**(1): 83–96.

Patil, A. S., R. B. Sable and R. M. Kothari (2012). "Occurrence, biochemical profile of vascular endothelial growth factor (VEGF) isoforms and their functions in endochondral ossification." *J Cell Physiol* **227**(4): 1298–308.

Patsalo, V., M. A. Yondola, B. Luan, I. Shoshani, C. Kisker, D. F. Green, D. P. Raleigh and P. Hearing (2012). "Biophysical and functional analyses suggest that adenovirus

E4-ORF3 protein requires higher-order multimerization to function against promyelocytic leukemia protein nuclear bodies." *J Biol Chem* **287**(27): 22573–83.

Patterson, S. and W. C. Russell (1983). "Ultrastructural and immunofluorescence studies of early events in adenovirus-HeLa cell interactions." *J Gen Virol* **64**(Pt 5): 1091–9.

Paull, T. T. (2015). "Mechanisms of ATM Activation." *Annu Rev Biochem* **84**: 711–38.

Payet, V., C. Arnauld, J. P. Picault, A. Jestin and P. Langlois (1998). "Transcriptional organization of the avian adenovirus CELO." *J Virol* **72**(11): 9278–85.

Paz, I., M. Sachse, N. Dupont, J. Mounier, C. Cederfur, J. Enninga, H. Leffler, F. Poirier, M. C. Prevost, F. Lafont and P. Sansonetti (2010). "Galectin-3, a marker for vacuole lysis by invasive pathogens." *Cell Microbiol* **12**(4): 530–44.

Pecenkova, T. and V. Paces (1999). "Molecular phylogeny of phi29-like phages and their evolutionary relatedness to other protein-primed replicating phages and other phages hosted by gram-positive bacteria." *J Mol Evol* **48**(2): 197–208.

Pelka, P., J. N. Ablack, G. J. Fonseca, A. F. Yousef and J. S. Mymryk (2008). "Intrinsic structural disorder in adenovirus E1A: a viral molecular hub linking multiple diverse processes." *J Virol* **82**(15): 7252–63.

Pelka, P., J. N. Ablack, M. Shuen, A. F. Yousef, M. Rasti, R. J. Grand, A. S. Turnell and J. S. Mymryk (2009). "Identification of a second independent binding site for the pCAF acetyltransferase in adenovirus E1A." *Virology* **391**(1): 90–8.

Pelka, P., J. N. Ablack, J. Torchia, A. S. Turnell, R. J. Grand and J. S. Mymryk (2009). "Transcriptional control by adenovirus E1A conserved region 3 via p300/CBP." *Nucleic Acids Res* **37**(4): 1095–106.

Peng, Y., E. Falck-Pedersen and K. B. Elkon (2001). "Variation in adenovirus transgene expression between BALB/c and C57BL/6 mice is associated with differences in interleukin-12 and gamma interferon production and NK cell activation." *J Virol* **75**(10): 4540–50.

Pennella, M. A., Y. Liu, J. L. Woo, C. A. Kim and A. J. Berk (2010). "Adenovirus E1B 55-kilodalton protein is a p53-SUMO1 E3 ligase that represses p53 and stimulates its nuclear export through interactions with promyelocytic leukemia nuclear bodies." *J Virol* **84**(23): 12210–25.

Pereira, H. G. (1958). "A protein factor responsible for the early cytopathic effect of adenoviruses." *Virology* **6**(3): 601–11.

Perez, D. and E. White (2000). "TNF-alpha signals apoptosis through a bid-dependent conformational change in Bax that is inhibited by E1B 19K." *Mol Cell* **6**(1): 53–63.

Perez-Berna, A. J., W. F. Mangel, W. J. McGrath, V. Graziano, J. Flint and C. San Martin (2014). "Processing of the L1 52/55k protein by the adenovirus protease: a new substrate and new insights into virion maturation." *J Virol* **88**(3): 1513–24.

Perez-Berna, A. J., R. Marabini, S. H. Scheres, R. Menendez-Conejero, I. P. Dmitriev, D. T. Curiel, W. F. Mangel, S. J. Flint and C. San Martin (2009). "Structure and uncoating of immature adenovirus." *J Mol Biol* **392**(2): 547–57.

Perez-Berna, A. J., A. Ortega-Esteban, R. Menendez-Conejero, D. C. Winkler, M. Menendez, A. C. Steven, S. J. Flint, P. J. de Pablo and C. San Martin (2012).

"The role of capsid maturation on adenovirus priming for sequential uncoating." *J Biol Chem* **287**(37): 31582–95.

Perez-Romero, P., K. E. Gustin and M. J. Imperiale (2006). "Dependence of the encapsidation function of the adenovirus L1 52/55-kilodalton protein on its ability to bind the packaging sequence." *J Virol* **80**(4): 1965–71.

Perez-Romero, P., R. E. Tyler, J. R. Abend, M. Dus and M. J. Imperiale (2005). "Analysis of the interaction of the adenovirus L1 52/55-kilodalton and IVa2 proteins with the packaging sequence in vivo and in vitro." *J Virol* **79**(4): 2366–74.

Perez-Vargas, J., R. C. Vaughan, C. Houser, K. M. Hastie, C. C. Kao and G. R. Nemerow (2014). "Isolation and characterization of the DNA and protein binding activities of adenovirus core protein V." *J Virol* **88**(16): 9287–96.

Perreau, M., G. Pantaleo and E. J. Kremer (2008). "Activation of a dendritic cell-T cell axis by Ad5 immune complexes creates an improved environment for replication of HIV in T cells." *J Exp Med* **205**(12): 2717–25.

Perricaudet, M., G. Akusjärvi, A. Virtanen and U. Pettersson (1979). "Structure of two spliced mRNAs from the transforming region of human subgroup C adenoviruses." *Nature* **281**: 694–696.

Persson, B. D., D. M. Reiter, M. Marttila, Y. F. Mei, J. M. Casasnovas, N. Arnberg and T. Stehle (2007). "Adenovirus type 11 binding alters the conformation of its receptor CD46." *Nat Struct Mol Biol* **14**(2): 164–6.

Pesonen, S., I. Diaconu, V. Cerullo, S. Escutenaire, M. Raki, L. Kangasniemi, P. Nokisalmi, G. Dotti, K. Guse, L. Laasonen, K. Partanen, E. Karli, E. Haavisto, M. Oksanen, A. Karioja-Kallio, P. Hannuksela, S. L. Holm, S. Kauppinen, T. Joensuu, A. Kanerva and A. Hemminki (2012). "Integrin targeted oncolytic adenoviruses Ad5-D24-RGD and Ad5-RGD-D24-GMCSF for treatment of patients with advanced chemotherapy refractory solid tumors." *Int J Cancer* **130**(8): 1937–47.

Peters, L. and G. Meister (2007). "Argonaute proteins: mediators of RNA silencing." *Mol Cell* **26**(5): 611–23.

Peters, W., J. R. Brandl, J. D. Lindbloom, C. J. Martinez, C. D. Scallan, G. R. Trager, D. W. Tingley, M. L. Kabongo and S. N. Tucker (2013). "Oral administration of an adenovirus vector encoding both an avian influenza A hemagglutinin and a TLR3 ligand induces antigen specific granzyme B and IFN-gamma T cell responses in humans." *Vaccine* **31**(13): 1752–8.

Pettersson, U. and L. Philipson (1975). "Location of sequences on the adenovirus genome coding for the 5.5S RNA." *Cell* **6**: 1–4.

Pettit, S. C., M. S. Horwitz and J. A. Engler (1989). "Mutations of the precursor to the terminal protein of adenovirus serotypes 2 and 5." *J Virol* **63**(12): 5244–50.

Phelps, W. C., S. Bagchi, J. A. Barnes, P. Raychaudhuri, V. Kraus, K. Munger, P. M. Howley and J. R. Nevins (1991). "Analysis of trans activation by human papillomavirus type 16 E7 and adenovirus 12S E1A suggests a common mechanism." *J Virol* **65**(12): 6922–30.

Philipson, L. (1967). "Attachment and eclipse of adenovirus." *J Virol* **1**(5): 868–75.

Philipson, L., K. Lonberg-Holm and U. Pettersson (1968). "Virus-receptor interaction in an adenovirus system." *J Virol* **2**(10): 1064–75.

Piao, Y., H. Jiang, R. Alemany, V. Krasnykh, F. C. Marini, J. Xu, M. M. Alonso, C. A. Conrad, K. D. Aldape, C. Gomez-Manzano and J. Fueyo (2009). "Oncolytic adenovirus retargeted to Delta-EGFR induces selective antiglioma activity." *Cancer Gene Ther* **16**(3): 256–65.

Pichla-Gollon, S. L., S. W. Lin, S. E. Hensley, M. O. Lasaro, L. Herkenhoff-Haut, M. Drinker, N. Tatsis, G. P. Gao, J. M. Wilson, H. C. Ertl and J. M. Bergelson (2009). "Effect of preexisting immunity on an adenovirus vaccine vector: in vitro neutralization assays fail to predict inhibition by antiviral antibody in vivo." *J Virol* **83**(11): 5567–73.

Pickles, R. J., D. McCarty, H. Matsui, P. J. Hart, S. H. Randell and R. C. Boucher (1998). "Limited entry of adenovirus vectors into well-differentiated airway epithelium is responsible for inefficient gene transfer." *J Virol* **72**(7): 6014–23.

Pilder, S., M. Moore, J. Logan and T. Shenk (1986). "The adenovirus E1B-55kd transforming polypeptide modulates transport or cytoplasmic stablization of viral and host cell mRNAs." *Mol Cell Biol* **6**: 470–476.

Pina, M. and M. Green (1969). "Biochemical studies on adenovirus multiplication. XIV. Macromolecule and enzyme synthesis in cells replicating oncogenic and nononcogenic human adenovirus." *Virology* **38**(4): 573–86.

Pincus, S., W. Robertson and D. Rekosh (1981). "Characterization of the effect of aphidicolin on adenovirus DNA replication: evidence in support of a protein primer model of initiation." *Nucleic Acids Res* **9**(19): 4919–38.

Pine, S. O., J. G. Kublin, S. M. Hammer, J. Borgerding, Y. Huang, D. R. Casimiro and M. J. McElrath (2011). "Pre-existing adenovirus immunity modifies a complex mixed Th1 and Th2 cytokine response to an Ad5/HIV-1 vaccine candidate in humans." *PLoS One* **6**(4): e18526.

Piya, S., E. J. White, S. R. Klein, H. Jiang, T. J. McDonnell, C. Gomez-Manzano and J. Fueyo (2011). "The E1B19K oncoprotein complexes with Beclin 1 to regulate autophagy in adenovirus-infected cells." *PLoS One* **6**(12): e29467.

Poderycki, M. J., V. A. Kickhoefer, C. S. Kaddis, S. Raval-Fernandes, E. Johansson, J. I. Zink, J. A. Loo and L. H. Rome (2006). "The vault exterior shell is a dynamic structure that allows incorporation of vault-associated proteins into its interior." *Biochemistry* **45**(39): 12184–93.

Pombo, A., J. Ferreira, E. Bridge and M. Carmo-Fonseca (1994). "Adenovirus replication and transcription sites are spatially separated in the nucleus of infected cells." *EMBO J* **13**(21): 5075–85.

Poulin, K. L., R. M. Lanthier, A. C. Smith, C. Christou, M. Risco Quiroz, K. L. Powell, R. W. O'Meara, R. Kothary, I. A. Lorimer and R. J. Parks (2010). "Retargeting of adenovirus vectors through genetic fusion of a single-chain or single-domain antibody to capsid protein IX." *J Virol* **84**(19): 10074–86.

Pozzuto, T., C. Roger, J. Kurreck and H. Fechner (2015). "Enhanced suppression of adenovirus replication by triple combination of anti-adenoviral siRNAs, soluble adenovirus receptor trap sCAR-Fc and cidofovir." *Antiviral Res* **120**: 72–78.

Prage, L., U. Pettersson, S. Hoglund, K. Lonberg-Holm and L. Philipson (1970). "Structural proteins of adenoviruses. IV. Sequential degradation of the adenovirus type 2 virion." *Virology* **42**(2): 341–58.

Prescott, J. and E. Falck-Pedersen (1994). "Sequence elements upstream of the 3′ cleavage site confer substrate strength to the adenovirus L1 and L3 polyadenylation sites." *Mol Cell Biol* **14**(7): 4682–93.

Prescott, J. C., L. Liu and E. Falck-Pedersen (1997). "Sequence-mediated regulation of adenovirus gene expression by repression of mRNA accumulation." *Mol Cell Biol* **17**(4): 2207–16.

Price, R. and S. Penman (1972). "A distinct RNA polymerase activity, synthesizing 5–5 s, 5 s and 4 s RNA in nuclei from adenovirus 2-infected HeLa cells." *J Mol Biol* **70**(3): 435–50.

Priddy, F. H., D. Brown, J. Kublin, K. Monahan, D. P. Wright, J. Lalezari, S. Santiago, M. Marmor, M. Lally, R. M. Novak, S. J. Brown, P. Kulkarni, S. A. Dubey, L. S. Kierstead, D. R. Casimiro, R. Mogg, M. J. DiNubile, J. W. Shiver, R. Y. Leavitt, M. N. Robertson, D. V. Mehrotra, E. Quirk and V. S. G. Merck (2008). "Safety and immunogenicity of a replication-incompetent adenovirus type 5 HIV-1 clade B gag/pol/nef vaccine in healthy adults." *Clin Infect Dis* **46**(11): 1769–81.

Proffitt, J. A. and G. E. Blair (1997). "The MHC-encoded TAP1/LMP2 bidirectional promoter is down-regulated in highly oncogenic adenovirus type 12 transformed cells." *FEBS Lett* **400**(2): 141–4.

Pronk, R. and P. C. van der Vliet (1993). "The adenovirus terminal protein influences binding of replication proteins and changes the origin structure." *Nucleic Acids Res* **21**(10): 2293–300.

Pronk, R., W. Van Driel and P. C. Van der Vliet (1994). "Replication of adenovirus DNA in vitro is ATP-independent." *FEBS Lett* **337**(1): 33–8.

Pruzan, R., P. K. Chatterjee and S. J. Flint (1992). "Specific transcription from the adenovirus E2E promoter by RNA polymerase III requires a subpopulation of TFIID." *Nucleic Acids Res.* **20**: 5705–5712.

Pruzan, R. and S. J. Flint (1995). "Transcription of adenovirus RNA polymerase III genes." *Curr. Top. Microbiol. Immunol.* **199**: 201–226.

Puig-Saus, C., A. Gros, R. Alemany and M. Cascallo (2012). "Adenovirus i-leader truncation bioselected against cancer-associated fibroblasts to overcome tumor stromal barriers." *Mol Ther* **20**(1): 54–62.

Puig-Saus, C., L. A. Rojas, E. Laborda, A. Figueras, R. Alba, C. Fillat and R. Alemany (2014). "iRGD tumor-penetrating peptide-modified oncolytic adenovirus shows enhanced tumor transduction, intratumoral dissemination and antitumor efficacy." *Gene Ther* **21**(8): 767–74.

Punga, T. and G. Akusjarvi (2000). "The adenovirus-2 E1B-55K protein interacts with a mSin3A/histone deacetylase 1 complex." *FEBS Lett* **476**(3): 248–52.

Puntener, D., M. F. Engelke, Z. Ruzsics, S. Strunze, C. Wilhelm and U. F. Greber (2011). "Stepwise loss of fluorescent core protein V from human adenovirus during entry into cells." *J Virol* **85**(1): 481–96.

Putzer, B. M., M. Hitt, W. J. Muller, P. Emtage, J. Gauldie and F. L. Graham (1997). "Interleukin 12 and B7-1 costimulatory molecule expressed by an adenovirus vector act synergistically to facilitate tumor regression." *Proc Natl Acad Sci U S A* **94**(20): 10889–94.

Puvion-Dutilleul, F., M. K. Chelbi-Alix, M. Koken, F. Quignon, E. Puvion and H. de The (1995). "Adenovirus infection induces rearrangements in the intranuclear distribution of the nuclear body-associated PML protein." *Exp Cell Res* **218**(1): 9–16.

Puvion-Dutilleul, F. and E. Puvion (1990). "Analysis by in situ hybridization and autoradiography of sites of replication and storage of single- and double-stranded adenovirus type 5 DNA in lytically infected HeLa cells." *J Struct Biol* **103**(3): 280–9.

Puvion-Dutilleul, F. and E. Puvion (1991). "Sites of transcription of adenovirus type 5 genomes in relation to early viral DNA replication in infected HeLa cells. A high resolution in situ hybridization and autoradiographical study." *Biol Cell* **71**(1–2): 135–47.

Puvion-Dutilleul, F., R. Roussev and E. Puvion (1992). "Distribution of viral RNA molecules during the adenovirus type 5 infectious cycle in HeLa cells." *J Struct Biol* **108**(3): 209–20.

Pyronnet, S. (2000). "Phosphorylation of the cap-binding protein eIF4E by the MAPK-activated protein kinase Mnk1." *Biochem Pharmacol* **60**(8): 1237–43.

Qiu, Q., Z. Xu, J. Tian, R. Moitra, S. Gunti, A. L. Notkins and A. P. Byrnes (2015). "Impact of natural IgM concentration on gene therapy with adenovirus type 5 vectors." *J Virol* **89**(6): 3412–6.

Qualmann, B., M. M. Kessels and R. B. Kelly (2000). "Molecular links between endocytosis and the actin cytoskeleton." *J Cell Biol* **150**(5): F111–6.

Querido, E., P. Blanchette, Q. Yan, T. Kamura, M. Morrison, D. Boivin, W. G. Kaelin, R. C. Conaway, J. W. Conaway and P. E. Branton (2001). "Degradation of p53 by adenovirus E4orf6 and E1B55K proteins occurs via a novel mechanism involving a Cullin-containing complex." *Genes Dev* **15**(23): 3104–17.

Querido, E., R. C. Marcellus, A. Lai, R. Charbonneau, J. G. Teodoro, G. Ketner and P. E. Branton (1997). "Regulation of p53 levels by the E1B 55-kilodalton protein and E4orf6 in adenovirus-infected cells." *J. Virol.* **71**(5): 3788–3798.

Querido, E., M. R. Morrison, H. Chu-Pham-Dang, S. W. Thirlwell, D. Boivin and P. E. Branton (2001). "Identification of three functions of the adenovirus e4orf6 protein that mediate p53 degradation by the E4orf6-E1B55K complex." *J Virol* **75**(2): 699–709.

Querido, E., J. G. Teodoro and P. E. Branton (1997). "Accumulation of p53 induced by the adenovirus E1A protein requires regions involved in the stimulation of DNA synthesis." *J Virol* **71**(5): 3526–33.

**Quinn, K. M., A. Da Costa, A. Yamamoto, D. Berry, R. W. Lindsay, P. A. Darrah, L. Wang, C. Cheng, W. P. Kong, J. G. Gall, A. Nicosia, A. Folgori, S. Colloca, R. Cortese, E. Gostick, D. A. Price, C. E. Gomez, M. Esteban, L. S. Wyatt,

B. Moss, C. Morgan, M. Roederer, R. T. Bailer, G. J. Nabel, R. A. Koup and R. A. Seder (2013). "Comparative analysis of the magnitude, quality, phenotype, and protective capacity of simian immunodeficiency virus gag-specific CD8+ T cells following human-, simian-, and chimpanzee-derived recombinant adenoviral vector immunization." *J Immunol* **190**(6): 2720–35.

Rabbitts, T. H. (1978). "Evidence for splicing of interrupted immunoglobulin variable and constant region sequences in nuclear RNA." *Nature* **275**(5678): 291–6.

Radin, J. M., A. W. Hawksworth, P. J. Blair, D. J. Faix, R. Raman, K. L. Russell and G. C. Gray (2014). "Dramatic decline of respiratory illness among US military recruits after the renewed use of adenovirus vaccines." *Clin Infect Dis* **59**(7): 962–8.

Radke, J. R., F. Grigera, D. S. Ucker and J. L. Cook (2014). "Adenovirus E1B 19-kilodalton protein modulates innate immunity through apoptotic mimicry." *J Virol* **88**(5): 2658–69.

Radko, S., M. Koleva, K. M. James, R. Jung, J. S. Mymryk and P. Pelka (2014). "Adenovirus E1A targets the DREF nuclear factor to regulate virus gene expression, DNA replication, and growth." *J Virol* **88**(22): 13469–81.

Radosevic, K., C. W. Wieland, A. Rodriguez, G. J. Weverling, R. Mintardjo, G. Gillissen, R. Vogels, Y. A. Skeiky, D. M. Hone, J. C. Sadoff, T. van der Poll, M. Havenga and J. Goudsmit (2007). "Protective immune responses to a recombinant adenovirus type 35 tuberculosis vaccine in two mouse strains: CD4 and CD8 T-cell epitope mapping and role of gamma interferon." *Infect Immun* **75**(8): 4105–15.

Radtke, K., D. Kieneke, A. Wolfstein, K. Michael, W. Steffen, T. Scholz, A. Karger and B. Sodeik (2010). "Plus- and minus-end directed microtubule motors bind simultaneously to herpes simplex virus capsids using different inner tegument structures." *PLoS Pathog* **6**(7): e1000991.

Rahman, S., I. Magalhaes, J. Rahman, R. K. Ahmed, D. R. Sizemore, C. A. Scanga, F. Weichold, F. Verreck, I. Kondova, J. Sadoff, R. Thorstensson, M. Spangberg, M. Svensson, J. Andersson, M. Maeurer and S. Brighenti (2012). "Prime-boost vaccination with rBCG/rAd35 enhances CD8(+) cytolytic T-cell responses in lesions from Mycobacterium tuberculosis-infected primates." *Mol Med* **18**: 647–58.

Rajsbaum, R., A. Garcia-Sastre and G. A. Versteeg (2014). "TRIMmunity: the roles of the TRIM E3-ubiquitin ligase family in innate antiviral immunity." *J Mol Biol* **426**(6): 1265–84.

Raman, S., T. H. Hsu, S. L. Ashley and K. R. Spindler (2009). "Usage of integrin and heparan sulfate as receptors for mouse adenovirus type 1." *J Virol* **83**(7): 2831–8.

Rampling, T., K. Ewer, G. Bowyer, D. Wright, E. B. Imoukhuede, R. Payne, F. Hartnell, M. Gibani, C. Bliss, A. Minhinnick, M. Wilkie, N. Venkatraman, I. Poulton, N. Lella, R. Roberts, K. Sierra-Davidson, V. Krahling, E. Berrie, F. Roman, I. De Ryck, A. Nicosia, N. J. Sullivan, D. A. Stanley, J. E. Ledgerwood, R. M. Schwartz, L. Siani, S. Colloca, A. Folgori, S. Di Marco, R. Cortese, S. Becker, B. S. Graham, R. A. Koup, M. M. Levine, V. Moorthy, A. J. Pollard, S. J. Draper, W. R. Ballou, A. Lawrie, S. C. Gilbert and A. V. Hill (2015).

"A Monovalent Chimpanzee Adenovirus Ebola Vaccine — Preliminary Report." *N Engl J Med.*

Rancourt, C., H. Keyvani-Amineh, S. Sircar, P. Labrecque and J. M. Weber (1995). "Proline 137 is critical for adenovirus protease encapsidation and activation but not enzyme activity." *Virology* **209**(1): 167–73.

Rancourt, C., K. Tihanyi, M. Bourbonniere and J. M. Weber (1994). "Identification of active-site residues of the adenovirus endopeptidase." *Proc Natl Acad Sci U S A* **91**(3): 844–7.

Rao, L., M. Debbas, P. Sabbatini, D. Hockenbery, S. Korsmeyer and E. White (1992). "The adenovirus E1A proteins induce apoptosis, which is inhibited by the E1B 19-kDa and Bcl-2 proteins." *Proc. Natl. Acad. Sci. U S A* **89**(16): 7742–7746.

Rao, V. B. and M. Feiss (2008). "The bacteriophage DNA packaging motor." *Annu Rev Genet* **42**: 647–81.

Raper, S. E., N. Chirmule, F. S. Lee, N. A. Wivel, A. Bagg, G. P. Gao, J. M. Wilson and M. L. Batshaw (2003). "Fatal systemic inflammatory response syndrome in a ornithine transcarbamylase deficient patient following adenoviral gene transfer." *Mol Genet Metab* **80**(1–2): 148–58.

Raper, S. E., Z. J. Haskal, X. Ye, C. Pugh, E. E. Furth, G. P. Gao and J. M. Wilson (1998). "Selective gene transfer into the liver of non-human primates with E1-deleted, E2A-defective, or E1-E4 deleted recombinant adenoviruses." *Hum Gene Ther* **9**(5): 671–9.

Raper, S. E., M. Yudkoff, N. Chirmule, G. P. Gao, F. Nunes, Z. J. Haskal, E. E. Furth, K. J. Propert, M. B. Robinson, S. Magosin, H. Simoes, L. Speicher, J. Hughes, J. Tazelaar, N. A. Wivel, J. M. Wilson and M. L. Batshaw (2002). "A pilot study of in vivo liver-directed gene transfer with an adenoviral vector in partial ornithine transcarbamylase deficiency." *Hum Gene Ther* **13**(1): 163–75.

Rasti, M., R. J. Grand, A. F. Yousef, M. Shuen, J. S. Mymryk, P. H. Gallimore and A. S. Turnell (2006). "Roles for APIS and the 20S proteasome in adenovirus E1A-dependent transcription." *EMBO J* **25**(12): 2710–22.

Rauma, T., J. Tuukkanen, J. M. Bergelson, G. Denning and T. Hautala (1999). "rab5 GTPase regulates adenovirus endocytosis." *J Virol* **73**(11): 9664–8.

Rawlins, D. R., P. J. Rosenfeld, R. J. Wides, M. D. Challberg and T. Kelly, Jr. (1984). "Structure and function of the adenovirus origin of replication." *Cell* **37**: 309–319.

Reach, M., L. E. Babiss and C. S. H. Young (1990). "The upstream factor-binding site is not essential for activation of transcription from the adenovirus major late promoter." *J. Virol.* **64**: 5851–5860.

Reach, M., L. X. Xu and C. S. Young (1991). "Transcription from the adenovirus major late promoter uses redundant activating elements." *Embo J* **10**(11): 3439–46.

Reddy, P. S., S. Ganesh and D. C. Yu (2006). "Enhanced gene transfer and oncolysis of head and neck cancer and melanoma cells by fiber chimeric oncolytic adenoviruses." *Clin Cancer Res* **12**(9): 2869–78.

Reddy, V. S., S. K. Natchiar, L. Gritton, T. M. Mullen, P. L. Stewart and G. R. Nemerow (2010). "Crystallization and preliminary X-ray diffraction analysis of human adenovirus." *Virology* **402**(1): 209–14.

Reddy, V. S., S. K. Natchiar, P. L. Stewart and G. R. Nemerow (2010). "Crystal structure of human adenovirus at 3.5 A resolution." *Science* **329**(5995): 1071–5.

Reddy, V. S. and G. R. Nemerow (2014). "Reply to Campos: Revised structures of adenovirus cement proteins represent a consensus model for understanding virus assembly and disassembly." *Proc Natl Acad Sci U S A* **111**(43): E4544–5.

Reddy, V. S. and G. R. Nemerow (2014). "Structures and organization of adenovirus cement proteins provide insights into the role of capsid maturation in virus entry and infection." *Proc Natl Acad Sci U S A* **111**(32): 11715–20.

Reed, R. and H. Cheng (2005). "TREX, SR proteins and export of mRNA." *Curr Opin Cell Biol* **17**(3): 269–73.

Reed, S. I. (1997). "Control of the G1/S transition." *Cancer Surv* **29**: 7–23.

Regad, T. and M. K. Chelbi-Alix (2001). "Role and fate of PML nuclear bodies in response to interferon and viral infections." *Oncogene* **20**(49): 7274–86.

Reich, N., R. Pine, D. Levy and J. E. Darnell, Jr. (1988). "Transcription of interferon-stimulated genes is induced by adenovirus particles but is suppressed by E1A gene products." *J Virol* **62**(1): 114–9.

Reich, P. R., B. G. Forget and S. M. Weissman (1966). "RNA of low molecular weight in KB cells infected with adenovirus type 2." *J Mol Biol* **17**(2): 428–39.

Reichel, P. A., W. C. Merrick, J. Siekierka and M. B. Mathews (1985). "Regulation of a protein synthesis initiation factor by adenovirus virus-associated RNA." *Nature* **313**(5999): 196–200.

Reid, T., R. Warren and D. Kirn (2002). "Intravascular adenoviral agents in cancer patients: lessons from clinical trials." *Cancer Gene Ther* **9**(12): 979–86.

Rekosh, D., W. C. Russell, A. Bellett and A. J. Robinson (1977). "Identification of a protein linked to the ends of adenovirus DNA." *Cell* **11**: 283–0.

Reyes-Sandoval, A., T. Berthoud, N. Alder, L. Siani, S. C. Gilbert, A. Nicosia, S. Colloca, R. Cortese and A. V. Hill (2010). "Prime-boost immunization with adenoviral and modified vaccinia virus Ankara vectors enhances the durability and polyfunctionality of protective malaria CD8+ T-cell responses." *Infect Immun* **78**(1): 145–53.

Reyes-Sandoval, A., S. Sridhar, T. Berthoud, A. C. Moore, J. T. Harty, S. C. Gilbert, G. Gao, H. C. Ertl, J. C. Wilson and A. V. Hill (2008). "Single-dose immunogenicity and protective efficacy of simian adenoviral vectors against Plasmodium berghei." *Eur J Immunol* **38**(3): 732–41.

Ribbeck, K. and D. Gorlich (2001). "Kinetic analysis of translocation through nuclear pore complexes." *EMBO J* **20**(6): 1320–30.

Rice, S. A. and D. F. Klessig (1984). "The function(s) provided by the adenovirus-specified, DNA-binding protein required for viral late gene expression is independent of the role of the protein in viral DNA replication." *J Virol* **49**(1): 35–49.

Riordan, J. R., J. M. Rommens, B. Kerem, N. Alon, R. Rozmahel, Z. Grzelczak, J. Zielenski, S. Lok, N. Plavsic, J. L. Chou and et al. (1989). "Identification of the cystic fibrosis gene: cloning and characterization of complementary DNA." *Science* **245**(4922): 1066–73.

Rivera, A. A., J. Davydova, S. Schierer, M. Wang, V. Krasnykh, M. Yamamoto, D. T. Curiel and D. M. Nettelbeck (2004). "Combining high selectivity of replication with fiber chimerism for effective adenoviral oncolysis of CAR-negative melanoma cells." *Gene Ther* **11**(23): 1694–702.

Roberts, D. M., A. Nanda, M. J. Havenga, P. Abbink, D. M. Lynch, B. A. Ewald, J. Liu, A. R. Thorner, P. E. Swanson, D. A. Gorgone, M. A. Lifton, A. A. Lemckert, L. Holterman, B. Chen, A. Dilraj, A. Carville, K. G. Mansfield, J. Goudsmit and D. H. Barouch (2006). "Hexon-chimaeric adenovirus serotype 5 vectors circumvent pre-existing anti-vector immunity." *Nature* **441**(7090): 239–43.

Roberts, M. M., J. L. White, M. G. Grutter and R. M. Burnett (1986). "Three-dimensional structure of the adenovirus major coat protein hexon." *Science* **232**(4754): 1148–51.

Robinson, A. J., H. B. Younghusband and A. J. D. Bellett (1973). "A circular DNA-protein complex from adenoviruses." *Virology* **56**: 54–69.

Robinson, C. M., G. Singh, C. Henquell, M. P. Walsh, H. Peigue-Lafeuille, D. Seto, M. S. Jones, D. W. Dyer and J. Chodosh (2011). "Computational analysis and identification of an emergent human adenovirus pathogen implicated in a respiratory fatality." *Virology* **409**(2): 141–7.

Robinson, C. M., G. Singh, J. Y. Lee, S. Dehghan, J. Rajaiya, E. B. Liu, M. A. Yousuf, R. A. Betensky, M. S. Jones, D. W. Dyer, D. Seto and J. Chodosh (2013). "Molecular evolution of human adenoviruses." *Sci Rep* **3**: 1812.

Robinson, C. M., X. Zhou, J. Rajaiya, M. A. Yousuf, G. Singh, J. J. DeSerres, M. P. Walsh, S. Wong, D. Seto, D. W. Dyer, J. Chodosh and M. S. Jones (2013). "Predicting the next eye pathogen: analysis of a novel adenovirus." *MBio* **4**(2): e00595–12.

Robinson, J. A. (2011). "Protein epitope mimetics as anti-infectives." *Curr Opin Chem Biol* **15**(3): 379–86.

Rocconi, R. P., Z. B. Zhu, M. Stoff-Khalili, A. A. Rivera, B. Lu, M. Wang, R. D. Alvarez, D. T. Curiel and S. K. Makhija (2007). "Treatment of ovarian cancer with a novel dual targeted conditionally replicative adenovirus (CRAd)." *Gynecol Oncol* **105**(1): 113–21.

Rodriguez-Rocha, H., J. G. Gomez-Gutierrez, A. Garcia-Garcia, X. M. Rao, L. Chen, K. M. McMasters and H. S. Zhou (2011). "Adenoviruses induce autophagy to promote virus replication and oncolysis." *Virology* **416**(1–2): 9–15.

Rodriguez, A., R. Mintardjo, D. Tax, G. Gillissen, J. Custers, M. G. Pau, J. Klap, S. Santra, H. Balachandran, N. L. Letvin, J. Goudsmit and K. Radosevic (2009). "Evaluation of a prime-boost vaccine schedule with distinct adenovirus vectors against malaria in rhesus monkeys." *Vaccine* **27**(44): 6226–33.

Rodriguez, E. and E. Everitt (1996). "Adenovirus uncoating and nuclear establishment are not affected by weak base amines." *J Virol* **70**(6): 3470–7.

Rodriguez, I., J. M. Lazaro, L. Blanco, S. Kamtekar, A. J. Berman, J. Wang, T. A. Steitz, M. Salas and M. de Vega (2005). "A specific subdomain in phi29 DNA polymerase confers both processivity and strand-displacement capacity." *Proc Natl Acad Sci U S A* **102**(18): 6407–12.

Roelvink, P. W., A. Lizonova, J. G. Lee, Y. Li, J. M. Bergelson, R. W. Finberg, D. E. Brough, I. Kovesdi and T. J. Wickham (1998). "The coxsackievirus-adenovirus receptor protein can function as a cellular attachment protein for adenovirus serotypes from subgroups A, C, D, E, and F." *J Virol* **72**(10): 7909–15.

Roelvink, P. W., G. Mi Lee, D. A. Einfeld, I. Kovesdi and T. J. Wickham (1999). "Identification of a conserved receptor-binding site on the fiber proteins of CAR-recognizing adenoviridae." *Science* **286**(5444): 1568–71.

Rogulski, K. R., M. S. Wing, D. L. Paielli, J. D. Gilbert, J. H. Kim and S. O. Freytag (2000). "Double suicide gene therapy augments the antitumor activity of a replication-competent lytic adenovirus through enhanced cytotoxicity and radiosensitization." *Hum Gene Ther* **11**(1): 67–76.

Rojas, J. J., M. Cascallo, S. Guedan, A. Gros, J. Martinez-Quintanilla, A. Hemminki and R. Alemany (2009). "A modified E2F-1 promoter improves the efficacy to toxicity ratio of oncolytic adenoviruses." *Gene Ther* **16**(12): 1441–51.

Rojas, J. J., S. Guedan, P. F. Searle, J. Martinez-Quintanilla, R. Gil-Hoyos, F. Alcayaga-Miranda, M. Cascallo and R. Alemany (2010). "Minimal RB-responsive E1A promoter modification to attain potency, selectivity, and trans-gene-arming capacity in oncolytic adenoviruses." *Mol Ther* **18**(11): 1960–71.

Romanowski, E. G., K. A. Yates and Y. J. Gordon (2001). "Antiviral prophylaxis with twice daily topical cidofovir protects against challenge in the adenovirus type 5/New Zealand rabbit ocular model." *Antiviral Res* **52**(3): 275–80.

Rommens, J. M., M. C. Iannuzzi, B. Kerem, M. L. Drumm, G. Melmer, M. Dean, R. Rozmahel, J. L. Cole, D. Kennedy, N. Hidaka et al. (1989). "Identification of the cystic fibrosis gene: chromosome walking and jumping." *Science* **245**(4922): 1059–65.

Roos, W. H. (2011). "How to perform a nanoindentation experiment on a virus." *Methods Mol Biol* **783**: 251–64.

Roos, W. H., M. M. Gibbons, A. Arkhipov, C. Uetrecht, N. R. Watts, P. T. Wingfield, A. C. Steven, A. J. Heck, K. Schulten, W. S. Klug and G. J. Wuite (2010). "Squeezing protein shells: how continuum elastic models, molecular dynamics simulations, and experiments coalesce at the nanoscale." *Biophys J* **99**(4): 1175–81.

Roovers, D. J., F. M. van der Lee, J. van der Wees and J. S. Sussenbach (1993). "Analysis of the adenovirus type 5 terminal protein precursor and DNA polymerase by linker insertion mutagenesis." *J Virol* **67**(1): 265–76.

Rosa, M. D., E. Gottlieb, M. R. Lerner and J. A. Steitz (1981). "Striking similarities are exhibited by two small Epstein-Barr virus-encoded ribonucleic acids and the adenovirus-associated ribonucleic acids VAI and VAII." *Mol Cell Biol* **1**(9): 785–96.

Rosenfeld, M. A., C. S. Chu, P. Seth, C. Danel, T. Banks, K. Yoneyama, K. Yoshimura and R. G. Crystal (1994). "Gene transfer to freshly isolated human respiratory epithelial cells in vitro using a replication-deficient adenovirus containing the human cystic fibrosis transmembrane conductance regulator cDNA." *Hum Gene Ther* **5**(3): 331–42.

Rosenfeld, M. A., W. Siegfried, K. Yoshimura, K. Yoneyama, M. Fukayama, L. E. Stier, P. K. Paakko, P. Gilardi, L. D. Stratford-Perricaudet, M. Perricaudet et al.

(1991). "Adenovirus-mediated transfer of a recombinant alpha 1-antitrypsin gene to the lung epithelium in vivo." *Science* **252**(5004): 431–4.

Rosenfeld, M. A., K. Yoshimura, B. C. Trapnell, K. Yoneyama, E. R. Rosenthal, W. Dalemans, M. Fukayama, J. Bargon, L. E. Stier, L. Stratford-Perricaudet *et al.* (1992). "In vivo transfer of the human cystic fibrosis transmembrane conductance regulator gene to the airway epithelium." *Cell* **68**(1): 143–55.

Ross, J. F., X. Liu and B. D. Dynlacht (1999). "Mechanism of transcriptional repression of E2F by the retinoblastoma tumor suppressor protein." *Mol Cell* **3**(2): 195–205.

Rotem-Yehudar, R., M. Groettrup, A. Soza, P. M. Kloetzel and R. Ehrlich (1996). "LMP-associated proteolytic activities and TAP-dependent peptide transport for class 1 MHC molecules are suppressed in cell lines transformed by the highly oncogenic adenovirus 12." *J Exp Med* **183**(2): 499–514.

Rotem-Yehudar, R., S. Winograd, S. Sela, J. E. Coligan and R. Ehrlich (1994). "Downregulation of peptide transporter genes in cell lines transformed with the highly oncogenic adenovirus 12." *J Exp Med* **180**(2): 477–88.

Rothmann, T., A. Hengstermann, N. J. Whitaker, M. Scheffner and H. zur Hausen (1998). "Replication of ONYX-015, a potential anticancer adenovirus, is independent of p53 status in tumor cells." *J Virol* **72**(12): 9470–8.

Rout, M. P., J. D. Aitchison, M. O. Magnasco and B. T. Chait (2003). "Virtual gating and nuclear transport: the hole picture." *Trends Cell Biol* **13**(12): 622–8.

Routes, J. M., S. Ryan, K. Morris, R. Takaki, A. Cerwenka and L. L. Lanier (2005). "Adenovirus serotype 5 E1A sensitizes tumor cells to NKG2D-dependent NK cell lysis and tumor rejection." *J Exp Med* **202**(11): 1477–82.

Rowe, W. F., R. J. Huebner, L. K. Gilmore, R. H. Parrott and T. G. Ward (1953). "Isolation of a cytogenic agent from human adenoids undergoing spontaneous degradation in tissue culture." *Proc. Soc. Exp. Biol. Med.* **84**: 570–573.

Roy, S., G. Gao, D. S. Clawson, L. H. Vandenberghe, S. F. Farina and J. M. Wilson (2004). "Complete nucleotide sequences and genome organization of four chimpanzee adenoviruses." *Virology* **324**(2): 361–72.

Roy, S., G. Gao, Y. Lu, X. Zhou, M. Lock, R. Calcedo and J. M. Wilson (2004). "Characterization of a family of chimpanzee adenoviruses and development of molecular clones for gene transfer vectors." *Hum Gene Ther* **15**(5): 519–30.

Rusinova, I., S. Forster, S. Yu, A. Kannan, M. Masse, H. Cumming, R. Chapman and P. J. Hertzog (2013). "Interferome v2.0: an updated database of annotated interferon-regulated genes." *Nucleic Acids Res* **41**(Database issue): D1040–6.

Russell, W. C. and G. D. Kemp (1995). "Role of adenovirus structural components in the regulation of adenovirus infection." *Curr Top Microbiol Immunol* **199** (Pt 1): 81–98.

Rux, J. J. and R. M. Burnett (2000). "Type-specific epitope locations revealed by X-ray crystallographic study of adenovirus type 5 hexon." *Mol Ther* **1**(1): 18–30.

Ruzek, M. C., B. F. Kavanagh, A. Scaria, S. M. Richards and R. D. Garman (2002). "Adenoviral vectors stimulate murine natural killer cell responses and demonstrate antitumor activities in the absence of transgene expression." *Mol Ther* **5**(2): 115–24.

Ruzindana-Umunyana, A., L. Imbeault and J. M. Weber (2002). "Substrate specificity of adenovirus protease." *Virus Res* **89**(1): 41–52.

Saban, S. D., M. Silvestry, G. R. Nemerow and P. L. Stewart (2006). "Visualization of alpha-helices in a 6-angstrom resolution cryoelectron microscopy structure of adenovirus allows refinement of capsid protein assignments." *J Virol* **80**(24): 12049–59.

Sala, V. and T. Crepaldi (2011). "Novel therapy for myocardial infarction: can HGF/Met be beneficial?" *Cell Mol Life Sci* **68**(10): 1703–17.

Salas, M. (1991). "Protein-priming of DNA replication." *Annu Rev Biochem* **60**: 39–71.

Salcedo, R., K. Wasserman, H. A. Young, M. C. Grimm, O. M. Howard, M. R. Anver, H. K. Kleinman, W. J. Murphy and J. J. Oppenheim (1999). "Vascular endothelial growth factor and basic fibroblast growth factor induce expression of CXCR4 on human endothelial cells: In vivo neovascularization induced by stromal-derived factor-1alpha." *Am J Pathol* **154**(4): 1125–35.

Salinas, S., L. G. Bilsland, D. Henaff, A. E. Weston, A. Keriel, G. Schiavo and E. J. Kremer (2009). "CAR-associated vesicular transport of an adenovirus in motor neuron axons." *PLoS Pathog* **5**(5): e1000442.

Salinas, S., C. Zussy, F. Loustalot, D. Henaff, G. Menendez, P. E. Morton, M. Parsons, G. Schiavo and E. J. Kremer (2014). "Disruption of the coxsackievirus and adenovirus receptor-homodimeric interaction triggers lipid microdomain- and dynamin-dependent endocytosis and lysosomal targeting." *J Biol Chem* **289**(2): 680–95.

Sambrook, J., J. Williams, P. A. Sharp and T. Grodzicker (1975). "Physical mapping of temperature-sensitive mutations of adenoviruses." *J Mol Biol* **97**(3): 369–90.

Sanchez-Cespedes, J., C. L. Moyer, L. R. Whitby, D. L. Boger and G. R. Nemerow (2014). "Inhibition of adenovirus replication by a trisubstituted piperazin-2-one derivative." *Antiviral Res* **108**: 65–73.

Sandkovsky, U., L. Vargas and D. F. Florescu (2014). "Adenovirus: current epidemiology and emerging approaches to prevention and treatment." *Curr Infect Dis Rep* **16**(8): 416.

Sano, M., Y. Kato and K. Taira (2006). "Sequence-specific interference by small RNAs derived from adenovirus VAI RNA." *FEBS Lett* **580**(6): 1553–64.

Santra, S., M. S. Seaman, L. Xu, D. H. Barouch, C. I. Lord, M. A. Lifton, D. A. Gorgone, K. R. Beaudry, K. Svehla, B. Welcher, B. K. Chakrabarti, Y. Huang, Z. Y. Yang, J. R. Mascola, G. J. Nabel and N. L. Letvin (2005). "Replication-defective adenovirus serotype 5 vectors elicit durable cellular and humoral immune responses in nonhuman primates." *J Virol* **79**(10): 6516–22.

Santra, S., Y. Sun, B. Korioth-Schmitz, J. Fitzgerald, C. Charbonneau, G. Santos, M. S. Seaman, S. J. Ratcliffe, D. C. Montefiori, G. J. Nabel, H. C. Ertl and N. L. Letvin (2009). "Heterologous prime/boost immunizations of rhesus monkeys using chimpanzee adenovirus vectors." *Vaccine* **27**(42): 5837–45.

Sarnow, P., C. A. Sullivan and A. J. Levine (1982). "A monoclonal antibody detecting the Ad5 E1B-58K tumor antigen in adenovirus-infected and transformed cells." *Virology* **120**: 387–394.

Sathish, N., F. X. Zhu and Y. Yuan (2009). "Kaposi's sarcoma-associated herpesvirus ORF45 interacts with kinesin-2 transporting viral capsid-tegument complexes along microtubules." *PLoS Pathog* **5**(3): e1000332.

Sauthoff, H., T. Pipiya, S. Heitner, S. Chen, B. Bleck, J. Reibman, W. Chang, R. G. Norman, W. N. Rom and J. G. Hay (2004). "Impact of E1a modifications on tumor-selective adenoviral replication and toxicity." *Mol Ther* **10**(4): 749–57.

Sawada, Y., B. Fohring, T. E. Shenk and K. Raska, Jr. (1985). "Tumorigenicity of adenovirus-transformed cells: region E1A of adenovirus 12 confers resistance to natural killer cells." *Virology* **147**(2): 413–21.

Sawadogo, M. and R. G. Roeder (1985). "Interaction of a gene-specific transcription factor with the adenovirus major late promoter upstream of the TATA box region." *Cell* **43**: 165–175.

Schaack, J., W. J.-W. Ho, P. Freimuth and T. Shenk (1990). "Adenovirus terminal protein mediates both nuclear matrix attachment and efficient transcription of adenovirus DNA." *Genes Dev.* **4**: 1197–1208.

Schaeper, U., T. Subramanian, L. Lim, J. M. Boyd and G. Chinnadurai (1998). "Interaction between a cellular protein that binds to the C-terminal region of adenovirus E1A (CtBP) and a novel cellular protein is disrupted by E1A through a conserved PLDLS motif." *J Biol Chem* **273**(15): 8549–52.

Scherer, F., M. Anton, U. Schillinger, J. Henke, C. Bergemann, A. Kruger, B. Gansbacher and C. Plank (2002). "Magnetofection: enhancing and targeting gene delivery by magnetic force in vitro and in vivo." *Gene Ther* **9**(2): 102–9.

Scherer, J. and R. B. Vallee (2015). "Conformational changes in the adenovirus hexon subunit responsible for regulating cytoplasmic dynein recruitment." *J Virol* **89**(2): 1013–23.

Scherer, J., J. Yi and R. B. Vallee (2014). "PKA-dependent dynein switching from lysosomes to adenovirus: a novel form of host-virus competition." *J Cell Biol* **205**(2): 163–77.

Schiedner, G., S. Hertel, M. Johnston, V. Biermann, V. Dries and S. Kochanek (2002). "Variables affecting in vivo performance of high-capacity adenovirus vectors." *J Virol* **76**(4): 1600–9.

Schiedner, G., N. Morral, R. J. Parks, Y. Wu, S. C. Koopmans, C. Langston, F. L. Graham, A. L. Beaudet and S. Kochanek (1998). "Genomic DNA transfer with a high-capacity adenovirus vector results in improved in vivo gene expression and decreased toxicity." *Nat Genet* **18**(2): 180–3.

Schipper, H., V. Alla, C. Meier, D. M. Nettelbeck, O. Herchenroder and B. M. Putzer (2014). "Eradication of metastatic melanoma through cooperative expression of RNA-based HDAC1 inhibitor and p73 by oncolytic adenovirus." *Oncotarget* **5**(15): 5893–907.

Schirmbeck, R., J. Reimann, S. Kochanek and F. Kreppel (2008). "The immunogenicity of adenovirus vectors limits the multispecificity of CD8 T-cell responses to vector-encoded transgenic antigens." *Mol Ther* **16**(9): 1609–16.

Schneider-Brachert, W., V. Tchikov, O. Merkel, M. Jakob, C. Hallas, M. L. Kruse, P. Groitl, A. Lehn, E. Hildt, J. Held-Feindt, T. Dobner, D. Kabelitz, M. Kronke and S. Schutze (2006). "Inhibition of TNF receptor 1 internalization by adenovirus 14.7K as a novel immune escape mechanism." *J Clin Invest* **116**(11): 2901–13.

Schneider, R. J., B. Safer, S. M. Munemitsu, C. E. Samuel and T. Shenk (1985). "Adenovirus VAI RNA prevents phosphorylation of the eukaryotic initiation factor 2 alpha subunit subsequent to infection." *Proc Natl Acad Sci U S A* **82**(13): 4321–5.

Schneider, W. M., M. D. Chevillotte and C. M. Rice (2014). "Interferon-stimulated genes: a complex web of host defenses." *Annu Rev Immunol* **32**: 513–45.

Schoehn, G., M. El Bakkouri, C. M. Fabry, O. Billet, L. F. Estrozi, L. Le, D. T. Curiel, A. V. Kajava, R. W. Ruigrok and E. J. Kremer (2008). "Three-dimensional structure of canine adenovirus serotype 2 capsid." *J Virol* **82**(7): 3192–203.

Schoehn, G., P. Fender, J. Chroboczek and E. A. Hewat (1996). "Adenovirus 3 penton dodecahedron exhibits structural changes of the base on fibre binding." *EMBO J* **15**(24): 6841–6.

Schoggins, J. W. (2014). "Interferon-stimulated genes: roles in viral pathogenesis." *Curr Opin Virol* **6**: 40–6.

Schoggins, J. W. and C. M. Rice (2011). "Interferon-stimulated genes and their antiviral effector functions." *Curr Opin Virol* **1**(6): 519–25.

Schreiner, S., S. Kinkley, C. Burck, A. Mund, P. Wimmer, T. Schubert, P. Groitl, H. Will and T. Dobner (2013). "SPOC1-mediated antiviral host cell response is antagonized early in human adenovirus type 5 infection." *PLoS Pathog* **9**(11): e1003775.

Schuldt, N. J., Y. A. Aldhamen, S. Godbehere-Roosa, S. S. Seregin, Y. A. Kousa and A. Amalfitano (2012). "Immunogenicity when utilizing adenovirus serotype 4 and 5 vaccines expressing circumsporozoite protein in naive and adenovirus (Ad5) immune mice." *Malar J* **11**: 209.

Sebestyen, Z., J. de Vrij, M. Magnusson, R. Debets and R. Willemsen (2007). "An oncolytic adenovirus redirected with a tumor-specific T-cell receptor." *Cancer Res* **67**(23): 11309–16.

Sedegah, M., M. R. Hollingdale, F. Farooq, H. Ganeshan, M. Belmonte, Y. Kim, B. Peters, A. Sette, J. Huang, S. McGrath, E. Abot, K. Limbach, M. Shi, L. Soisson, C. Diggs, I. Chuang, C. Tamminga, J. E. Epstein, E. Villasante and T. L. Richie (2014). "Sterile immunity to malaria after DNA prime/adenovirus boost immunization is associated with effector memory CD8+T cells targeting AMA1 class I epitopes." *PLoS One* **9**(9): e106241.

Segerman, A., J. P. Atkinson, M. Marttila, V. Dennerquist, G. Wadell and N. Arnberg (2003). "Adenovirus type 11 uses CD46 as a cellular receptor." *J Virol* **77**(17): 9183–91.

Seguin, L., J. S. Desgrosellier, S. M. Weis and D. A. Cheresh (2015). "Integrins and cancer: regulators of cancer stemness, metastasis, and drug resistance." *Trends Cell Biol* **25**(4): 234–40.

Segura, M. M., R. Alba, A. Bosch and M. Chillon (2008). "Advances in helper-dependent adenoviral vector research." *Curr Gene Ther* **8**(4): 222–35.

Seiradake, E. and S. Cusack (2005). "Crystal structure of enteric adenovirus serotype 41 short fiber head." *J Virol* **79**(22): 14088–94.

Seiradake, E., D. Henaff, H. Wodrich, O. Billet, M. Perreau, C. Hippert, F. Mennechet, G. Schoehn, H. Lortat-Jacob, H. Dreja, S. Ibanes, V. Kalatzis, J. P. Wang, R. W. Finberg, S. Cusack and E. J. Kremer (2009). "The cell adhesion molecule "CAR" and sialic acid on human erythrocytes influence adenovirus in vivo biodistribution." *PLoS Pathog* **5**(1): e1000277.

Seiradake, E., H. Lortat-Jacob, O. Billet, E. J. Kremer and S. Cusack (2006). "Structural and mutational analysis of human Ad37 and canine adenovirus 2 fiber heads in complex with the D1 domain of coxsackie and adenovirus receptor." *J Biol Chem* **281**(44): 33704–16.

Sekaly, R. P. (2008). "The failed HIV Merck vaccine study: a step back or a launching point for future vaccine development?" *J Exp Med* **205**(1): 7–12.

Selsted, M. E. and A. J. Ouellette (2005). "Mammalian defensins in the antimicrobial immune response." *Nat Immunol* **6**(6): 551–7.

Serangeli, C., O. Bicanic, M. H. Scheible, D. Wernet, P. Lang, H. G. Rammensee, S. Stevanovic, R. Handgretinger and T. Feuchtinger (2010). "Ex vivo detection of adenovirus specific CD4+ T-cell responses to HLA-DR-epitopes of the Hexon protein show a contracted specificity of T(HELPER) cells following stem cell transplantation." *Virology* **397**(2): 277–84.

Sergeant, A., M. A. Tigges and H. J. Raskas (1979). "Nucleosome-like structural subunits of intranuclear parental adenovirus type 2 DNA." *J Virol* **29**(3): 888–98.

Sester, M. and H. G. Burgert (1994). "Conserved cysteine residues within the E3/19K protein of adenovirus type 2 are essential for binding to major histocompatibility complex antigens." *J Virol* **68**(9): 5423–32.

Sester, M., K. Koebernick, D. Owen, M. Ao, Y. Bromberg, E. May, E. Stock, L. Andrews, V. Groh, T. Spies, A. Steinle, B. Menz and H. G. Burgert (2010). "Conserved amino acids within the adenovirus 2 E3/19K protein differentially affect downregulation of MHC class I and MICA/B proteins." *J Immunol* **184**(1): 255–67.

Sester, M., Z. Ruszics, E. Mackley and H. G. Burgert (2013). "The transmembrane domain of the adenovirus E3/19K protein acts as an endoplasmic reticulum retention signal and contributes to intracellular sequestration of major histocompatibility complex class I molecules." *J Virol* **87**(11): 6104–17.

Seth, P., D. J. Fitzgerald, M. C. Willingham and I. Pastan (1984). "Role of a low-pH environment in adenovirus enhancement of the toxicity of a Pseudomonas exotoxin-epidermal growth factor conjugate." *J Virol* **51**(3): 650–5.

Seth, P., M. C. Willingham and I. Pastan (1984). "Adenovirus-dependent release of 51Cr from KB cells at an acidic pH." *J Biol Chem* **259**(23): 14350–3.

Seto, D., J. Chodosh, J. R. Brister and M. S. Jones (2011). "Using the whole-genome sequence to characterize and name human adenoviruses." *J Virol* **85**(11): 5701–2.

Seto, D., M. S. Jones, D. W. Dyer and J. Chodosh (2013). "Characterizing, typing, and naming human adenovirus type 55 in the era of whole genome data." *J Clin Virol* **58**(4): 741–2.

Seto, J., M. P. Walsh, P. Mahadevan, A. Purkayastha, J. M. Clark, C. Tibbetts and D. Seto (2009). "Genomic and bioinformatics analyses of HAdV-14p, reference strain of a re-emerging respiratory pathogen and analysis of B1/B2." *Virus Res* **143**(1): 94–105.

Sha, J., M. K. Ghosh, K. Zhang and M. L. Harter (2010). "E1A interacts with two opposing transcriptional pathways to induce quiescent cells into S phase." *J Virol* **84**(8): 4050–9.

Shah, D. P., S. S. Ghantoji, V. E. Mulanovich, E. J. Ariza-Heredia and R. F. Chemaly (2012). "Management of respiratory viral infections in hematopoietic cell transplant recipients." *Am J Blood Res* **2**(4): 203–18.

Shah, G. A. and C. C. O'Shea (2015). "Viral and Cellular Genomes Activate Distinct DNA Damage Responses." *Cell* **162**(5): 987–1002.

Sharp, P. A., C. Moore and J. Haverty (1976). "The infectivity of adenovirus 5 DNA protein complex." *Virology* **75**: 442–456.

Shaw, A. R. and E. B. Ziff (1980). "Transcripts from the adenovirus-2 major late promoter yield a single early family of 3' coterminal mRNAs and five late families." *Cell* **22**(3): 905–16.

Shayakhmetov, D. M., A. Gaggar, S. Ni, Z. Y. Li and A. Lieber (2005). "Adenovirus binding to blood factors results in liver cell infection and hepatotoxicity." *J Virol* **79**(12): 7478–91.

Shayakhmetov, D. M., Z. Y. Li, V. Ternovoi, A. Gaggar, H. Gharwan and A. Lieber (2003). "The interaction between the fiber knob domain and the cellular attachment receptor determines the intracellular trafficking route of adenoviruses." *J Virol* **77**(6): 3712–23.

Shayakhmetov, D. M. and A. Lieber (2000). "Dependence of adenovirus infectivity on length of the fiber shaft domain." *J Virol* **74**(22): 10274–86.

Shen, W., R. Falahati, R. Stark, D. Leitenberg and S. Ladisch (2005). "Modulation of CD4 Th cell differentiation by ganglioside GD1a in vitro." *J Immunol* **175**(8): 4927–34.

Shen, W. and S. Ladisch (2002). "Ganglioside GD1a impedes lipopolysaccharide-induced maturation of human dendritic cells." *Cell Immunol* **220**(2): 125–33.

Shepard, R. N. and D. A. Ornelles (2004). "Diverse roles for E4orf3 at late times of infection revealed in an E1B 55-kilodalton protein mutant background." *J Virol* **78**(18): 9924–35.

Sherr, C. J. (1994). "Growth factor-regulated G1 cyclins." *Stem Cells* **12 Suppl 1**: 47–55; discussion 55–7.

Sherr, C. J. and F. McCormick (2002). "The RB and p53 pathways in cancer." *Cancer Cell* **2**(2): 103–12.

Sherwood, V., E. King, S. Totemeyer, I. Connerton and K. H. Mellits (2012). "Interferon treatment suppresses enteric adenovirus infection in a model gastrointestinal cell-culture system." *J Gen Virol* **93**(Pt 3): 618–23.

Shi, L., M. Ramaswamy, L. J. Manzel and D. C. Look (2007). "Inhibition of Jak1-dependent signal transduction in airway epithelial cells infected with adenovirus." *Am J Respir Cell Mol Biol* **37**(6): 720–8.

Shin, S. P., H. H. Seo, J. H. Shin, H. B. Park, D. P. Lim, H. S. Eom, Y. S. Bae, I. H. Kim, K. Choi and S. J. Lee (2013). "Adenovirus expressing both thymidine kinase and soluble PD1 enhances antitumor immunity by strengthening CD8 T-cell response." *Mol Ther* **21**(3): 688–95.

Shisler, J., C. Yang, B. Walter, C. F. Ware and L. R. Gooding (1997). "The adenovirus E3–10.4K/14.5K complex mediates loss of cell surface Fas (CD95) and resistance to Fas-induced apoptosis." *J Virol* **71**(11): 8299–306.

Shiver, J. W. and E. A. Emini (2004). "Recent advances in the development of HIV-1 vaccines using replication-incompetent adenovirus vectors." *Annu Rev Med* **55**: 355–72.

Shiver, J. W., T. M. Fu, L. Chen, D. R. Casimiro, M. E. Davies, R. K. Evans, Z. Q. Zhang, A. J. Simon, W. L. Trigona, S. A. Dubey, L. Huang, V. A. Harris, R. S. Long, X. Liang, L. Handt, W. A. Schleif, L. Zhu, D. C. Freed, N. V. Persaud, L. Guan, K. S. Punt, A. Tang, M. Chen, K. A. Wilson, K. B. Collins, G. J. Heidecker, V. R. Fernandez, H. C. Perry, J. G. Joyce, K. M. Grimm, J. C. Cook, P. M. Keller, D. S. Kresock, H. Mach, R. D. Troutman, L. A. Isopi, D. M. Williams, Z. Xu, K. E. Bohannon, D. B. Volkin, D. C. Montefiori, A. Miura, G. R. Krivulka, M. A. Lifton, M. J. Kuroda, J. E. Schmitz, N. L. Letvin, M. J. Caulfield, A. J. Bett, R. Youil, D. C. Kaslow and E. A. Emini (2002). "Replication-incompetent adenoviral vaccine vector elicits effective anti-immunodeficiency-virus immunity." *Nature* **415**(6869): 331–5.

Siekierka, J., T. M. Mariano, P. A. Reichel and M. B. Mathews (1985). "Translational control by adenovirus: lack of virus-associated RNAI during adenovirus infection results in phosphorylation of initiation factor eIF-2 and inhibition of protein synthesis." *Proc Natl Acad Sci U S A* **82**(7): 1959–63.

Silva, A. C., P. Fernandes, M. F. Sousa and P. M. Alves (2014). "Scalable production of adenovirus vectors." *Methods Mol Biol* **1089**: 175–96.

Silvestry, M., S. Lindert, J. G. Smith, O. Maier, C. M. Wiethoff, G. R. Nemerow and P. L. Stewart (2009). "Cryo-electron microscopy structure of adenovirus type 2 temperature-sensitive mutant 1 reveals insight into the cell entry defect." *J Virol* **83**(15): 7375–83.

Sims, B., L. Gu, A. Krendelchtchikov and Q. L. Matthews (2014). "Neural stem cell-derived exosomes mediate viral entry." *Int J Nanomedicine* **9**: 4893–7.

Singh, A. K., R. Menendez-Conejero, C. San Martin and M. J. van Raaij (2014). "Crystal structure of the fibre head domain of the Atadenovirus Snake Adenovirus 1." *PLoS One* **9**(12): e114373.

Singh, G., C. M. Robinson, S. Dehghan, M. S. Jones, D. W. Dyer, D. Seto and J. Chodosh (2013). "Homologous recombination in E3 genes of human adenovirus species D." *J Virol* **87**(22): 12481–8.

Singh, G., X. Zhou, J. Y. Lee, M. A. Yousuf, M. Ramke, A. M. Ismail, J. S. Lee, C. M. Robinson, D. Seto, D. W. Dyer, M. S. Jones, J. Rajaiya and J. Chodosh (2015). "Recombination of the epsilon determinant and corneal tropism: Human adenovirus species D types 15, 29, 56, and 69." *Virology* **485**: 452–459.

Singh, S., S. Vedi, W. Li, S. K. Samrat, R. Kumar and B. Agrawal (2014). "Recombinant adenoviral vector expressing HCV NS4 induces protective immune responses in a mouse model of Vaccinia-HCV virus infection: a dose and route conundrum." *Vaccine* **32**(23): 2712–21.

Singleton, D. C., D. Li, S. Y. Bai, S. P. Syddall, J. B. Smaill, Y. Shen, W. A. Denny, W. R. Wilson and A. V. Patterson (2007). "The nitroreductase prodrug SN 28343 enhances the potency of systemically administered armed oncolytic adenovirus ONYX-411(NTR)." *Cancer Gene Ther* **14**(12): 953–67.

Sinha, S., B. Medhi and R. Sehgal (2014). "Challenges of drug-resistant malaria." *Parasite* **21**: 61.

Sirena, D., B. Lilienfeld, M. Eisenhut, S. Kalin, K. Boucke, R. R. Beerli, L. Vogt, C. Ruedl, M. F. Bachmann, U. F. Greber and S. Hemmi (2004). "The human membrane cofactor CD46 is a receptor for species B adenovirus serotype 3." *J Virol* **78**(9): 4454–62.

Small, E. J., M. A. Carducci, J. M. Burke, R. Rodriguez, L. Fong, L. van Ummersen, D. C. Yu, J. Aimi, D. Ando, P. Working, D. Kirn and G. Wilding (2006). "A phase I trial of intravenous CG7870, a replication-selective, prostate-specific antigen-targeted oncolytic adenovirus, for the treatment of hormone-refractory, metastatic prostate cancer." *Mol Ther* **14**(1): 107–17.

Small, J. C., L. H. Haut, A. Bian and H. C. Ertl (2014). "The effect of adenovirus-specific antibodies on adenoviral vector-induced, transgene product-specific T cell responses." *J Leukoc Biol* **96**(5): 821–31.

Smart, J. E. and B. W. Stillman (1982). "Adenovirus terminal protein precursor. Partial amino acid sequence and the site of covalent linkage to virus DNA." *J Biol Chem* **257**(22): 13499–506.

Smirnov, D. A., S. Hou, X. Liu, E. Claudio, U. K. Siebenlist and R. P. Ricciardi (2001). "Coup-TFII is up-regulated in adenovirus type 12 tumorigenic cells and is a repressor of MHC class I transcription." *Virology* **284**(1): 13–9.

Smirnov, D. A., S. Hou and R. P. Ricciardi (2000). "Association of histone deacetylase with COUP-TF in tumorigenic Ad12-transformed cells and its potential role in shut-off of MHC class I transcription." *Virology* **268**(2): 319–28.

Smith T. A., N. Idamakanti, J. Marshall-Neff, M. L. Rolence, P. Wright, C. Mech, L. Dinges, W.O. Iverson, A.D. Sherer, J.E. Markovits, R. M. Lyons, M. Kaleko and S. C. Stevenson (2003). "Receptor interactions involved in adenoviral-mediated gene delivery after systemic administration in non-human primates." *Hum Gene Ther* **14**(17): 15905–604.

Smith, J. G., A. Cassany, L. Gerace, R. Ralston and G. R. Nemerow (2008). "Neutralizing antibody blocks adenovirus infection by arresting microtubule-dependent cytoplasmic transport." *J Virol* **82**(13): 6492–500.

Smith, J. G. and G. R. Nemerow (2008). "Mechanism of adenovirus neutralization by Human alpha-defensins." *Cell Host Microbe* **3**(1): 11–9.

Smith, J. G., M. Silvestry, S. Lindert, W. Lu, G. R. Nemerow and P. L. Stewart (2010). "Insight into the mechanisms of adenovirus capsid disassembly from studies of defensin neutralization." *PLoS Pathog* **6**(6): e1000959.

Smith, J. S., Z. Xu and A. P. Byrnes (2008). "A quantitative assay for measuring clearance of adenovirus vectors by Kupffer cells." *J Virol Methods* **147**(1): 54–60.

Smith, J. S., Z. Xu, J. Tian, D. J. Palmer, P. Ng and A. P. Byrnes (2011). "The role of endosomal escape and mitogen-activated protein kinases in adenoviral activation of the innate immune response." *PLoS One* **6**(10): e26755.

Smith, J. S., Z. Xu, J. Tian, S. C. Stevenson and A. P. Byrnes (2008). "Interaction of systemically delivered adenovirus vectors with Kupffer cells in mouse liver." *Hum Gene Ther* **19**(5): 547–54.

Smith, T. A., N. Idamakanti, M. L. Rollence, J. Marshall-Neff, J. Kim, K. Mulgrew, G. R. Nemerow, M. Kaleko and S. C. Stevenson (2003). "Adenovirus serotype 5 fiber shaft influences in vivo gene transfer in mice." *Hum Gene Ther* **14**(8): 777–87.

Snijder, J., M. Benevento, C. L. Moyer, V. Reddy, G. R. Nemerow and A. J. Heck (2014). "The cleaved N-terminus of pVI binds peripentonal hexons in mature adenovirus." *J Mol Biol* **426**(9): 1971–9.

Snijder, J., V. S. Reddy, E. R. May, W. H. Roos, G. R. Nemerow and G. J. Wuite (2013). "Integrin and defensin modulate the mechanical properties of adenovirus." *J Virol* **87**(5): 2756–66.

Söderlund, H., U. Pettersson, B. Vennström, L. Philipson and M. B. Mathews (1976). "A new species of virus-coded, low molecular weight RNA from cells infected with adenovirus type 2." *Cell* **7**: 585–593.

Sohn, S. Y., R. G. Bridges and P. Hearing (2015). "Proteomic analysis of ubiquitin-like posttranslational modifications induced by the adenovirus E4-ORF3 protein." *J Virol* **89**(3): 1744–55.

Sohn, S. Y. and P. Hearing (2011). "Adenovirus sequesters phosphorylated STAT1 at viral replication centers and inhibits STAT dephosphorylation." *J Virol* **85**(15): 7555–62.

Sohn, S. Y. and P. Hearing (2012). "Adenovirus regulates sumoylation of Mre11-Rad50-Nbs1 components through a paralog-specific mechanism." *J Virol* **86**(18): 9656–65.

Soria, C., F. E. Estermann, K. C. Espantman and C. C. O'Shea (2010). "Heterochromatin silencing of p53 target genes by a small viral protein." *Nature* **466**(7310): 1076–81.

Soudais, C., C. Laplace-Builhe, K. Kissa and E. J. Kremer (2001). "Preferential transduction of neurons by canine adenovirus vectors and their efficient retrograde transport in vivo." *FASEB J* **15**(12): 2283–5.

Soudais, C., N. Skander and E. J. Kremer (2004). "Long-term in vivo transduction of neurons throughout the rat CNS using novel helper-dependent CAV-2 vectors." *FASEB J* **18**(2): 391–3.

Sova, P., X. W. Ren, S. Ni, K. M. Bernt, J. Mi, N. Kiviat and A. Lieber (2004). "A tumor-targeted and conditionally replicating oncolytic adenovirus vector expressing TRAIL for treatment of liver metastases." *Mol Ther* **9**(4): 496–509.

Spanggord, R. J. and P. A. Beal (2001). "Selective binding by the RNA binding domain of PKR revealed by affinity cleavage." *Biochemistry* **40**(14): 4272–80.

Spanggord, R. J., M. Vuyisich and P. A. Beal (2002). "Identification of binding sites for both dsRBMs of PKR on kinase-activating and kinase-inhibiting RNA ligands." *Biochemistry* **41**(14): 4511–20.

Spindler, K. R., C. Y. Eng and A. J. Berk (1985). "An adenovirus early region 1A protein is required for maximal viral DNA replication in growth-arrested human cells." *J Virol* **53**: 742–750.

Spjut, S., W. Qian, J. Bauer, R. Storm, L. Frangsmyr, T. Stehle, N. Arnberg and M. Elofsson (2011). "A potent trivalent sialic acid inhibitor of adenovirus type 37 infection of human corneal cells." *Angew Chem Int Ed Engl* **50**(29): 6519–21.

Stanley, D. A., A. N. Honko, C. Asiedu, J. C. Trefry, A. W. Lau-Kilby, J. C. Johnson, L. Hensley, V. Ammendola, A. Abbate, F. Grazioli, K. E. Foulds, C. Cheng, L. Wang, M. M. Donaldson, S. Colloca, A. Folgori, M. Roederer, G. J. Nabel, J. Mascola, A. Nicosia, R. Cortese, R. A. Koup and N. J. Sullivan (2014). "Chimpanzee adenovirus vaccine generates acute and durable protective immunity against ebolavirus challenge." *Nat Med* **20**(10): 1126–9.

Steegenga, W. T., N. Riteco, A. G. Jochemsen, F. J. Fallaux and J. L. Bos (1998). "The large E1B protein together with the E4orf6 protein target p53 for active degradation in adenovirus infected cells." *Oncogene* **16**(3): 349–357.

Stehle, T. and T. S. Dermody (2003). "Structural evidence for common functions and ancestry of the reovirus and adenovirus attachment proteins." *Rev Med Virol* **13**(2): 123–32.

Stein, R. W., M. Corrigan, P. Yaciuk, J. Whelan and E. Moran (1990). "Analysis of E1A-mediated growth regulation functions: binding of the 300-kilodalton cellular product correlates with E1A enhancer repression function and DNA synthesis-inducing activity." *J Virol* **64**(9): 4421–7.

Stein, S. C. and E. Falck-Pedersen (2012). "Sensing adenovirus infection: activation of interferon regulatory factor 3 in RAW 264.7 cells." *J Virol* **86**(8): 4527–37.

Steitz, J., J. Bruck, J. Knop and T. Tuting (2001). "Adenovirus-transduced dendritic cells stimulate cellular immunity to melanoma via a CD4(+) T cell-dependent mechanism." *Gene Ther* **8**(16): 1255–63.

Steitz, T. A. (1998). "A mechanism for all polymerases." *Nature* **391**(6664): 231–2.

Stephen, A. G., S. Raval-Fernandes, T. Huynh, M. Torres, V. A. Kickhoefer and L. H. Rome (2001). "Assembly of vault-like particles in insect cells expressing only the major vault protein." *J Biol Chem* **276**(26): 23217–20.

Stephens, C. and E. Harlow (1987). "Differential splicing yields novel adenovirus 5 E1A mRNAs that encode 30 kd and 35 kd proteins." *EMBO J* **6**(7): 2027–35.

Stephenson, K. E., J. Hural, S. P. Buchbinder, F. Sinangil and D. H. Barouch (2012). "Preexisting adenovirus seropositivity is not associated with increased HIV-1 acquisition in three HIV-1 vaccine efficacy trials." *J Infect Dis* **205**(12): 1806–10.

Stevens, J. L., G. T. Cantin, G. Wang, A. Shevchenko, A. Shevchenko and A. J. Berk (2002). "Transcription control by E1A and MAP kinase pathway via Sur2 mediator subunit." *Science* **296**(5568): 755–8.

Stewart, P. L., R. M. Burnett, M. Cyrklaff and S. D. Fuller (1991). "Image reconstruction reveals the complex molecular organization of adenovirus." *Cell* **67**(1): 145–54.

Stewart, P. L., C. Y. Chiu, S. Huang, T. Muir, Y. Zhao, B. Chait, P. Mathias and G. R. Nemerow (1997). "Cryo-EM visualization of an exposed RGD epitope on adenovirus that escapes antibody neutralization." *EMBO J* **16**(6): 1189–98.

Stewart, P. L. and G. R. Nemerow (2007). "Cell integrins: commonly used receptors for diverse viral pathogens." *Trends Microbiol* **15**(11): 500–7.

Stillman, B. W., J. B. Lewis, L. T. Chow, M. B. Mathews and J. E. Smart (1981). "Identification of the gene and mRNA for the adenovirus terminal protein precursor." *Cell* **23**(2): 497–508.

Stillman, B. W., F. Tamanoi and M. B. Mathews (1982). "Purification of an adenovirus-coded DNA polymerase that is required for initiation of DNA replication." *Cell* **31**: 613–632.

Stinchcombe, J. C. and G. M. Griffiths (2003). "The role of the secretory immunological synapse in killing by CD8+ CTL." *Semin Immunol* **15**(6): 301–5.

Stoff-Khalili, M. A., A. A. Rivera, A. Nedeljkovic-Kurepa, A. DeBenedetti, X. L. Li, Y. Odaka, J. Podduturi, D. A. Sibley, G. P. Siegal, A. Stoff, S. Young, Z. B. Zhu, D. T. Curiel and J. M. Mathis (2008). "Cancer-specific targeting of a conditionally replicative adenovirus using mRNA translational control." *Breast Cancer Res Treat* **108**(1): 43–55.

Stoff-Khalili, M. A., A. A. Rivera, A. Stoff, J. Michael Mathis, R. P. Rocconi, Q. L. Matthews, M. T. Numnum, I. Herrmann, P. Dall, D. E. Eckhoff, J. T. Douglas, G. P. Siegal, Z. B. Zhu and D. T. Curiel (2007). "Combining high selectivity of replication via CXCR4 promoter with fiber chimerism for effective adenoviral oncolysis in breast cancer." *Int J Cancer* **120**(4): 935–41.

Stone, D., S. Ni, Z. Y. Li, A. Gaggar, N. DiPaolo, Q. Feng, V. Sandig and A. Lieber (2005). "Development and assessment of human adenovirus type 11 as a gene transfer vector." *J Virol* **79**(8): 5090–104.

Stonebraker, J. R., D. Wagner, R. W. Lefensty, K. Burns, S. J. Gendler, J. M. Bergelson, R. C. Boucher, W. K. O'Neal and R. J. Pickles (2004). "Glycocalyx restricts adenoviral vector access to apical receptors expressed on respiratory epithelium in vitro and in vivo: role for tethered mucins as barriers to lumenal infection." *J Virol* **78**(24): 13755–68.

Stracker, T. H., C. T. Carson and M. D. Weitzman (2002). "Adenovirus oncoproteins inactivate the Mre11-Rad50-NBS1 DNA repair complex." *Nature* **418**(6895): 348–52.

Stracker, T. H., D. V. Lee, C. T. Carson, F. D. Araujo, D. A. Ornelles and M. D. Weitzman (2005). "Serotype-specific reorganization of the Mre11 complex by adenoviral E4orf3 proteins." *J Virol* **79**(11): 6664–73.

Strom, A. C., P. Ohlsson and G. Akusjarvi (1998). "AR1 is an integral part of the adenovirus type 2 E1A-CR3 transactivation domain." *J Virol* **72**(7): 5978–83.

Stromsten, N. J., D. H. Bamford and J. K. Bamford (2003). "The unique vertex of bacterial virus PRD1 is connected to the viral internal membrane." *J Virol* **77**(11): 6314–21.

Strunze, S., M. F. Engelke, I. H. Wang, D. Puntener, K. Boucke, S. Schleich, M. Way, P. Schoenenberger, C. J. Burckhardt and U. F. Greber (2011). "Kinesin-1-mediated capsid disassembly and disruption of the nuclear pore complex promote virus infection." *Cell Host Microbe* **10**(3): 210–23.

Stucki, M. and S. P. Jackson (2006). "gammaH2AX and MDC1: anchoring the DNA-damage-response machinery to broken chromosomes." *DNA Repair (Amst)* **5**(5): 534–43.

Stuiver, M. H., W. G. Bergsma, A. C. Arnberg, H. van Amerongen, R. van Grondelle and P. C. van der Vliet (1992). "Structural alterations of double-stranded DNA in complex with the adenovirus DNA-binding protein. Implications for its function in DNA replication." *J Mol Biol* **225**(4): 999–1011.

Stuiver, M. H. and P. C. van der Vliet (1990). "Adenovirus DNA-binding protein forms a multimeric protein complex with double-stranded DNA and enhances binding of nuclear factor I." *J Virol* **64**(1): 379–86.

Stunnenberg, H. G., H. Lange, L. Philipson, R. T. van Miltenburg and P. C. van der Vliet (1988). "High expression of functional adenovirus DNA polymerase and precursor terminal protein using recombinant vaccinia virus." *Nucleic Acids Res* **16**(6): 2431–44.

Stupack, D. G., E. Li, S. A. Silletti, J. A. Kehler, R. L. Geahlen, K. Hahn, G. R. Nemerow and D. A. Cheresh (1999). "Matrix valency regulates integrin-mediated lymphoid adhesion via Syk kinase." *J Cell Biol* **144**(4): 777–88.

Subramanian, T., M. Kuppuswamy, J. Gysbers, S. Mak and G. Chinnadurai (1984). "19-kDa tumor antigen coded by early region E1b of adenovirus 2 is required for efficient synthesis and for protection of viral DNA." *J Biol Chem* **259**(19): 11777–83.

Subramanian, T., M. Kuppuswamy, S. Mak and G. Chinnadurai (1984). "Adenovirus cyt+ locus, which controls cell transformation and tumorigenicity, is an allele of lp+ locus, which codes for a 19-kilodalton tumor antigen." *J Virol* **52**(2): 336–43.

Subramanian, T., B. Tarodi and G. Chinnadurai (1995). "p53-independent apoptotic and necrotic cell deaths induced by adenovirus infection: suppression by E1B 19K and Bcl-2 proteins." *Cell Growth Differ* **6**(2): 131–7.

Subramanian, T., S. Vijayalingam, E. Lomonosova, L. J. Zhao and G. Chinnadurai (2007). "Evidence for involvement of BH3-only proapoptotic members in adenovirus-induced apoptosis." *J Virol* **81**(19): 10486–95.

Subramanian, T., L. J. Zhao and G. Chinnadurai (2013). "Interaction of CtBP with adenovirus E1A suppresses immortalization of primary epithelial cells and enhances virus replication during productive infection." *Virology* **443**(2): 313–20.

Sugawara, K., Z. Gilead, W. S. Wold and M. Green (1977). "Immunofluorescence study of the adenovirus type 2 single-stranded DNA binding protein in infected and transformed cells." *J Virol* **22**(2): 527–39.

Sullivan, N. J., T. W. Geisbert, J. B. Geisbert, L. Xu, Z. Y. Yang, M. Roederer, R. A. Koup, P. B. Jahrling and G. J. Nabel (2003). "Accelerated vaccination for Ebola virus haemorrhagic fever in non-human primates." *Nature* **424**(6949): 681–4.

Sumida, S. M., D. M. Truitt, M. G. Kishko, J. C. Arthur, S. S. Jackson, D. A. Gorgone, M. A. Lifton, W. Koudstaal, M. G. Pau, S. Kostense, M. J. Havenga, J. Goudsmit, N. L. Letvin and D. H. Barouch (2004). "Neutralizing antibodies and CD8+ T lymphocytes both contribute to immunity to adenovirus serotype 5 vaccine vectors." *J Virol* **78**(6): 2666–73.

Sumida, S. M., D. M. Truitt, A. A. Lemckert, R. Vogels, J. H. Custers, M. M. Addo, S. Lockman, T. Peter, F. W. Peyerl, M. G. Kishko, S. S. Jackson, D. A. Gorgone, M. A. Lifton, M. Essex, B. D. Walker, J. Goudsmit, M. J. Havenga and D. H. Barouch (2005). "Neutralizing antibodies to adenovirus serotype 5 vaccine vectors are directed primarily against the adenovirus hexon protein." *J Immunol* **174**(11): 7179–85.

Sundararajan, R. and E. White (2001). "E1B 19K blocks Bax oligomerization and tumor necrosis factor alpha-mediated apoptosis." *J Virol* **75**(16): 7506–16.

Sundquist, B., E. Everitt, L. Philipson and S. Hoglund (1973). "Assembly of adenoviruses." *J Virol* **11**(3): 449–59.

Suomalainen, M., M. Y. Nakano, S. Keller, K. Boucke, R. P. Stidwill and U. F. Greber (1999). "Microtubule-dependent plus- and minus end-directed motilities are competing processes for nuclear targeting of adenovirus." *J Cell Biol* **144**(4): 657–72.

Suprenant, K. A. (2002). "Vault ribonucleoprotein particles: sarcophagi, gondolas, or safety deposit boxes?" *Biochemistry* **41**(49): 14447–54.

Svensson, U. and R. Persson (1984). "Entry of adenovirus 2 into HeLa cells." *J Virol* **51**(3): 687–94.

Swadling, L., S. Capone, R. D. Antrobus, A. Brown, R. Richardson, E. W. Newell, J. Halliday, C. Kelly, D. Bowen, J. Fergusson, A. Kurioka, V. Ammendola, M. Del Sorbo, F. Grazioli, M. L. Esposito, L. Siani, C. Traboni, A. Hill, S. Colloca, M. Davis, A. Nicosia, R. Cortese, A. Folgori, P. Klenerman and E. Barnes (2014). "A human vaccine strategy based on chimpanzee adenoviral and MVA vectors that primes, boosts, and sustains functional HCV-specific T cell memory." *Sci Transl Med* **6**(261): 261ra153.

Swaminathan, S. and B. Thimmapaya (1995). "Regulation of adenovirus E2 transcription unit." *Curr. Top. Microbiol. Immunol.* **199**((Pt 3)): 177–194.

Taatjes, D. J. (2010). "The human Mediator complex: a versatile, genome-wide regulator of transcription." *Trends Biochem Sci* **35**(6): 315–22.

Taipale, K., I. Liikanen, J. Juhila, R. Turkki, S. Tahtinen, M. Kankainen, L. Vassilev, A. Ristimaki, A. Koski, A. Kanerva, I. Diaconu, V. Cerullo, M. Vaha-Koskela, M. Oksanen, N. Linder, T. Joensuu, J. Lundin and A. Hemminki (2016). "Chronic Activation of Innate Immunity Correlates With Poor Prognosis in Cancer Patients Treated With Oncolytic Adenovirus." *Mol Ther* **24**(1): 175–83.

Takagi-Kimura, M., T. Yamano, A. Tamamoto, N. Okamura, H. Okamura, T. Hashimoto-Tamaoki, M. Tagawa, N. Kasahara and S. Kubo (2013). "Enhanced antitumor efficacy of fiber-modified, midkine promoter-regulated oncolytic adenovirus in human malignant mesothelioma." *Cancer Sci* **104**(11): 1433–9.

Takemori, N. (1972). "Genetic studies with tumorigenic adenoviruses. 3. Recombination in adenovirus type 12." *Virology* **47**(1): 157–67.

Tam, C., V. Idone, C. Devlin, M. C. Fernandes, A. Flannery, X. He, E. Schuchman, I. Tabas and N. W. Andrews (2010). "Exocytosis of acid sphingomyelinase by wounded cells promotes endocytosis and plasma membrane repair." *J Cell Biol* **189**(6): 1027–38.

Tam, J. C., S. R. Bidgood, W. A. McEwan and L. C. James (2014). "Intracellular sensing of complement C3 activates cell autonomous immunity." *Science* **345**(6201): 1256070.

Tamanini, A., E. Nicolis, A. Bonizzato, V. Bezzerri, P. Melotti, B. M. Assael and G. Cabrini (2006). "Interaction of adenovirus type 5 fiber with the coxsackievirus and adenovirus receptor activates inflammatory response in human respiratory cells." *J Virol* **80**(22): 11241–54.

Tamminga, C., M. Sedegah, S. Maiolatesi, C. Fedders, S. Reyes, A. Reyes, C. Vasquez, Y. Alcorta, I. Chuang, M. Spring, M. Kavanaugh, H. Ganeshan, J. Huang, M. Belmonte, E. Abot, A. Belmonte, J. Banania, F. Farooq, J. Murphy, J. Komisar, N. O. Richie, J. Bennett, K. Limbach, N. B. Patterson, J. T. Bruder, M. Shi, E. Miller, S. Dutta, C. Diggs, L. A. Soisson, M. R. Hollingdale, J. E. Epstein and T. L. Richie (2013). "Human adenovirus 5-vectored Plasmodium falciparum NMRC-M3V-Ad-PfCA vaccine encoding CSP and AMA1 is safe, well-tolerated and immunogenic but does not protect against controlled human malaria infection." *Hum Vaccin Immunother* **9**(10): 2165–77.

Tang, J., M. Olive, R. Pulmanausahakul, M. Schnell, N. Flomenberg, L. Eisenlohr and P. Flomenberg (2006). "Human CD8+ cytotoxic T cell responses to adenovirus capsid proteins." *Virology* **350**(2): 312–22.

Tao, N., G. P. Gao, M. Parr, J. Johnston, T. Baradet, J. M. Wilson, J. Barsoum and S. E. Fawell (2001). "Sequestration of adenoviral vector by Kupffer cells leads to a nonlinear dose response of transduction in liver." *Mol Ther* **3**(1): 28–35.

Tapia, M. D., S. O. Sow, K. E. Lyke, F. C. Haidara, F. Diallo, M. Doumbia, A. Traore, F. Coulibaly, M. Kodio, U. Onwuchekwa, M. B. Sztein, R. Wahid, J. D. Campbell, M. P. Kieny, V. Moorthy, E. B. Imoukhuede, T. Rampling, F. Roman, I. De Ryck, A. R. Bellamy, L. Dally, O. T. Mbaya, A. Ploquin, Y. Zhou, D. A. Stanley, R. Bailer, R. A. Koup, M. Roederer, J. Ledgerwood, A. V. Hill, W. R. Ballou, N. Sullivan, B. Graham and M. M. Levine (2016). "Use of ChAd3-EBO-Z Ebola virus vaccine in Malian and US adults, and boosting of Malian adults with MVA-BN-Filo: a phase 1, single-blind, randomised trial, a phase 1b, open-label and double-blind, dose-escalation trial, and a nested, randomised, double-blind, placebo-controlled trial." *Lancet Infect Dis* **16**(1): 31–42.

Tarodi, B., T. Subramanian and G. Chinnadurai (1993). "Functional similarity between adenovirus e1b 19k gene and bcl2 oncogene — mutant complementation and suppression of cell-death induced by DNA-damaging agents." *Int J Oncol* **3**(3): 467–72.

Tate, V. E. and L. Philipson (1979). "Parental adenovirus DNA accumulates in nucleo-some-like structures in infected cells." *Nucleic Acids Res* **6**(8): 2769–85.

Tatsis, N., M. O. Lasaro, S. W. Lin, L. H. Haut, Z. Q. Xiang, D. Zhou, L. Dimenna, H. Li, A. Bian, S. Abdulla, Y. Li, W. Giles-Davis, J. Engram, S. J. Ratcliffe, G. Silvestri, H. C. Ertl and M. R. Betts (2009). "Adenovirus vector-induced immune responses in nonhuman primates: responses to prime boost regimens." *J Immunol* **182**(10): 6587–99.

Teigler, J. E., J. C. Kagan and D. H. Barouch (2014). "Late endosomal trafficking of alternative serotype adenovirus vaccine vectors augments antiviral innate immunity." *J Virol* **88**(18): 10354–63.

Temperley, S. M. and R. T. Hay (1992). "Recognition of the adenovirus type 2 origin of DNA replication by the virally encoded DNA polymerase and preterminal proteins." *EMBO J* **11**(2): 761–8.

Tenge, V. R., A. P. Gounder, M. E. Wiens, W. Lu and J. G. Smith (2014). "Delineation of interfaces on human alpha-defensins critical for human adenovirus and human papillomavirus inhibition." *PLoS Pathog* **10**(9): e1004360.

Teodoro, J. G. and P. E. Branton (1997). "Regulation of p53-dependent apoptosis, transcriptional repression, and cell transformation by phosphorylation of the 55-kilodalton E1B protein of human adenovirus type 5." *J. Virol.* **71**(5): 3620–3627.

Teodoro, J. G., T. Halliday, S. G. Whalen, D. Takayesu, F. L. Graham and P. E. Branton (1994). "Phosphorylation at the carboxy terminus of the 55-kilodalton adenovirus type 5 E1B protein regulates transforming activity." *J. Virol.* **68**(2): 776–786.

Teodoro, J. G., G. C. Shore and P. E. Branton (1995). "Adenovirus E1A proteins induce apoptosis by both p53-dependent and p53-independent mechanisms." *Oncogene* **11**(3): 467–74.

Thimmappaya, B., C. Weinberger, R. J. Schneider and T. Shenk (1982). "Adenovirus VAI RNA is required for efficient translation of viral mRNAs at late times after infection." *Cell* **31**(3 Pt 2): 543–51.

Thomas, G. B. and M. B. Mathews (1980). "DNA replication and the early to late transition in adenovirus infection." *Cell* **22**: 523–533.

Thorne, S. H., B. Y. Tam, D. H. Kirn, C. H. Contag and C. J. Kuo (2006). "Selective intratumoral amplification of an antiangiogenic vector by an oncolytic virus produces enhanced antivascular and anti-tumor efficacy." *Mol Ther* **13**(5): 938–46.

Thorner, A. R., A. A. Lemckert, J. Goudsmit, D. M. Lynch, B. A. Ewald, M. Denholtz, M. J. Havenga and D. H. Barouch (2006). "Immunogenicity of heterologous recombinant adenovirus prime-boost vaccine regimens is enhanced by circumventing vector cross-reactivity." *J Virol* **80**(24): 12009–16.

Thurston, T. L., M. P. Wandel, N. von Muhlinen, A. Foeglein and F. Randow (2012). "Galectin 8 targets damaged vesicles for autophagy to defend cells against bacterial invasion." *Nature* **482**(7385): 414–8.

Tian, J., Z. Xu, J. S. Smith, S. E. Hofherr, M. A. Barry and A. P. Byrnes (2009). "Adenovirus activates complement by distinctly different mechanisms in vitro and in vivo: indirect complement activation by virions in vivo." *J Virol* **83**(11): 5648–58.

Tian, X., Q. Ma, Z. Jiang, J. Huang, Q. Liu, X. Lu, Q. Luo and R. Zhou (2015). "Identification and Application of Neutralizing Epitopes of Human Adenovirus Type 55 Hexon Protein." *Viruses* **7**(10): 5632–42.

Tibbetts, C. (1977). "Viral DNA sequences from incomplete particles of human adenovirus type 7." *Cell* **12**(1): 243–9.

Tiemessen, C. T. and A. H. Kidd (1993). "Sensitivity of subgroup F adenoviruses to interferon." *Arch Virol* **128**(1–2): 1–13.

Tilghman, S. M., P. J. Curtis, D. C. Tiemeier, P. Leder and C. Weissmann (1978). "The intervening sequence of a mouse beta-globin gene is transcribed within the 15S beta-globin mRNA precursor." *Proc Natl Acad Sci U S A* **75**(3): 1309–13.

Tilghman, S. M., D. C. Tiemeier, J. G. Seidman, B. M. Peterlin, M. Sullivan, J. V. Maizel and P. Leder (1978). "Intervening sequence of DNA identified in the structural portion of a mouse beta-globin gene." *Proc Natl Acad Sci U S A* **75**(2): 725–9.

Toietta, G., V. P. Mane, W. S. Norona, M. J. Finegold, P. Ng, A. F. McDonagh, A. L. Beaudet and B. Lee (2005). "Lifelong elimination of hyperbilirubinemia in the Gunn rat with a single injection of helper-dependent adenoviral vector." *Proc Natl Acad Sci U S A* **102**(11): 3930–5.

Tollefson, A. E., T. W. Hermiston, D. L. Lichtenstein, C. F. Colle, R. A. Tripp, T. Dimitrov, K. Toth, C. E. Wells, P. C. Doherty and W. S. Wold (1998). "Forced degradation of Fas inhibits apoptosis in adenovirus-infected cells." *Nature* **392**(6677): 726–30.

Tollefson, A. E., A. Scaria, T. W. Hermiston, J. S. Ryerse, L. J. Wold and W. S. Wold (1996). "The adenovirus death protein (E3–11.6K) is required at very late stages of infection for efficient cell lysis and release of adenovirus from infected cells." *J Virol* **70**(4): 2296–306.

Tollefson, A. E., J. F. Spencer, B. Ying, R. M. Buller, W. S. Wold and K. Toth (2014). "Cidofovir and brincidofovir reduce the pathology caused by systemic infection with human type 5 adenovirus in immunosuppressed Syrian hamsters, while ribavirin is largely ineffective in this model." *Antiviral Res* **112**: 38–46.

Tollefson, A. E., K. Toth, K. Doronin, M. Kuppuswamy, O. A. Doronina, D. L. Lichtenstein, T. W. Hermiston, C. A. Smith and W. S. Wold (2001). "Inhibition of TRAIL-induced apoptosis and forced internalization of TRAIL receptor 1 by adenovirus proteins." *J Virol* **75**(19): 8875–87.

Tollefson, A. E., B. Ying, K. Doronin, P. D. Sidor and W. S. Wold (2007). "Identification of a new human adenovirus protein encoded by a novel late l-strand transcription unit." *J Virol* **81**(23): 12918–26.

Tomasec, P., E. C. Wang, V. Groh, T. Spies, B. P. McSharry, R. J. Aicheler, R. J. Stanton and G. W. Wilkinson (2007). "Adenovirus vector delivery stimulates natural killer cell recognition." *J Gen Virol* **88**(Pt 4): 1103–8.

Tomko, R. P., R. Xu and L. Philipson (1997). "HCAR and MCAR: the human and mouse cellular receptors for subgroup C adenoviruses and group B coxsackieviruses." *Proc Natl Acad Sci U S A* **94**(7): 3352–6.

Tomson, B. N. and K. M. Arndt (2013). "The many roles of the conserved eukaryotic Paf1 complex in regulating transcription, histone modifications, and disease states." *Biochim Biophys Acta* **1829**(1): 116–26.

Topham, N. J. and E. W. Hewitt (2009). "Natural killer cell cytotoxicity: how do they pull the trigger?" *Immunology* **128**(1): 7–15.

Tordo, N., A. Foumier, C. Jallet, M. Szelechowski, B. Klonjkowski and M. Eloit (2008). "Canine adenovirus based rabies vaccines." *Dev Biol (Basel)* **131**: 467–76.

Tormanen, H., E. Backstrom, A. Carlsson and G. Akusjarvi (2006). "L4-33K, an adenovirus-encoded alternative RNA splicing factor." *J Biol Chem* **281**(48): 36510–7.

Tormanen Persson, H., A. K. Aksaas, A. K. Kvissel, T. Punga, A. Engstrom, B. S. Skalhegg and G. Akusjarvi (2012). "Two cellular protein kinases, DNA-PK and PKA, phosphorylate the adenoviral L4-33K protein and have opposite effects on L1 alternative RNA splicing." *PLoS One* **7**(2): e31871.

Toth, K., H. Djeha, B. Ying, A. E. Tollefson, M. Kuppuswamy, K. Doronin, P. Krajcsi, K. Lipinski, C. J. Wrighton and W. S. Wold (2004). "An oncolytic adenovirus vector combining enhanced cell-to-cell spreading, mediated by the ADP cytolytic protein, with selective replication in cancer cells with deregulated wnt signaling." *Cancer Res* **64**(10): 3638–44.

Toth, K., S. R. Lee, B. Ying, J. F. Spencer, A. E. Tollefson, J. E. Sagartz, I. K. Kong, Z. Wang and W. S. Wold (2015). "STAT2 Knockout Syrian Hamsters Support Enhanced Replication and Pathogenicity of Human Adenovirus, Revealing an Important Role of Type I Interferon Response in Viral Control." *PLoS Pathog* **11**(8): e1005084.

Toth, K., J. F. Spencer, D. Dhar, J. E. Sagartz, R. M. Buller, G. R. Painter and W. S. Wold (2008). "Hexadecyloxypropyl-cidofovir, CMX001, prevents adenovirus-induced mortality in a permissive, immunosuppressed animal model." *Proc Natl Acad Sci U S A* **105**(20): 7293–7.

Trentin, J. J., Y. Yabe and G. Taylor (1962). "The quest for human cancer viruses." *Science* **137**: 835–841.

Trevejo, J. M., M. W. Marino, N. Philpott, R. Josien, E. C. Richards, K. B. Elkon and E. Falck-Pedersen (2001). "TNF-alpha -dependent maturation of local dendritic cells is critical for activating the adaptive immune response to virus infection." *Proc Natl Acad Sci U S A* **98**(21): 12162–7.

Tribouley, C., P. Lutz, A. Staub and C. Kedinger (1994). "The product of the adenovirus intermediate gene IVa2 is a transcription activator of the major late promoter." *J. Virol.* **68**: 4450–4457.

Trinh, H. V., G. Lesage, V. Chennamparampil, B. Vollenweider, C. J. Burckhardt, S. Schauer, M. Havenga, U. F. Greber and S. Hemmi (2012). "Avidity binding of human adenovirus serotypes 3 and 7 to the membrane cofactor CD46 triggers infection." *J Virol* **86**(3): 1623–37.

Trotman, L. C., N. Mosberger, M. Fornerod, R. P. Stidwill and U. F. Greber (2001). "Import of adenovirus DNA involves the nuclear pore complex receptor CAN/Nup214 and histone H1." *Nat Cell Biol* **3**(12): 1092–100.

Tuboly, T. and E. Nagy (2001). "Construction and characterization of recombinant porcine adenovirus serotype 5 expressing the transmissible gastroenteritis virus spike gene." *J Gen Virol* **82**(Pt 1): 183–90.

Tucker, P. A., D. Tsernoglou, A. D. Tucker, F. E. Coenjaerts, H. Leenders and P. C. van der Vliet (1994). "Crystal structure of the adenovirus DNA binding protein reveals a hook-on model for cooperative DNA binding." *EMBO J* **13**(13): 2994–3002.

Tuettenberg, A., H. Jonuleit, T. Tuting, J. Bruck, J. Knop and A. H. Enk (2003). "Priming of T cells with Ad-transduced DC followed by expansion with peptide-pulsed DC significantly enhances the induction of tumor-specific CD8+ T cells: implications for an efficient vaccination strategy." *Gene Ther* **10**(3): 243–50.

Turnell, A. S., R. J. Grand and P. H. Gallimore (1999). "The replicative capacities of large E1B-null group A and group C adenoviruses are independent of host cell p53 status." *J Virol* **73**(3): 2074–83.

Turner, R. L., J. C. Wilkinson and D. A. Ornelles (2014). "E1B and E4 oncoproteins of adenovirus antagonize the effect of apoptosis inducing factor." *Virology* **456–457**: 205–19.

Tuve, S., H. Wang, C. Ware, Y. Liu, A. Gaggar, K. Bernt, D. Shayakhmetov, Z. Li, R. Strauss, D. Stone and A. Lieber (2006). "A new group B adenovirus receptor is expressed at high levels on human stem and tumor cells." *J Virol* **80**(24): 12109–20.

Uchio, E., H. Inoue, A. Fuchigami and K. Kadonosono (2011). "Anti-adenoviral effect of interferon-beta and interferon-gamma in serotypes that cause acute keratoconjunctivitis." *Clin Experiment Ophthalmol* **39**(4): 358–63.

Ueyama, K., K. Mori, T. Shoji, H. Omata, P. L. Gehlbach, D. E. Brough, L. L. Wei and S. Yoneya (2014). "Ocular localization and transduction by adenoviral vectors are serotype-dependent and can be modified by inclusion of RGD fiber modifications." *PLoS One* **9**(9): e108071.

Uil, T. G., J. Vellinga, J. de Vrij, S. K. van den Hengel, M. J. Rabelink, S. J. Cramer, J. J. Eekels, Y. Ariyurek, M. van Galen and R. C. Hoeben (2011). "Directed adenovirus evolution using engineered mutator viral polymerases." *Nucleic Acids Res* **39**(5): e30.

Ulasov, I., A. V. Borovjagin, N. Kaverina, B. Schroeder, N. Shah, B. Lin, A. Baryshnikov and C. Cobbs (2015). "MT1-MMP silencing by an shRNA-armed glioma-targeted conditionally replicative adenovirus (CRAd) improves its anti-glioma efficacy in vitro and in vivo." *Cancer Lett* **365**(2): 240–50.

Ulasov, I. V., M. A. Tyler, A. A. Rivera, D. M. Nettlebeck, J. T. Douglas and M. S. Lesniak (2008). "Evaluation of E1A double mutant oncolytic adenovectors in anti-glioma gene therapy." *J Med Virol* **80**(9): 1595–603.

Ulfendahl, P. J., S. Linder, J. P. Kreivi, K. Nordqvist, C. Sevensson, H. Hultberg and G. Akusjärvi (1987). "A novel adenovirus-2 E1A mRNA encoding a protein with transcription activation properties." *EMBO J* **6**: 2037–2044.

Ullman, A. J. and P. Hearing (2008). "Cellular proteins PML and Daxx mediate an innate antiviral defense antagonized by the adenovirus E4 ORF3 protein." *J Virol* **82**(15): 7325–35.

Ullman, A. J., N. C. Reich and P. Hearing (2007). "Adenovirus E4 ORF3 protein inhibits the interferon-mediated antiviral response." *J Virol* **81**(9): 4744–52.

Uriel, S., E. M. Brey and H. P. Greisler (2006). "Sustained low levels of fibroblast growth factor-1 promote persistent microvascular network formation." *Am J Surg* **192**(5): 604–9.

Uusi-Kerttula, H., M. Legut, J. Davies, R. Jones, E. Hudson, L. Hanna, R. J. Stanton, J. D. Chester and A. L. Parker "Incorporation of Peptides Targeting EGFR and FGFR1 into the Adenoviral Fiber Knob Domain and Their Evaluation as Targeted Cancer Therapies." *Hum Gene Ther* **26**(5): 320–9.

Vachon, V. K. and G. L. Conn (2015). "Adenovirus VA RNA: An essential pro-viral non-coding RNA." *Virus Res* **212**: 39–52.

Vales, L. D. and J. E. Darnell (1989). "Promoter occlusion prevents transcription of adenovirus polypeptide 1X mRNA until after DNA replication." *Genes Dev.* **3**: 49–59.

van Bergen, B. G. and P. C. van der Vliet (1983). "Temperature-sensitive initiation and elongation of adenovirus DNA replication in vitro with nuclear extracts from H5ts36-, H5ts149-, and H5ts125-infected HeLa cells." *J Virol* **46**(2): 642–8.

van Breukelen, B., P. N. Kanellopoulos, P. A. Tucker and P. C. van der Vliet (2000). "The formation of a flexible DNA-binding protein chain is required for efficient DNA unwinding and adenovirus DNA chain elongation." *J Biol Chem* **275**(52): 40897–903.

van den Bosch, M., R. T. Bree and N. F. Lowndes (2003). "The MRN complex: coordinating and mediating the response to broken chromosomes." *EMBO Rep* **4**(9): 844–9.

van der Poel, H. G., B. Molenaar, V. W. van Beusechem, H. J. Haisma, R. Rodriguez, D. T. Curiel and W. R. Gerritsen (2002). "Epidermal growth factor receptor targeting of replication competent adenovirus enhances cytotoxicity in bladder cancer." *J Urol* **168**(1): 266–72.

Van der Vliet, P. C. (1995). "Adenovirus DNA replication." *Curr Top Microbiol Immunol* **199** (Pt 2): 1–30.

Van der Vliet, P. C. and R. C. Hoeben (2006). Adenovirus. In: *DNA Replication and Human Disease*. M. L. DePamphilis (ed). Cold Spring Harbor, NY, Cold Spring Harbor Laboratory Press: 645–661.

van der Vliet, P. C. and A. J. Levine (1973). "DNA-binding proteins specific for cells infected by adenovirus." *Nat New Biol* **246**(154): 170–4.

van Dyke, M. W. and R. G. Roeder (1987). "Multiple proteins bind to VA RNA genes of adenovirus type 2." *Mol. Cell. Biol.* **7**: 1021–1031.

van Leeuwen, H. C., M. Rensen and P. C. van der Vliet (1997). "The Oct-1 POU homeodomain stabilizes the adenovirus preinitiation complex via a direct interaction with the priming protein and is displaced when the replication fork passes." *J Biol Chem* **272**(6): 3398–405.

van Oostrum, J. and R. M. Burnett (1985). "Molecular composition of the adenovirus type 2 virion." *J Virol* **56**(2): 439–48.

van Raaij, M. J., E. Chouin, H. van der Zandt, J. M. Bergelson and S. Cusack (2000). "Dimeric structure of the coxsackievirus and adenovirus receptor D1 domain at 1.7 A resolution." *Structure* **8**(11): 1147–55.

van Raaij, M. J., N. Louis, J. Chroboczek and S. Cusack (1999). "Structure of the human adenovirus serotype 2 fiber head domain at 1.5 A resolution." *Virology* **262**(2): 333–43.

van Raaij, M. J., A. Mitraki, G. Lavigne and S. Cusack (1999). "A triple beta-spiral in the adenovirus fibre shaft reveals a new structural motif for a fibrous protein." *Nature* **401**(6756): 935–8.

van Tol, M. J., A. C. Kroes, J. Schinkel, W. Dinkelaar, E. C. Claas, C. M. Jol-van der Zijde and J. M. Vossen (2005). "Adenovirus infection in paediatric stem cell transplant recipients: increased risk in young children with a delayed immune recovery." *Bone Marrow Transplant* **36**(1): 39–50.

Varghese, R., Y. Mikyas, P. L. Stewart and R. Ralston (2004). "Postentry neutralization of adenovirus type 5 by an antihexon antibody." *J Virol* **78**(22): 12320–32.

Vasavada, R., K. B. Eager, G. Barbanti-Brodano, A. Caputo and R. P. Ricciardi (1986). "Adenovirus type 12 early region 1A proteins repress class I HLA expression in transformed human cells." *Proc Natl Acad Sci U S A* **83**(14): 5257–61.

Veesler, D., K. Cupelli, M. Burger, P. Graber, T. Stehle and J. E. Johnson (2014). "Single-particle EM reveals plasticity of interactions between the adenovirus penton base and integrin alphaVbeta3." *Proc Natl Acad Sci U S A* **111**(24): 8815–9.

Veesler, D. and J. E. Johnson (2012). "Virus maturation." *Annu Rev Biophys* **41**: 473–96.

Veesler, D., T. S. Ng, A. K. Sendamarai, B. J. Eilers, C. M. Lawrence, S. M. Lok, M. J. Young, J. E. Johnson and C. Y. Fu (2013). "Atomic structure of the 75 MDa extremophile Sulfolobus turreted icosahedral virus determined by CryoEM and X-ray crystallography." *Proc Natl Acad Sci U S A* **110**(14): 5504–9.

Veltrop-Duits, L. A., B. Heemskerk, C. C. Sombroek, T. van Vreeswijk, S. Gubbels, R. E. Toes, C. J. Melief, K. L. Franken, M. Havenga, M. J. van Tol and M. W. Schilham (2006). "Human CD4+ T cells stimulated by conserved adenovirus 5 hexon peptides recognize cells infected with different species of human adenovirus." *Eur J Immunol* **36**(9): 2410–23.

Venkataraman, S., S. P. Reddy, J. Loo, N. Idamakanti, P. L. Hallenbeck and V. S. Reddy (2008). "Structure of Seneca Valley Virus-001: an oncolytic picornavirus representing a new genus." *Structure* **16**(10): 1555–61.

Vertegaal, A. C., H. B. Kuiperij, A. Houweling, M. Verlaan, A. J. van der Eb and A. Zantema (2003). "Differential expression of tapasin and immunoproteasome subunits in adenovirus type 5- versus type 12-transformed cells." *J Biol Chem* **278**(1): 139–46.

Vigant, F., D. Descamps, B. Jullienne, S. Esselin, E. Connault, P. Opolon, T. Tordjmann, E. Vigne, M. Perricaudet and K. Benihoud (2008). "Substitution of hexon hypervariable region 5 of adenovirus serotype 5 abrogates blood factor binding and limits gene transfer to liver." *Mol Ther* **16**(8): 1474–80.

Vijayalingam, S. and G. Chinnadurai (2013). "Adenovirus L-E1A activates transcription through mediator complex-dependent recruitment of the super elongation complex." *J Virol* **87**(6): 3425–34.

Vink, E. I., M. A. Yondola, K. Wu and P. Hearing (2012). "Adenovirus E4-ORF3-dependent relocalization of TIF1alpha and TIF1gamma relies on access to the Coiled-Coil motif." *Virology* **422**(2): 317–25.

Vink, E. I., Y. Zheng, R. Yeasmin, T. Stamminger, L. T. Krug and P. Hearing (2015). "Impact of Adenovirus E4-ORF3 Oligomerization and Protein Localization on Cellular Gene Expression." *Viruses* **7**(5): 2428–49.

Voelkerding, K. and D. F. Klessig (1986). "Identification of two nuclear subclasses of the adenovirus type 5- encoded DNA-binding protein." *J. Virol.* **60**(2): 353–362.

Vogels, R., D. Zuijdgeest, R. van Rijnsoever, E. Hartkoorn, I. Damen, M. P. de Bethune, S. Kostense, G. Penders, N. Helmus, W. Koudstaal, M. Cecchini, A. Wetterwald, M. Sprangers, A. Lemckert, O. Ophorst, B. Koel, M. van Meerendonk, P. Quax, L. Panitti, J. Grimbergen, A. Bout, J. Goudsmit and M. Havenga (2003). "Replication-deficient human adenovirus type 35 vectors for gene transfer and vaccination: efficient human cell infection and bypass of preexisting adenovirus immunity." *J Virol* **77**(15): 8263–71.

Von Seggern, D. J., C. Y. Chiu, S. K. Fleck, P. L. Stewart and G. R. Nemerow (1999). "A helper-independent adenovirus vector with E1, E3, and fiber deleted: structure and infectivity of fiberless particles." *J Virol* **73**(2): 1601–8.

Vos, H. L., F. M. van der Lee, A. M. Reemst, A. E. van Loon and J. S. Sussenbach (1988). "The genes encoding the DNA binding protein and the 23K protease of adenovirus types 40 and 41." *Virology* **163**(1): 1–10.

Vousden, K. H. (1995). "Regulation of the cell cycle by viral oncoproteins." *Semin Cancer Biol* **6**(2): 109–16.

Waddington, S. N., J. H. McVey, D. Bhella, A. L. Parker, K. Barker, H. Atoda, R. Pink, S. M. Buckley, J. A. Greig, L. Denby, J. Custers, T. Morita, I. M. Francischetti, R. Q. Monteiro, D. H. Barouch, N. van Rooijen, C. Napoli, M. J. Havenga, S. A. Nicklin and A. H. Baker (2008). "Adenovirus serotype 5 hexon mediates liver gene transfer." *Cell* **132**(3): 397–409.

Waddington, S. N., A. L. Parker, M. Havenga, S. A. Nicklin, S. M. Buckley, J. H. McVey and A. H. Baker (2007). "Targeting of adenovirus serotype 5 (Ad5) and 5/47 pseudotyped vectors in vivo: fundamental involvement of coagulation factors and redundancy of CAR binding by Ad5." *J Virol* **81**(17): 9568–71.

Wadell, G. (1972). "Sensitization and neutralization of adenovirus by specific sera against capsid subunits." *J Immunol* **108**(3): 622–32.

Waga, S. and B. Stillman (1998). "Cyclin-dependent kinase inhibitor p21 modulates the DNA primer-template recognition complex." *Mol Cell Biol* **18**(7): 4177–87.

Wahid, A. M., V. K. Coventry and G. L. Conn (2008). "Systematic deletion of the adenovirus-associated RNAI terminal stem reveals a surprisingly active RNA inhibitor of double-stranded RNA-activated protein kinase." *J Biol Chem* **283**(25): 17485–93.

Wahid, A. M., V. K. Coventry and G. L. Conn (2009). "The PKR-binding domain of adenovirus VA RNAI exists as a mixture of two functionally non-equivalent structures." *Nucleic Acids Res* **37**(17): 5830–7.

Wakayama, M., M. Abei, R. Kawashima, E. Seo, K. Fukuda, H. Ugai, T. Murata, N. Tanaka, I. Hyodo, H. Hamada and K. K. Yokoyama (2007). "E1A, E1B double-restricted adenovirus with RGD-fiber modification exhibits enhanced oncolysis for CAR-deficient biliary cancers." *Clin Cancer Res* **13**(10): 3043–50.

Wallach, D., E. E. Varfolomeev, N. L. Malinin, Y. V. Goltsev, A. V. Kovalenko and M. P. Boldin (1999). "Tumor necrosis factor receptor and Fas signaling mechanisms." *Annu Rev Immunol* **17**: 331–67.

Walport, M. J. (2001). "Complement. First of two parts." *N Engl J Med* **344**(14): 1058–66.

Walsh, M. P., J. Seto, D. Tirado, J. Chodosh, D. Schnurr, D. Seto and M. S. Jones (2010). "Computational analysis of human adenovirus serotype 18." *Virology* **404**(2): 284–92.

Walters, R. W., P. Freimuth, T. O. Moninger, I. Ganske, J. Zabner and M. J. Welsh (2002). "Adenovirus fiber disrupts CAR-mediated intercellular adhesion allowing virus escape." *Cell* **110**(6): 789–99.

Walters, R. W., T. Grunst, J. M. Bergelson, R. W. Finberg, M. J. Welsh and J. Zabner (1999). "Basolateral localization of fiber receptors limits adenovirus infection from the apical surface of airway epithelia." *J Biol Chem* **274**(15): 10219–26.

Wang, B. X. and E. N. Fish (2012). "The yin and yang of viruses and interferons." *Trends Immunol* **33**(4): 190–7.

Wang, G., G. T. Cantin, J. L. Stevens and A. J. Berk (2001). "Characterization of mediator complexes from HeLa cell nuclear extract." *Mol Cell Biol* **21**(14): 4604–13.

Wang, H., I. Beyer, J. Persson, H. Song, Z. Li, M. Richter, H. Cao, R. van Rensburg, X. Yao, K. Hudkins, R. Yumul, X. B. Zhang, M. Yu, P. Fender, A. Hemminki and A. Lieber (2012). "A new human DSG2-transgenic mouse model for studying the tropism and pathology of human adenoviruses." *J Virol* **86**(11): 6286–302.

Wang, H., Z. Y. Li, Y. Liu, J. Persson, I. Beyer, T. Moller, D. Koyuncu, M. R. Drescher, R. Strauss, X. B. Zhang, J. K. Wahl, 3rd, N. Urban, C. Drescher, A. Hemminki, P. Fender and A. Lieber (2011). "Desmoglein 2 is a receptor for adenovirus serotypes 3, 7, 11 and 14." *Nat Med* **17**(1): 96–104.

Wang, H., Y. Liu, Z. Y. Li, X. Fan, A. Hemminki and A. Lieber (2010). "A recombinant adenovirus type 35 fiber knob protein sensitizes lymphoma cells to rituximab therapy." *Blood* **115**(3): 592–600.

Wang, H., R. Yumul, H. Cao, L. Ran, X. Fan, M. Richter, F. Epstein, J. Gralow, C. Zubieta, P. Fender and A. Lieber (2013). "Structural and functional studies on the interaction of adenovirus fiber knobs and desmoglein 2." *J Virol* **87**(21): 11346–62.

Wang, H. G., Y. Rikitake, M. C. Carter, P. Yaciuk, S. E. Abraham, B. Zerler and E. Moran (1993). "Identification of specific adenovirus E1A N-terminal residues critical to the binding of cellular proteins and to the control of cell growth." *J Virol* **67**(1): 476–88.

Wang, K., S. Huang, A. Kapoor-Munshi and G. Nemerow (1998). "Adenovirus internalization and infection require dynamin." *J Virol* **72**(4): 3455–8.

Wang, Q., T. Matsui, T. Domitrovic, Y. Zheng, P. C. Doerschuk and J. E. Johnson (2013). "Dynamics in cryo EM reconstructions visualized with maximum-likelihood derived variance maps." *J Struct Biol* **181**(3): 195–206.

Wang, Y., S. A. Xue, G. Hallden, J. Francis, M. Yuan, B. E. Griffin and N. R. Lemoine (2005). "Virus-associated RNA I-deleted adenovirus, a potential oncolytic agent targeting EBV-associated tumors." *Cancer Res* **65**(4): 1523–31.

Watkins, D. I., D. R. Burton, E. G. Kallas, J. P. Moore and W. C. Koff (2008). "Nonhuman primate models and the failure of the Merck HIV-1 vaccine in humans." *Nat Med* **14**(6): 617–21.

Watkinson, R. E., W. A. McEwan, J. C. Tam, M. Vaysburd and L. C. James (2015). "TRIM21 Promotes cGAS and RIG-I Sensing of Viral Genomes during Infection by Antibody-Opsonized Virus." *PLoS Pathog* **11**(10): e1005253.

Weaver, E. A. (2014). "Vaccines within vaccines: the use of adenovirus types 4 and 7 as influenza vaccine vectors." *Hum Vaccin Immunother* **10**(3): 544–56.

Weber, F., V. Wagner, N. Kessler and O. Haller (2006). "Induction of interferon synthesis by the PKR-inhibitory VA RNAs of adenoviruses." *J Interferon Cytokine Res* **26**(1): 1–7.

Weber, J. (1976). "Genetic analysis of adenovirus type 2 III. Temperature sensitivity of processing viral proteins." *J Virol* **17**(2): 462–71.

Webster, A., I. R. Leith and R. T. Hay (1994). "Activation of adenovirus-coded protease and processing of preterminal protein." *J Virol* **68**(11): 7292–300.

Webster, A., I. R. Leith, J. Nicholson, J. Hounsell and R. T. Hay (1997). "Role of preterminal protein processing in adenovirus replication." *J Virol* **71**(9): 6381–9.

Webster, A., S. Russell, P. Talbot, W. C. Russell and G. D. Kemp (1989). "Characterization of the adenovirus proteinase: substrate specificity." *J Gen Virol* **70** (Pt 12): 3225–34.

Webster, K. (1999). "Molecular switches for regulating therapeutic genes." *Gene Ther* **6**(6): 951–3.

Webster, L. C., K. Zhang, B. Chance, I. Ayene, J. S. Culp, W. J. Huang, F. Y. Wu and R. P. Ricciardi (1991). "Conversion of the E1A Cys4 zinc finger to a nonfunctional His2,Cys2 zinc finger by a single point mutation." *Proc Natl Acad Sci U S A* **88**(22): 9989–93.

Weiden, M. D. and H. S. Ginsberg (1994). "Deletion of the E4 region of the genome produces adenovirus DNA concatemers." *Proc Natl Acad Sci U S A* **91**(1): 153–7.

Weil, P. A., D. S. Luse, J. Segall and R. G. Roeder (1979). "Selective and accurate initiation of transcription at the Ad2 major late promotor in a soluble system dependent on purified RNA polymerase II and DNA." *Cell* **18**(2): 469–84.

Weinmann, R., H. J. Raskas and R. G. Roeder (1974). "Role of DNA-dependent RNA polymerases II and III in transcription of the adenovirus genome late in productive infection." *Proc Natl Acad Sci U S A* **71**(9): 3426–39.

Whitby, L. R. and D. L. Boger (2012). "Comprehensive peptidomimetic libraries targeting protein-protein interactions." *Acc Chem Res* **45**(10): 1698–709.

White, E. (1995). "Regulation of p53-dependent apoptosis by E1A and E1B." *Curr Top Microbiol Immunol* **199 (Pt 3)**: 34–58.

White, E., P. Sabbatini, M. Debbas, W. S. Wold, D. I. Kusher and L. R. Gooding (1992). "The 19-kilodalton adenovirus E1B transforming protein inhibits programmed cell death and prevents cytolysis by tumor necrosis factor alpha." *Mol Cell Biol* **12**(6): 2570–80.

Whittaker, G. R., M. Kann and A. Helenius (2000). "Viral entry into the nucleus." *Annu Rev Cell Dev Biol* **16**: 627–51.

Whyte, P., N. M. Williamson and E. Harlow (1989). "Cellular targets for transformation by the adenovirus E1A proteins." *Cell* **56**(1): 67–75.

Wickham, T. J., P. Mathias, D. A. Cheresh and G. R. Nemerow (1993). "Integrins alpha v beta 3 and alpha v beta 5 promote adenovirus internalization but not virus attachment." *Cell* **73**(2): 309–19.

Wiethoff, C. M. and G. R. Nemerow (2015). "Adenovirus membrane penetration: Tickling the tail of a sleeping dragon." *Virology* **479–480**: 591–9.

Wiethoff, C. M., H. Wodrich, L. Gerace and G. R. Nemerow (2005). "Adenovirus protein VI mediates membrane disruption following capsid disassembly." *J Virol* **79**(4): 1992–2000.

Williams, G. J., S. P. Lees-Miller and J. A. Tainer (2010). "Mre11-Rad50-Nbs1 conformations and the control of sensing, signaling, and effector responses at DNA double-strand breaks." *DNA Repair (Amst)* **9**(12): 1299–306.

Williams, J., B. D. Karger, Y. S. Ho, C. L. Castiglia, T. Mann and S. J. Flint (1986). "The adenovirus E1B 495R protein plays a role in regulating the transport and stability of the viral late messages." *Cancer Cells* **4**: 275–284.

Williams, J. F., Y. Zhang, M. A. Williams, S. Hou, D. Kushner and R. P. Ricciardi (2004). "E1A-based determinants of oncogenicity in human adenovirus groups A and C." *Curr Top Microbiol Immunol* **273**: 245–88.

Williams, R. S., J. S. Williams and J. A. Tainer (2007). "Mre11-Rad50-Nbs1 is a keystone complex connecting DNA repair machinery, double-strand break signaling, and the chromatin template." *Biochem Cell Biol* **85**(4): 509–20.

Wilson, J. L., V. K. Vachon, S. Sunita, S. L. Schwartz and G. L. Conn (2014). "Dissection of the adenoviral VA RNAI central domain structure reveals minimum requirements for RNA-mediated inhibition of PKR." *J Biol Chem* **289**(33): 23233–45.

Wilson, J. M. (2009). "Lessons learned from the gene therapy trial for ornithine transcarbamylase deficiency." *Mol Genet Metab* **96**(4): 151–7.

Wilson, S. S., A. Tocchi, M. K. Holly, W. C. Parks and J. G. Smith (2015). "A small intestinal organoid model of non-invasive enteric pathogen-epithelial cell interactions." *Mucosal Immunol* **8**(2): 352–61.

Wilson, S. S., M. E. Wiens and J. G. Smith (2013). "Antiviral mechanisms of human defensins." *J Mol Biol* **425**(24): 4965–80.

Windheim, M. and H. G. Burgert (2002). "Characterization of E3/49K, a novel, highly glycosylated E3 protein of the epidemic keratoconjunctivitis-causing adenovirus type 19a." *J Virol* **76**(2): 755–66.

Windheim, M., A. Hilgendorf and H. G. Burgert (2004). "Immune evasion by adenovirus E3 proteins: exploitation of intracellular trafficking pathways." *Curr Top Microbiol Immunol* **273**: 29–85.

Windheim, M., J. H. Southcombe, E. Kremmer, L. Chaplin, D. Urlaub, C. S. Falk, M. Claus, J. Mihm, M. Braithwaite, K. Dennehy, H. Renz, M. Sester, C. Watzl and H. G. Burgert (2013). "A unique secreted adenovirus E3 protein binds to the leukocyte common antigen CD45 and modulates leukocyte functions." *Proc Natl Acad Sci U S A* **110**(50): E4884–93.

Wirth, T., L. Zender, B. Schulte, B. Mundt, R. Plentz, K. L. Rudolph, M. Manns, S. Kubicka and F. Kuhnel (2003). "A telomerase-dependent conditionally replicating adenovirus for selective treatment of cancer." *Cancer Res* **63**(12): 3181–8.

Wodrich, H., A. Cassany, M. A. D'Angelo, T. Guan, G. Nemerow and L. Gerace (2006). "Adenovirus core protein pVII is translocated into the nucleus by multiple import receptor pathways." *J Virol* **80**(19): 9608–18.

Wodrich, H., T. Guan, G. Cingolani, D. Von Seggern, G. Nemerow and L. Gerace (2003). "Switch from capsid protein import to adenovirus assembly by cleavage of nuclear transport signals." *EMBO J* **22**(23): 6245–55.

Wodrich, H., D. Henaff, B. Jammart, C. Segura-Morales, S. Seelmeir, O. Coux, Z. Ruzsics, C. M. Wiethoff and E. J. Kremer (2010). "A capsid-encoded PPxY-motif facilitates adenovirus entry." *PLoS Pathog* **6**(3): e1000808.

Wohlfart, C. (1988). "Neutralization of adenoviruses: kinetics, stoichiometry, and mechanisms." *J Virol* **62**(7): 2321–8.

Wohlfart, C. E., U. K. Svensson and E. Everitt (1985). "Interaction between HeLa cells and adenovirus type 2 virions neutralized by different antisera." *J Virol* **56**(3): 896–903.

Wold, W. S., C. Cladaras, S. L. Deutscher and Q. S. Kapoor (1985). "The 19-kDa glycoprotein coded by region E3 of adenovirus. Purification, characterization, and structural analysis." *J Biol Chem* **260**(4): 2424–31.

Wold, W. S. and M. Horwitz (2006). Adenoviruses. In: *Fields Virology*. D. M. Knipe and P. M. Howley (eds). Philadelphia, PA, Wolters-Kluwer/Lippincott Williams and Wilkins.

Wold, W. S. and M. G. Ison (2013). Adenoviruses. In: *Fields Virology*. D. M. Knipe and P. M. Howley (eds). Philadelphia, PA, Wolters Kluwer/Lippincott, Williams and Wilkins.

Wolff, G., S. Worgall, N. van Rooijen, W. R. Song, B. G. Harvey and R. G. Crystal (1997). "Enhancement of in vivo adenovirus-mediated gene transfer and expression by prior depletion of tissue macrophages in the target organ." *J Virol* **71**(1): 624–9.

Wolfrum, N. and U. F. Greber (2013). "Adenovirus signalling in entry." *Cell Microbiol* **15**(1): 53–62.

Worgall, S., P. L. Leopold, G. Wolff, B. Ferris, N. Van Roijen and R. G. Crystal (1997). "Role of alveolar macrophages in rapid elimination of adenovirus vectors administered to the epithelial surface of the respiratory tract." *Hum Gene Ther* **8**(14): 1675–84.

Woo, J. L. and A. J. Berk (2007). "Adenovirus protein ubiquitin ligase stimlates late mRNA nuclear export." *J Virol* **81**(2): 575–87.

Wright, J., Z. Atwan, S. J. Morris and K. N. Leppard (2015). "The Human Adenovirus Type 5 L4 Promoter Is Negatively Regulated by TFII-I and L4-33K." *J Virol* **89**(14): 7053–63.

Wright, J. and K. N. Leppard (2013). "The human adenovirus 5 L4 promoter is activated by cellular stress response protein p53." *J Virol* **87**(21): 11617–25.

Wu, C., L. Bai, Z. Li, C. E. Samuel, G. Akusjarvi and C. Svensson (2015). "Poor growth of human adenovirus-12 compared to adenovirus-2 correlates with a failure to impair PKR activation during the late phase of infection." *Virology* **475**: 120–8.

Wu, E., J. Fernandez, S. K. Fleck, D. J. Von Seggern, S. Huang and G. R. Nemerow (2001). "A 50-kDa membrane protein mediates sialic acid-independent binding and infection of conjunctival cells by adenovirus type 37." *Virology* **279**(1): 78–89.

Wu, E. and G. R. Nemerow (2004). "Virus yoga: the role of flexibility in virus host cell recognition." *Trends Microbiol* **12**(4): 162–9.

Wu, E., L. Pache, D. J. Von Seggern, T. M. Mullen, Y. Mikyas, P. L. Stewart and G. R. Nemerow (2003). "Flexibility of the adenovirus fiber is required for efficient receptor interaction." *J Virol* **77**(13): 7225–35.

Wu, E., S. A. Trauger, L. Pache, T. M. Mullen, D. J. von Seggern, G. Siuzdak and G. R. Nemerow (2004). "Membrane cofactor protein is a receptor for adenoviruses associated with epidemic keratoconjunctivitis." *J Virol* **78**(8): 3897–905.

Wu, K., D. Orozco and P. Hearing (2012). "The adenovirus L4-22K protein is multifunctional and is an integral component of crucial aspects of infection." *J Virol* **86**(19): 10474–83.

Wu, L., P. Zhou, X. Ge, L. F. Wang, M. L. Baker and Z. Shi (2013). "Deep RNA sequencing reveals complex transcriptional landscape of a bat adenovirus." *J Virol* **87**(1): 503–11.

Xi, Q., R. Cuesta and R. J. Schneider (2004). "Tethering of eIF4G to adenoviral mRNAs by viral 100k protein drives ribosome shunting." *Genes Dev.* **18**: 1997–2009.

Xi, Q., R. Cuesta and R. J. Schneider (2005). "Regulation of translation by ribosome shunting through phosphotyrosine-dependent coupling of adenovirus protein 100k to viral mRNAs." *J Virol* **79**(9): 5676–83.

Xia, D., L. J. Henry, R. D. Gerard and J. Deisenhofer (1994). "Crystal structure of the receptor-binding domain of adenovirus type 5 fiber protein at 1.7 A resolution." *Structure* **2**(12): 1259–70.

Xia, Z. J., J. H. Chang, L. Zhang, W. Q. Jiang, Z. Z. Guan, J. W. Liu, Y. Zhang, X. H. Hu, G. H. Wu, H. Q. Wang, Z. C. Chen, J. C. Chen, Q. H. Zhou, J. W. Lu, Q. X. Fan, J. J. Huang and X. Zheng (2004). "Phase III randomized clinical trial of intratumoral injection of E1B gene-deleted adenovirus (H101) combined with cisplatin-based chemotherapy in treating squamous cell cancer of head and neck or esophagus." *Ai Zheng* **23**(12): 1666–70.

Xiang, Z., G. Gao, A. Reyes-Sandoval, C. J. Cohen, Y. Li, J. M. Bergelson, J. M. Wilson and H. C. Ertl (2002). "Novel, chimpanzee serotype 68-based adenoviral vaccine carrier for induction of antibodies to a transgene product." *J Virol* **76**(6): 2667–75.

Xiang, Z., Y. Li, A. Cun, W. Yang, S. Ellenberg, W. M. Switzer, M. L. Kalish and H. C. Ertl (2006). "Chimpanzee adenovirus antibodies in humans, sub-Saharan Africa." *Emerg Infect Dis* **12**(10): 1596–9.

Xiang, Z. Q., Y. Yang, J. M. Wilson and H. C. Ertl (1996). "A replication-defective human adenovirus recombinant serves as a highly efficacious vaccine carrier." *Virology* **219**(1): 220–7.

Xiao, T., J. K. Fan, H. L. Huang, J. F. Gu, L. Y. Li and X. Y. Liu (2010). "VEGI-armed oncolytic adenovirus inhibits tumor neovascularization and directly induces mitochondria-mediated cancer cell apoptosis." *Cell Res* **20**(3): 367–78.

Xiao, T., C. F. Kao, N. J. Krogan, Z. W. Sun, J. F. Greenblatt, M. A. Osley and B. D. Strahl (2005). "Histone H2B ubiquitylation is associated with elongating RNA polymerase II." *Mol Cell Biol* **25**(2): 637–51.

Xiao, T., J. Takagi, B. S. Coller, J. H. Wang and T. A. Springer (2004). "Structural basis for allostery in integrins and binding to fibrinogen-mimetic therapeutics." *Nature* **432**(7013): 59–67.

Xu, N., B. Segerman, X. Zhou and G. Akusjarvi (2007). "Adenovirus virus-associated RNAII-derived small RNAs are efficiently incorporated into the rna-induced silencing complex and associate with polyribosomes." *J Virol* **81**(19): 10540–9.

Xu, X., H. G. Zhang, Z. Y. Liu, Q. Wu, P. A. Yang, S. H. Sun, J. Chen, H. C. Hsu and J. D. Mountz (2004). "Defective clearance of adenovirus in IRF-1 mice associated with defects in NK and T cells but not macrophages." *Scand J Immunol* **60**(1–2): 89–99.

Xu, Z., Q. Qiu, J. Tian, J. S. Smith, G. M. Conenello, T. Morita and A. P. Byrnes (2013). "Coagulation factor X shields adenovirus type 5 from attack by natural antibodies and complement." *Nat Med* **19**(4): 452–7.

Xu, Z., J. Tian, J. S. Smith and A. P. Byrnes (2008). "Clearance of adenovirus by Kupffer cells is mediated by scavenger receptors, natural antibodies, and complement." *J Virol* **82**(23): 11705–13.

Xue, Y., J. S. Johnson, D. A. Ornelles, J. Lieberman and D. A. Engel (2005). "Adenovirus protein VII functions throughout early phase and interacts with cellular proteins SET and pp32." *J Virol* **79**(4): 2474–83.

Yamada, K., H. Moriyama, H. Yasuda, K. Hara, Y. Maniwa, H. Hamada, K. Yokono and M. Nagata (2007). "Modification of the Rb-binding domain of replication-competent adenoviral vector enhances cytotoxicity against human esophageal cancers via NF-kappaB activity." *Hum Gene Ther* **18**(5): 389–400.

Yamaguchi, T., K. Kawabata, N. Koizumi, F. Sakurai, K. Nakashima, H. Sakurai, T. Sasaki, N. Okada, K. Yamanishi and H. Mizuguchi (2007). "Role of MyD88 and TLR9 in the innate immune response elicited by serotype 5 adenoviral vectors." *Hum Gene Ther* **18**(8): 753–62.

Yamaguchi, T., K. Kawabata, E. Kouyama, K. J. Ishii, K. Katayama, T. Suzuki, S. Kurachi, F. Sakurai, S. Akira and H. Mizuguchi (2010). "Induction of type I interferon by adenovirus-encoded small RNAs." *Proc Natl Acad Sci U S A* **107**(40): 17286–91.

Yamamoto, Y., N. Hiraoka, N. Goto, Y. Rin, K. Miura, K. Narumi, H. Uchida, M. Tagawa and K. Aoki (2014). "A targeting ligand enhances infectivity and cytotoxicity of an oncolytic adenovirus in human pancreatic cancer tissues." *J Control Release* **192**: 284–93.

Yan, W., G. Kitzes, F. Dormishian, L. Hawkins, A. Sampson-Johannes, J. Watanabe, J. Holt, V. Lee, T. Dubensky, A. Fattaey, T. Hermiston, A. Balmain and Y. Shen (2003). "Developing novel oncolytic adenoviruses through bioselection." *J Virol* **77**(4): 2640–50.

Yang, S. W., J. J. Cody, A. A. Rivera, R. Waehler, M. Wang, K. J. Kimball, R. A. Alvarez, G. P. Siegal, J. T. Douglas and S. Ponnazhagan (2011). "Conditionally replicating adenovirus expressing TIMP2 for ovarian cancer therapy." *Clin Cancer Res* **17**(3): 538–49.

Yang, U.-C., W. Huang and S. J. Flint (1996). "mRNA export correlates with activation of transcription in human subgroup C adenovirus-infected cells." *J. Virol.* **70**: 4071–4080.

Yang, Y., K. U. Jooss, Q. Su, H. C. Ertl and J. M. Wilson (1996). "Immune responses to viral antigens versus transgene product in the elimination of recombinant adenovirus-infected hepatocytes in vivo." *Gene Ther* **3**(2): 137–44.

Yang, Y., Q. Li, H. C. Ertl and J. M. Wilson (1995). "Cellular and humoral immune responses to viral antigens create barriers to lung-directed gene therapy with recombinant adenoviruses." *J Virol* **69**(4): 2004–15.

Yang, Y., F. A. Nunes, K. Berencsi, E. E. Furth, E. Gonczol and J. M. Wilson (1994). "Cellular immunity to viral antigens limits E1-deleted adenoviruses for gene therapy." *Proc Natl Acad Sci U S A* **91**(10): 4407–11.

Yang, Y., H. Xu, W. Huang, M. Ding, J. Xiao, D. Yang, H. Li, X. Y. Liu and L. Chu (2015). "Targeting lung cancer stem-like cells with TRAIL gene armed oncolytic adenovirus." *J Cell Mol Med* **19**(5): 915–23.

Yang, Z. and D. J. Klionsky (2010). "Eaten alive: a history of macroautophagy." *Nat Cell Biol* **12**(9): 814–22.

Yang, Z. Y., L. S. Wyatt, W. P. Kong, Z. Moodie, B. Moss and G. J. Nabel (2003). "Overcoming immunity to a viral vaccine by DNA priming before vector boosting." *J Virol* **77**(1): 799–803.

Yao, W., G. Guo, Q. Zhang, L. Fan, N. Wu and Y. Bo (2014). "The application of multiple miRNA response elements enables oncolytic adenoviruses to possess specificity to glioma cells." *Virology* **458–459**: 69–82.

Yap, M. W. and J. P. Stoye (2012). "TRIM proteins and the innate immune response to viruses." *Adv Exp Med Biol* **770**: 93–104.

Yates, V. J. and D. E. Fry (1957). "Observations on a chicken embryo lethal orphan (CELO) virus." *Am J Vet Res* **18**(68): 657–60.

Yatherajam, G., W. Huang and S. J. Flint (2011). "Export of adenoviral late mRNA from the nucleus requires the Nxf1/Tap export receptor." *J Virol* **85**(4): 1429–38.

Ye, X., M. B. Robinson, M. L. Batshaw, E. E. Furth, I. Smith and J. M. Wilson (1996). "Prolonged metabolic correction in adult ornithine transcarbamylase-deficient mice with adenoviral vectors." *J Biol Chem* **271**(7): 3639–46.

Ye, X., M. B. Robinson, C. Pabin, T. Quinn, A. Jawad, J. M. Wilson and M. L. Batshaw (1997). "Adenovirus-mediated in vivo gene transfer rapidly protects ornithine transcarbamylase-deficient mice from an ammonium challenge." *Pediatr Res* **41**(4 Pt 1): 527–34.

Yeh-Kai, L., G. Akusjarvi, P. Alestrom, U. Pettersson, M. Tremblay and J. Weber (1983). "Genetic identification of an endoproteinase encoded by the adenovirus genome." *J Mol Biol* **167**(1): 217–22.

Yew, P. R. and A. J. Berk (1992). "Inhibition of p53 transactivation required for transformation by adenovirus early 1B protein." *Nature* **357**(6373): 82–85.

Yew, P. R., C. C. Kao and A. J. Berk (1990). "Dissection of functional domains in the adenovirus 2 early 1B 55k polypeptide by suppressor-linker-insertional mutagenesis." *Virol.* **179**: 795–805.

Yew, P. R., X. Liu and A. J. Berk (1994). "Adenovirus E1B oncoprotein tethers a transcriptional repression domain to p53." *Genes Dev* **8**(2): 190–202.

Yewdell, J. W., J. R. Bennink, K. B. Eager and R. P. Ricciardi (1988). "CTL recognition of adenovirus-transformed cells infected with influenza virus: lysis by anti-influenza CTL parallels adenovirus-12-induced suppression of class I MHC molecules." *Virology* **162**(1): 236–8.

Ying, B., A. E. Tollefson, J. F. Spencer, L. Balakrishnan, S. Dewhurst, C. Capella, R. M. Buller, K. Toth and W. S. Wold (2014). "Ganciclovir inhibits human adenovirus replication and pathogenicity in permissive immunosuppressed Syrian hamsters." *Antimicrob Agents Chemother* **58**(12): 7171–81.

Ying, B., A. E. Tollefson and W. S. Wold (2010). "Identification of a previously unrecognized promoter that drives expression of the UXP transcription unit in the human adenovirus type 5 genome." *J Virol* **84**(21): 11470–8.

Ylosmaki, E., T. Hakkarainen, A. Hemminki, T. Visakorpi, R. Andino and K. Saksela (2008). "Generation of a conditionally replicating adenovirus based on targeted destruction of E1A mRNA by a cell type-specific MicroRNA." *J Virol* **82**(22): 11009–15.

Yondola, M. A. and P. Hearing (2007). "The adenovirus E4 ORF3 protein binds and reorganizes the TRIM family member transcriptional intermediary factor 1 alpha." *J Virol* **81**(8): 4264–71.

Yoo, J. Y., J. H. Kim, Y. G. Kwon, E. C. Kim, N. K. Kim, H. J. Choi and C. O. Yun (2007). "VEGF-specific short hairpin RNA-expressing oncolytic adenovirus elicits potent inhibition of angiogenesis and tumor growth." *Mol Ther* **15**(2): 295–302.

Yoshimi, R., T. H. Chang, H. Wang, T. Atsumi, H. C. Morse, 3rd and K. Ozato (2009). "Gene disruption study reveals a nonredundant role for TRIM21/Ro52 in NF-kappaB-dependent cytokine expression in fibroblasts." *J Immunol* **182**(12): 7527–38.

Youil, R., T. J. Toner, Q. Su, M. Chen, A. Tang, A. J. Bett and D. Casimiro (2002). "Hexon gene switch strategy for the generation of chimeric recombinant adenovirus." *Hum Gene Ther* **13**(2): 311–20.

Youn, J. I., S. H. Park, H. T. Jin, C. G. Lee, S. H. Seo, M. Y. Song, C. W. Lee and Y. C. Sung (2008). "Enhanced delivery efficiency of recombinant adenovirus into tumor and mesenchymal stem cells by a novel PTD." *Cancer Gene Ther* **15**(11): 703–12.

Yu, W. and H. Fang (2007). "Clinical trials with oncolytic adenovirus in China." *Curr Cancer Drug Targets* **7**(2): 141–8.

Yuan, T. L. and L. C. Cantley (2008). "PI3K pathway alterations in cancer: variations on a theme." *Oncogene* **27**(41): 5497–510.

Yueh, A. and R. J. Schneider (1996). "Selective translation initiation by ribosome jumping in adenovirus-infected and heat-shocked cells." *Genes Dev* **10**(12): 1557–67.

Yueh, A. and R. J. Schneider (2000). "Translation by ribosome shunting on adenovirus and hsp70 mRNAs facilitated by complementarity to 18S rRNA." *Genes Dev* **14**(4): 414–21.

Zabner, J., L. A. Couture, R. J. Gregory, S. M. Graham, A. E. Smith and M. J. Welsh (1993). "Adenovirus-mediated gene transfer transiently corrects the chloride transport defect in nasal epithelia of patients with cystic fibrosis." *Cell* **75**(2): 207–16.

Zahn, R., G. Gillisen, A. Roos, M. Koning, E. van der Helm, D. Spek, M. Weijtens, M. Grazia Pau, K. Radosevic, G. J. Weverling, J. Custers, J. Vellinga, H. Schuitemaker, J. Goudsmit and A. Rodriguez (2012). "Ad35 and ad26 vaccine vectors induce potent and cross-reactive antibody and T-cell responses to multiple filovirus species." *PLoS One* **7**(12): e44115.

Zak, D. E., E. Andersen-Nissen, E. R. Peterson, A. Sato, M. K. Hamilton, J. Borgerding, A. T. Krishnamurty, J. T. Chang, D. J. Adams, T. R. Hensley, A. I. Salter, C. A. Morgan, A. C. Duerr, S. C. De Rosa, A. Aderem and M. J. McElrath (2012). "Merck Ad5/HIV induces broad innate immune activation that predicts CD8(+) T-cell responses but is attenuated by preexisting Ad5 immunity." *Proc Natl Acad Sci U S A* **109**(50): E3503–12.

Zakhartchouk, A. N., C. Pyne, G. K. Mutwiri, Z. Papp, M. E. Baca-Estrada, P. Griebel, L. A. Babiuk and S. K. Tikoo (1999). "Mucosal immunization of calves with recombinant bovine adenovirus-3: induction of protective immunity to bovine herpesvirus-1." *J Gen Virol* **80** (Pt 5): 1263–9.

Zeng, X. and C. R. Carlin (2013). "Host cell autophagy modulates early stages of adenovirus infections in airway epithelial cells." *J Virol* **87**(4): 2307–19.

Zerler, B., R. Roberts, M. B. Mathews and E. Moran (1987). "Different functional domains of the adenovirus E1A genes are involved in the regulation of host cell cycle products." *Mol. Cell. Biol.* **7**: 821–829.

Zhang, J. J., U. Vinkemeier, W. Gu, D. Chakravarti, C. M. Horvath and J. E. Darnell, Jr. (1996). "Two contact regions between Stat1 and CBP/p300 in interferon gamma signaling." *Proc Natl Acad Sci U S A* **93**(26): 15092–6.

Zhang, Q., G. Chen, L. Peng, X. Wang, Y. Yang, C. Liu, W. Shi, C. Su, H. Wu, X. Liu, M. Wu and Q. Qian (2006). "Increased safety with preserved antitumoral efficacy on hepatocellular carcinoma with dual-regulated oncolytic adenovirus." *Clin Cancer Res* **12**(21): 6523–31.

Zhang, S., W. Huang, X. Zhou, Q. Zhao, Q. Wang and B. Jia (2013). "Seroprevalence of neutralizing antibodies to human adenoviruses type-5 and type-26 and chimpanzee adenovirus type-68 in healthy Chinese adults." *J Med Virol* **85**(6): 1077–84.

Zhang, W. and M. J. Imperiale (2000). "Interaction of the adenovirus IVa2 protein with viral packaging sequences." *J Virol* **74**(6): 2687–93.

Zhang, W. and M. J. Imperiale (2003). "Requirement of the adenovirus IVa2 protein for virus assembly." *J Virol* **77**(6): 3586–94.

Zhang, Y., P. J. Dolph and R. J. Schneider (1989). "Secondary structure analysis of adenovirus tripartite leader." *J Biol Chem* **264**(18): 10679–84.

Zhang, Y., D. Feigenblum and R. J. Schneider (1994). "A late adenovirus factor induces e1F-4B dephosphorylation and inhibition of cellular protein synthesis." *J. Virol.* **68**: 7040–7050.

Zhang, Z., W. Zou, J. Wang, J. Gu, Y. Dang, B. Li, L. Zhao, C. Qian, Q. Qian and X. Liu (2005). "Suppression of tumor growth by oncolytic adenovirus-mediated delivery of an antiangiogenic gene, soluble Flt-1." *Mol Ther* **11**(4): 553–62.

Zhao, B., S. Hou and R. P. Ricciardi (2003). "Chromatin repression by COUP-TFII and HDAC dominates activation by NF-kappaB in regulating major histocompatibility complex class I transcription in adenovirus tumorigenic cells." *Virology* **306**(1): 68–76.

Zhao, B. and R. P. Ricciardi (2006). "E1A is the component of the MHC class I enhancer complex that mediates HDAC chromatin repression in adenovirus-12 tumorigenic cells." *Virology* **352**(2): 338–44.

Zhao, H., M. Chen and U. Pettersson (2013). "Identification of adenovirus-encoded small RNAs by deep RNA sequencing." *Virology* **442**(2): 148–55.

Zhao, H., M. Chen and U. Pettersson (2014). "A new look at adenovirus splicing." *Virology* **456–457**: 329–41.

Zhao, H., M. Dahlo, A. Isaksson, A. C. Syvanen and U. Pettersson (2012). "The transcriptome of the adenovirus infected cell." *Virology* **424**(2): 115–28.

Zheng, Y., T. Stamminger and P. Hearing (2016). "E2F/Rb Family Proteins Mediate Interferon Induced Repression of Adenovirus Immediate Early Transcription to Promote Persistent Viral Infection." *PLoS Pathog* **12**(1): e1005415.

Zhou, J., Q. Gao, G. Chen, X. Huang, Y. Lu, K. Li, D. Xie, L. Zhuang, J. Deng and D. Ma (2005). "Novel oncolytic adenovirus selectively targets tumor-associated polo-like kinase 1 and tumor cell viability." *Clin Cancer Res* **11**(23): 8431–40.

Zhu, J., X. Huang and Y. Yang (2007). "Innate immune response to adenoviral vectors is mediated by both Toll-like receptor-dependent and -independent pathways." *J Virol* **81**(7): 3170–80.

Zhu, J., X. Huang and Y. Yang (2007). "Type I IFN signaling on both B and CD4 T cells is required for protective antibody response to adenovirus." *J Immunol* **178**(6): 3505–10.

Zhu, J., X. Huang and Y. Yang (2008). "A critical role for type I IFN-dependent NK cell activation in innate immune elimination of adenoviral vectors in vivo." *Mol Ther* **16**(7): 1300–7.

Zhu, J., X. Huang and Y. Yang (2010). "NKG2D is required for NK cell activation and function in response to E1-deleted adenovirus." *J Immunol* **185**(12): 7480–6.

Zhu, T. X., B. Lan, L. Y. Meng, Y. L. Yang, R. X. Li, E. M. Li, S. Y. Zheng and L. Y. Xu (2013). "ECM-related gene expression profile in vascular smooth muscle cells from human saphenous vein and internal thoracic artery." *J Cardiothorac Surg* **8**: 155.

Zhu, Z. B., Y. Chen, S. K. Makhija, B. Lu, M. Wang, A. A. Rivera, M. Yamamoto, S. Wang, G. P. Siegal, D. T. Curiel and J. M. McDonald (2006). "Survivin promoter-based conditionally replicative adenoviruses target cholangiocarcinoma." *Int J Oncol* **29**(5): 1319–29.

Zhu, Z. B., B. Lu, M. Park, S. K. Makhija, T. M. Numnum, J. E. Kendrick, M. Wang, Y. Tsuruta, P. Fisher, R. D. Alvarez, F. Zhou, G. P. Siegal, H. Wu and D. T. Curiel (2008). "Development of an optimized conditionally replicative adenoviral agent for ovarian cancer." *Int J Oncol* **32**(6): 1179–88.

Ziff, E. B. and R. M. Evans (1978). "Coincidence of the promoter and capped 5' terminus of RNA from the adenovirus 2 major late transcription unit." *Cell* **15**(4): 1463–75.

Zijderveld, D. C. and P. C. van der Vliet (1994). "Helix-destabilizing properties of the adenovirus DNA-binding protein." *J Virol* **68**(2): 1158–64.

Zubieta, C., G. Schoehn, J. Chroboczek and S. Cusack (2005). "The structure of the human adenovirus 2 penton." *Mol Cell* **17**(1): 121–35.

Zuckerman, J. B., C. B. Robinson, K. S. McCoy, R. Shell, T. J. Sferra, N. Chirmule, S. A. Magosin, K. J. Propert, E. C. Brown-Parr, J. V. Hughes, J. Tazelaar, C. Baker, M. J. Goldman and J. M. Wilson (1999). "A phase I study of adenovirus-mediated transfer of the human cystic fibrosis transmembrane conductance regulator gene to a lung segment of individuals with cystic fibrosis." *Hum Gene Ther* **10**(18): 2973–85.

Index